DATE DUE

APR 18 '89	
JUL 05 1989	
OCT 26 1992	
NOV 11 1992	
DEC 02 1992	
MAR 31 1993	
JUN 22 1993	
AUG 11 1993	
AUG 29 1995	
APR 07 1997	

Second Edition

Topics from the Theory of Numbers

Emil Grosswald
Temple University

Topics
from the
Theory of Numbers

second edition

Birkhäuser
BOSTON • BASEL • STUTTGART

Emil Grosswald
Department of Mathematics
Temple University
Philadelphia, Pennsylvania 19122, USA

Library of Congress Cataloging in Publication Data

Grosswald, Emil.
   Topics from the theory of numbers.
   Includes bibliographies and index.
   1. Numbers, Theory of. I. Title.
QA241.G74   1984   512'.7   81-4318
ISBN 3-7643-3044-9   AACR2

CIP-Kurztitelaufnahme der Deutschen Bibliothek

CIP-Kurztitelaufnahme der Deutschen Bibliothek
Grosswald, Emil:
Topics from the theory of numbers / Emil
Grosswald. - 2. ed. - Boston ; Basel ; Stuttgart :
Birkhäuser, 1984
   ISBN 3-7643-3044-9

© Birkhäuser Boston, 1984
ISBN: 0-8176-3044-9
ISBN: 3-7643-3044-9
Printed in USA

A B C D E F G H I J

To the memory of my father
Paul Grosswald

Contents

Preface

There are now almost exactly 20 years since I wrote the lecture notes that eventually became this book. In the meantime, great progress has been made in virtually all branches of mathematics and number theory is no exception. In the elementary part (covered, essentially, in the first seven chapters), definitive theorems had been known since the time of Euler or even Fermat, and there the new results are of rather minor importance (new Mersenne primes, more stringent conditions for the highly unlikely existence of odd perfect numbers, better error terms in "elementary" proofs of the Prime Number Theorem, etc.). On more advanced topics, however, the progress has been spectacular. Among others, during this short time interval, the theory of Diophantine equations have been thoroughly altered; Hilbert's Tenth Problem was finally settled in 1970; effective methods were introduced from about 1965 (however, also see Roth's work from 1955); Weil's conjectures were proved one by one; progress was made in class field theory. These are a few of the topics that profoundly altered many aspects of number theory.

Not all these recent discoveries were relevant to the present text, which grew out of notes for undergraduate lectures. Nevertheless, I felt that, for instance, the chapter on Diophantine equations had to be rewritten. In it I made an effort to bring within the grasp of the prospective reader (I still have in mind the bright, eager undergraduate junior or senior) at least the relevance of the progress made, even if the proofs of many of the statements could not be fitted into the framework of the book. Nevertheless, the present volume should remain a mathematical book and not a book *about* mathematics. As many statements as possible are formalized as theorems and are proved. This in turn required some additional preparatory material. Much of this is so very interesting in its own right that I added it without regret. It consists, essentially, of one section on the Riemann Zeta function, a chapter on number fields (much of this material overlaps with Section 5 of Chapter 9 in the first edition) and one on L-functions and their use in connection with primes in arithmetic progressions. On the other hand, the non-consecutive chapters of the first edition that both treat the topic of Fermat's equation have now been collapsed into a single one, with the elimination of much material now covered in the above-mentioned chapter on number fields.

As some material had to be added to do justice to the progress made during the last 20 years, it had become desirable also to eliminate some parts of the text, so that the book should be kept to a reasonable size. This turned out to be rather easy. Indeed in 1962 many of the students who used the book had no course in the theory of functions of a complex variable, nor any in abstract algebra; for their benefit, all needed results from those two fields were listed in two sections of the text and two appendices were added for the same purpose. The classroom experience with the book seems to show, however, that both those sections and the appendices were of only limited usefulness, so they were deleted from the present edition.

In this second edition, the first six chapters (i.e., the old Part I and Part II) are essentially unchanged (except, hopefully, for the correction of an unusually large number of printing errors, which had marred the first edition) and can be read by anybody who is acquainted with the customary high school mathematics. This is the case also with the chapter on Partitions and, as this material is independent of the topics in Part III, the chapter has been renumbered as Chapter 7 and now follows naturally after Chapter 6 on Arithmetical Functions, as the last chapter of Part II. The first seven chapters offer ample materiàl for a one-semester course. It is hoped that during this time, the student will have taken the usual undergraduate junior-senior courses in analysis and algebra. This will be more than sufficient to enable him to study Part III of the book. A first course on functions of a complex variable will be all he needs for Chapter 8 (The distribution of Primes and Riemann's Zeta function), Chapter 9 (The Prime Number Theorem), and practically all of Chapter 12 (*L*-functions and Primes in Arithmetic Progressions).

Chapter 10 (The arithmetic of number fields) assumes knowledge of the *basic* concepts of groups, rings, fields, and ideals, but not, say of Noetherian rings, Dedekind domains, etc. This was necessary to maintain the basically *elementary* character of the book. For that reason, after some hesitation I decided to keep some 12 pages of the old Chapter 10 (Ideal Theory), renumbered now as Chapter 11, on the properties of ideals in rings of algebraic integers. Chapter 13 (Diophantine Equations) requires some algebra, but mainly a thorough understanding of the first six chapters and of Chapter 10. Chapter 14 on Fermat's Equation consists of portions of the old Chapters 9 and 11. Some material has been eliminated because it could be treated more systematically in the new Chapter 10. The present arrangement has the further advantage that the material can be presented in a logical sequence, rather than as before in two disjoint parts. Chapters 10, 11, 13, and 14 require no knowledge of complex variables.

Some new problems have been added in the old chapters and the bibliography has been updated, but otherwise the style and general approach remains the same as in the first edition, with as much emphasis on historical development as is consistent with a text that is not meant to be a history of number theory. Also, the general organization of the book remains the same as before and is described in detail in the Preface to the First Edition.

In addition to previous acknowledgements, I would like to mention my indebtedness to Professor S. Chowla for enlightening conversations, and to Professors W.W. Adams and L.J. Goldstein, whose book I read with great pleasure; some of Chapter 14 would have been different, had I not seen their book. Also to Professor J.M. Gandhi, whose untimely death we all deplore, thanks for an excellent suggestion were due.

My thanks go to Birkhäuser Boston, Inc., and in particular to Drs. Alice and Klaus Peters for their cooperation in bringing out this second edition. I also want to express my gratitude to the TECHNION in Haifa (Israel), where I found a congenial atmosphere and pleasant working conditions and where I did much of the work on this new edition.

Haifa, Israel, January 1981   E. Grosswald

Preface to the First Edition

In 1962 I gave a course in number theory for undergraduates at the University of Pennsylvania. To my pleasant surprise, the group of students turned out to be exceptionally eager and well prepared, and contained not only undergraduates, but also a sprinkling of graduate students. This led me to supplement the excellent text of Niven and Zuckerman with some notes of my own. These included among other topics a self-contained proof of the Prime Number Theorem. In the obvious references during the lectures to the discoverers of the first proofs of this theorem, I spoke of the then almost centenary Hadamard and of de la Vallée-Poussin as illustrious contemporary mathematicians; I may even have quoted a joke (first heard from P. Erdös) to the effect that proving such an important result seems to confer immortality. Between the time of those lectures and completion of the manuscript, both Hadamard and de la Vallée-Poussin have joined the ranks of the immortal mathematicians of the past. May this book help to keep alive the memory of their imperishable work!

Those first notes were rewritten and expanded as I had further opportunities to lecture on number theory; yet it was only with considerable doubt that I gave in to the insistence of some of my students and colleagues (and editors) and decided to "polish" those notes in order to write the present volume. Indeed, there existed in English several excellent books on number theory on a comparable level, and I was not convinced that there was a need for still another one, or that I could improve on those previous presentations. However, with the "finished product" in front of me, it seems that there is rather little overlap with any other text I know of, except, of course, for the most fundamental topics which are common to all books in this field. Also the style seems to be different from that of most other books because of both the emphasis given to the historical point of view and the inclusion of references to a large amount of source material (books and papers) quoted in the text and the bibliographies.

The book consists of three parts and two appendices. *Part I* is an introductory, mainly historic part, consisting of two chapters (Introduction, Notations). Most of my students have reacted extremely well to the classroom presentation of this past, which I would never omit. However, if the students (or their instructor) are not interested in the historic development, this part can be skipped without inconvenience, except, perhaps for a few pages in which symbols and notations are explained. In case the class is interested in the historic aspect but has little time available, *Part I* may be given as a supplementary reading assignment, and the instructor may begin with *Part II*.

Part II consists of four chapters in elementary number theory (Divisibility, Congruences, Quadratic Residue, Arithmetic Functions). While much of this material is standard fare, some of its is developed well beyond minimal requirements (for instance, intervals between consecutive primes, the number of solutions of higher congruences modulo prime powers, perfect numbers, Ramanujan function, and so on).

Part III consists of three divisions, which may be studied independently of each other and in any desired order. In a standard one-semester course on the junior-senior level. I was never able to cover more than three of the six chapters — that is, at most two or three divisions. These are:

(A) The Riemann Zeta Function (Chapter 7) and the Prime Number Theorem (Chapter 8);

(B) Diophantine Equations and Fermat's Conjecture (Chapter 9), Ideal Theory (Chapter 10), and the Proof of Fermat's Conjecture for Regular Primes (Chapter 11);
(C) The Theory of Partitions (Chapter 12).

In view of the title of this book it does not seem to me that any special explanation is due concerning the topics covered or any apology necessary for those topics which have been omitted. In order to make this book as self-contained as possible (consistent with a reasonable size), two appendices have been added, one on topics from advanced calculus and general analysis, the other on topics from algebra. Needless to say, these appendices are here mainly for reference purposes and are not meant to replace regular courses in their respective fields.

Throughout the book more difficult sections and problems are marked with a star (*) and a few particularly difficult ones by a double star (**). Within each chapter, the definitions, lemmas, theorems, and remarks have separate consecutive numbers; the same is true for the numbered formulas. The corollaries carry the number of the theorem from which they are derived, followed by a number of their own. Thus, the second corollary of Theorem 6 is called Corollary 6.2. References to definitions, lemmas, theorems, corollaries, or remarks of the same chapter carry only the corresponding number; however, if reference is made to one of them in another chapter, then the number of the statement is preceded by that of the chapter where it is stated. If a theorem, corollary, or definition contains several statements, these are numbered (1), (2), and so on. If one wants to refer to statement (3) of the first corollary to Theorem 1 of Chapter 12, one speaks of Corollary 1.1(3) in Chapter 12 and of Corollary 12.1.1.(3) in any other chapter. Sections within a chapter are numbered in a similar way. Section 3 of Chapter 7 and Section 5 of Chapter 9 contain miscellaneous definitions and theorems which are listed consecutively. Statement 12 of Section 3 in Chapter 7 will be quoted as 3.12 in Chapter 7 and as 7.3.12 in any other chapter, with similar notations for the statements of Section 5 in Chapter 9.

Each chapter in Parts II and III is followed by a number of problems. Some of these are simple exercises, others (occasionally marked by a star) are actually theorems of independent interest that did not fit into the main text; still others require the student to supply complete proofs for statements accepted in the text as "obvious" or which have been dismissed there because they follow by "trivial arguments." I sincerely hope that the students will agree with me and prove the "obviousness" to themselves by dashing off the needed arguments in no time at all! Should I be wrong, then I shall be glad to hear from irate students and instructors, in order to supply the not-so-obvious arguments in some future edition. But, on this subject of textbook problems, as well as on another, related one, I cannot express my own feelings better than by quoting from Ahlfors' *Complex Analysis*, ". . . the author has not had the inclination to relieve the teacher from making up more and better exercises . . ." and in particular, ". . . it is to be hoped that no teacher will follow this book page by page, for nothing could be more deadening." This, of course, holds true even for the author of a text and I myself could never use my own book without considerable (and each time different) modifications, suggested by the abilities and interests of the students.

Finally, a word of acknowledgement is in order. My interest in number theory was awakened by my revered teacher, Professor H. Rademacher. It is through him that I became aware of the classical works of Fermat, Euler, Gauss, Riemann, Hadamard, and de la Vallée-Poussin. Furthermore, within my generation, no student of number theory could fail to be influenced by the works of Landau, Hardy, Littlewood, Ramanujan, Hilbert, and Hecke. Also, some of the excellent textbooks by contemporary mathematicians such as those by Niven and Zuckerman, LeVeque, Vinogradov, or Harvey Cohn may have influenced my presentation of some of the topics. I owe, however, a special debt of gratitude to Professors M. Kac, P. Bateman, and R. Dixon. Without the strong encouragement of Professor Kac I might never have written this book. Professor Bateman read part of the manuscript and Professor Dixon, after an incredibly careful reading, made numerous and most pertinent suggestions for corrections of errors, improvements in presentation, and even for changes in the formulation of some problems. Also my students, Dr. D. Goldsmith and Dr. R. Alter, were helpful in the early stages of the preparation of the manuscript, and Mrs. S. Goldsmith did the excellent typing of its final version. I also am grateful to the University of Pennsylvania, which gave me the opportunity both to study and to teach the theory of numbers. The manuscript was completed during the first few weeks of a sabbatical leave, with support from the University of Pennsylvania and the National Science Foundation (Under Grant GP 3137).

E. G.

Paris

Part One

Introduction,
Historical Background,
and Notations

Chapter 1

Introduction and Historical Background

1 NUMBER THEORY — THE QUEEN OF MATHEMATICS

Numbers have exerted a fascination on the human mind since the beginnings of recorded history. Among the treasures of Egyptian antiquity we find the famous Rhind papyrus [6], which tells us about the mathematics practiced in Egypt almost 2000 years B.C. Still older cuneiform tablets (see [5]) show us that arithmetic, at least, was already quite sophisticated in Mesopotamia at the end of the third millenium B.C.

Of all the branches of mathematics, the one that seems to appeal most to our esthetic feelings is number theory, considered by many (by Euler (1707–1783), for instance) to be the queen of mathematics. Why is that so? Some people believe that this strong esthetic appeal is due to the very limited practical usefulness of number theory. However, there might be a better reason. In hardly any other branch of mathematics is it possible to ask really significant, non-trivial questions without preceding them by an annoyingly long list of definitions. In number theory, on the other hand, one can ask many questions in such simple terms that the famous "man in the street" can immediately understand—but generally not answer them! In fact, the answers to some of these "simple to ask" questions are so difficult that nobody has yet found them. Other questions, of course, have been answered already by the ingenious mathematicians of antiquity, or by those of medieval and modern times. In most cases (excepting, of course, the trivial questions) the answers *cannot* be formulated quite so easily as the questions themselves. In some other cases it is utterly impossible to convince the "man in the street" of the correctness of a certain answer, laboriously arrived at, which does not strike one as plausible at first sight. Indeed, it might be necessary first to teach him a considerable amount of mathematics and then to present him with a formal proof of the answer.

2 A PROBLEM

In order to illustrate these situations, consider, for instance, the following problem: A man has a debt of $10. He wants to repay it in bills of $1, $5, or $10. In how many ways can he pay his debt? Here the answer is clear; he may pay by

(i) one $10 bill;

(ii) two $5 bills;

(iii) one $5 bill and five $1 bills;

(iv) ten $1 bills.

In other words, there are four solutions to this problem. If we denote by x, y, and z the number of $10, $5, and $1 bills, respectively, used in the payment, then our problem may be formulated as follows: Find the solution of the equation

$$10x + 5y + z = 10 \qquad (1)$$

in non-negative integers x, y, and z.

Let us observe first that here the situation is different from that usually encountered in elementary algebra. Indeed, we are supposed to determine *three* quantities x, y, z and dispose of only *one* equation. Clearly, without additional conditions, the problem is not very interesting, as one may choose arbitrary values for two of the quantities, say, x and y, and then solve (1) for z, obtaining $z = 10 - 10x - 5y$. Here, however, we are *not* free to select *arbitrary* values for x and y, but must choose non-negative integers. (We cannot use a negative number of bills to pay a debt, nor can we tear up a bill and use part of it.) It is this added restriction that makes the problem more difficult—and at the same time more interesting and meaningful. Such equations, which must be solved in integers, sometimes with added side conditions (such as here that x, y, $z \geq 0$), were already known in antiquity and are called Diophantine equations, after Diophantus of Alexandria (who lived in the third or fourth century A.D.). So far, presumably, every person of normal intelligence will have understood both the problem and its solution. Let us now consider briefly the formal solution of (1). From $y \geq 0$, $z \geq 0$ it follows that $10 = 10x + 5y + z \geq 10x$. Hence, $x \leq 1$; as we also have $x \geq 0$, there are, clearly, only two possibilities:

(i) either $x = 1$, and (1) becomes $5y + z = 0$, so that (remembering $y \geq 0$, $z \geq 0$) $y = z = 0$; or else

(ii) $x = 0$ and (1) becomes $5y + z = 10$.

If $y = 0$, then $z = 10$; if $y = 1$, then $z = 5$; if $y = 2$, then $z = 0$; finally, if $y > 2$, then $z < 0$, contrary to our condition $z \geq 0$. These are precisely the answers already obtained informally. It seems likely that whoever understood the question in the first place will have no trouble also understanding the four possible solutions,

(i) $x = 1, y = 0, z = 0$;

(ii) $x = 0, y = 0, z = 10$;

(iii) $x = 0, y = 1, z = 5$;

(iv) $x = 0, y = 2, z = 0$,

even if the formal proof given is not clear to him in every detail. Let us now, however, consider once more equation (1), ignoring the original problem and, while still requiring that x, y, z be integers, drop the condition of nonnegativity.

Let us assume that x_0, y_0, z_0 are some integral values that satisfy (1) (e.g., $0, 2, 0$) and consider the expressions

$$x = x_0 + a_1 t + b_1 u,$$

$$y = y_0 + a_2 t + b_2 u,$$

$$z = z_0 + a_3 t + b_3 u.$$

Substituting these in (1), we obtain

$$10x_0 + 5y_0 + z_0 + (10a_1 + 5a_2 + a_3)t + (10b_1 + 5b_2 + b_3)u = 10,$$

or, because $10x_0 + 5y_0 + z_0 = 10$,

$$(10a_1 + 5a_2 + a_3)t + (10b_1 + 5b_2 + b_3)u = 0.$$

This equality will be satisfied for all values of t and u, provided that we select for (a_1, a_2, a_3) and for (b_1, b_2, b_3) sets of solutions (not necessarily the same) of the equation

$$10x + 5y + z = 0, \tag{1'}$$

obtained from (1) by dropping the "second member." Equation (1') is usually referred to as "the homogeneous equation corresponding to (1)." We may choose, for instance, $a_1 = 0$, $a_2 = 1$, $a_3 = -5$ and $b_1 = -1$, $b_2 = 1$, $b_3 = 5$; then, with above values $x_0 = 0$, $y_0 = 2$, $z_0 = 0$, we obtain

$$x = -u,$$

$$y = 2 + t + u, \tag{2}$$

$$z = -5t + 5u.$$

From the way in which the expressions (2) have been obtained, it is clear that if we substitute for u and t any integers, we obtain an integer-valued solution of (1). Actually, it may be shown that, at proper choice of a_1, a_2, a_3 and of b_1, b_2, b_3, we obtain in this way *all* integer-valued solutions of (1); here, however, we shall only verify that if we now insist that x, y, z be non-negative, we obtain once more the well-known solutions of the original problem. First,

$x \geq 0$ requires that $u \leq 0$. Next, $y \geq 0$ requires that $t + u \geq -2$, and $z \geq 0$ requires that $-t + u \geq 0$. Adding the last two shows that $2u \geq -2$ or $u \geq -1$. Hence, in view of $u \leq 0$, we may only have $u = -1$ or $u = 0$. In the first case, $x = 1$, $y = 1 + t$, $z = -5t - 5$. From $0 \leq y = 1 + t$ and $0 \leq z = -5t - 5$ we obtain that $0 \leq 1 + t \leq 0$, so that $t = -1$ and $y = z = 0$. If $u = 0$, then $x = 0$ and $y = 2 + t \geq 0$, $z = -5t \geq 0$. The last two inequalities require $-2 \leq t \leq 0$, so that $t = -2, -1$, or 0, leading to the solutions $x = 0$, $y = 0$, $z = 10$; $x = 0$, $y = 1$, $z = 5$; and $x = 0$, $y = 2$, $z = 0$, respectively. Adjoining to these three solutions for $u = 0$ the single solution obtained with $u = -1$, we have indeed the four already-known, non-negative, integer-valued solutions of (1). It would not be too surprising if the meaning of (2) as general solution of (1) could not be grasped without some effort by every person who understands perfectly the *problem* of solving equation (1) in integers.

3   SOMETHING ABOUT THE CONTENTS OF THIS BOOK

Often the simple-sounding problems asked within the context of number theory turned out to be too difficult even for the most powerful mathematicians who attacked them, but their efforts were not spent in vain. As a matter of fact, many flourishing branches of mathematics owe their very existence or their development to unsuccessful attempts to solve problems in number theory. In what follows, we shall mention a few instances and describe a few number-theoretic problems—some solved, some still open—all of which have greatly stimulated the development of mathematics as a whole. Incidentally, this description should prove more helpful in telling what number theory is than a formal definition of the term, which will not be given here. Next, we shall try to acquire some of the technical tools used in the attack on these problems and indicate in detail the solutions of some of them. For many other problems, however, the techniques needed are of a highly specialized nature and we shall have to forgo their detailed discussion in the present book.

4   NUMBER THEORY AND OTHER BRANCHES
OF MATHEMATICS

Among the many problems in the theory of numbers that had great impact upon the development of entire branches of mathematics, we quote as examples the following:

(a) The study of the distribution of primes sparked the development of the *theory of functions of a complex variable* and, in particular, that of the *theory of entire functions*. With the help of these theories, the original problem has been essentially solved, but in the process of finding the solution new problems arose, some of which are still wide open.

(b) The so-called "last theorem of Fermat" led to the creation of the *theory of algebraic numbers*—one of the most important and flourishing branches of modern number theory—and through it to that of much of modern algebra. Fermat's "last theorem" still has not been proven, but one is almost inclined to forget this fact as a rather irrelevant detail, because the successes of the theory of algebraic numbers completely overshadow the not terribly important statement of Fermat.

(c) The Theory of Partitions gave a strong impetus to the study (by Euler) of *generating functions*. Many of the original problems of Euler have been solved, but the theory of partitions continues to lead to new developments in such diverse fields as modular functions, the saddle-point method of integration, and combinatorial analysis.

We shall consider these problems in greater detail.

5   THE DISTRIBUTION OF PRIME NUMBERS

Among the oldest and most fascinating problems in number theory is that of the *distribution of prime numbers*. In order to discuss it, we need the following:

DEFINITION 1.   An integer $p > 1$ that is not the product of two other positive integers both smaller than p is called a *prime number*; an integer $a > 1$ that is not a prime is called *composite*.

The integer 11 is a prime, because there are no integers a, b such that $a \cdot b = 11, 1 \le a \le b < 11$. But 42 is not a prime; it is a composite number, because $6 \cdot 7 = 42$, and $1 \le 6 \le 7 < 42$ holds. Similarly, $25 = 5 \cdot 5$ is not a prime, and so on. It is convenient to agree that 1 is not a prime. If we list the primes in increasing order, the first few are:

$$2, 3, 5, 7, 11, 13, \ldots .$$

It is easy to write up all primes less than, say, 100 or 200, but to make a complete list of primes up to say, 10^7 is rather time consuming. Nevertheless, reliable printed lists of primes exist up to 10^7 (see [3]) and (apparently less reliable ones (see [2])) even up to 10^8; reliable lists up to 10^8 are available on microfilm (see [1]). If we study these lists in some detail we observe two contrasting features:

(i) A great irregularity in detail. For instance we observe again and again the occurrence of "twin primes," that is, primes p and q, with $q = p + 2$;

at the same time we meet with arbitrarily large "isolated primes," primes preceded and followed by a large number of composite numbers.

(ii) A certain regularity in the distribution of primes, in the sense that on the average the prime numbers seem to thin out steadily. This means, more precisely, that the number of primes out of, say, 1000 consecutive integers seems to decrease with a certain regularity. For instance, in the ten blocks of 1000 consecutive integers between 1 and 10,000 (i.e., in 1–1000, 1001–2000,...,9001–10,000) one finds as number of primes per block 168, 135, 127, 120, 119, 114, 117, 107, 110, and 112, respectively, and there are only 53 primes in the block of 1000 integers from 9,999,001 to 10,000,000. This observation may lead one to suspect that from some point on perhaps all numbers will turn out to be composite, or, in other words, that the total number of primes might be finite (even if, presumably, very large). That this is not so was already known in antiquity (Euclid, ca. 300 B.C.) and we shall soon see a very short proof of the fact that there are infinitely many primes (see Theorem 3.9).

We shall denote by $\pi(x)$ the number of primes up to but not larger than some given quantity x, or, in symbols, set $\pi(x) = \sum_{p \leq x} 1$. It has already been mentioned that if x increases, $\pi(x)$ also increases beyond any preassigned bound. In fact, on the basis of counting the primes, one may be led to suspect that $\pi(x)$ increases somewhat like $x/(\log x)$. Actually, Legendre (1752–1833) and Gauss (1777–1855; the corresponding statement is found in a notebook published only posthumously) stated the conjecture that the ratio between $\pi(x)$ and $x/(\log x)$ approaches unity as $x \to \infty$. This may be expressed symbolically by $\pi(x) \sim x/(\log x)$. An equivalent formulation is

$$\lim_{x \to \infty} \pi(x) \cdot (\log x)/x = L \qquad \text{exists, and} \quad L = 1. \tag{3}$$

Tchebycheff (1821–1894), in an attempt to prove (3), showed that there exist two positive constants c and C, such that $c \leq 1 \leq C$ and

$$c\frac{x}{\log x} < \pi(x) < C\frac{x}{\log x}$$

holds for all $x \geq 2$. He also showed that if the limit L exists at all, then $L = 1$ follows. Hence, if one could "only" show that the limit in (3) exists, the Gauss-Legendre conjecture would be completely proven. However, it turned out that to prove the existence of the limit in (3) is very hard and no direct approach seemed to work. In 1859 Riemann (1826–1866) undertook the study of this problem, in a famous memoir, by a very different, indirect approach. Following an idea that occurs already in Euler's work, he connected the problem of prime numbers with the properties of the function $\zeta(s) = \sum_{n=1}^{\infty} n^{-s}$. While Euler considered $\zeta(s)$ only for real values of s, Riemann let s take complex values. Riemann is one of the founders of the theory of functions of a complex variable and a case can be (and has been) made for the assertion that it was his interest in the study of primes that prompted him to investigate the general theory of functions of a complex variable.

In spite of his brilliant achievements, Riemann was not completely successful; his sketch of a proof of (3) had serious gaps. The most important of these could not be filled until properties of the class of functions called *entire functions* had been established. During the last decade of the 19th century, J. Hadamard (1865–1963) became interested in the problem of primes. Realizing the nature of the tool needed for its solution, he set out to systematize and complete the work previously done by Laguerre (1834–1886), Poincaré (1854–1912), Borel (1871–1956), Picard (1856–1941), and others. The result was his celebrated theory of entire functions. Using this theory, Hadamard, and almost simultaneously, de la Vallée Poussin (1866–1962), succeeded in proving (3), which since then has been known as the *prime number theorem*. Several gaps in Riemann's memoir still remained. Some of these were taken care of by the work of von Mangoldt (1854–1925), Landau (1877–1938), and others. But at least one conjecture, very important for a more precise formulation of the prime number theorem, has so far stubbornly defied all attempts at a proof (or at a refutation). This famous *Riemann hypothesis* states that $\zeta(s) \neq 0$ in the half plane $\operatorname{Re} s > \frac{1}{2}$. The attempts to prove it, while so far unsuccessful, led to such beautiful developments as the theory of almost periodic functions (Bohr, 1887–1951), and the end of this story is not yet in sight.

It should be added that in 1949 Selberg and Erdös succeeded in finding an elementary (but by no means easy) proof of (3), thus dispensing altogether with the use of the theory of functions, created largely in order to cope with this problem.

6 FERMAT'S "LAST THEOREM"

In 1637, Fermat (1601–1665) stated that the Diophantine equation $x^n + y^n = z^n$, with integral $n > 2$, has no solutions in positive integers x, y, z. For $n = 2$, such solutions of course exist, for instance $3^2 + 4^2 = 5^2$. Fermat asserted he had a "truly marvelous proof" of his statement, but today it is generally believed that his argument (which apparently was never revealed) must have been incomplete (see however Mordell [4]). A proof of Fermat's statement for the particular case $n = 4$ is known and is quite easy. It is also comparatively simple to prove the statement for $n = 3$ (Euler, ca. 1760). In fact, the statement has been proven at least for $n \leq 125,000$ (see [8]). But, trying to adapt the method of proof that works for $n = 3$ to the general case, Kummer (1810–1893) ran into a completely unexpected difficulty. We know (and shall prove it formally) that any integer can be factored into primes in essentially (i.e., except for the order of the factors) one way only. Most people feel that the uniqueness of factorization into primes is so obvious, that a formal proof is almost a waste of time. Yet Kummer (and Dirichlet (1805–1859) even before him) found that in an only slightly more general setting the statement (of essential uniqueness of factorization) is actually false. This of course points out once more that even the seemingly most obvious statements have to be proven (starting from some system of axioms) before they can be accepted. Kummer overcame the difficulty by introducing "ideal numbers"; these led, through Kummer's own work, Dedekind's (1831–1916), and that of their followers, to the development of the theory of algebraic numbers and much of modern algebra.

The study of other Diophantine equations led to the development of algebraic geometry, one of the most active branches of contemporary mathematics.

7   THE THEORY OF PARTITIONS

If an integer $n > 0$ is given, we may represent it in general, and in many ways, as a sum of positive integers. Taking for instance $n = 5$, we observe that

$$5 = 4 + 1 = 3 + 2 = 3 + 1 + 1 = 2 + 2 + 1$$
$$= 2 + 1 + 1 + 1 = 1 + 1 + 1 + 1 + 1.$$

Each such representation by a sum (including the one which involves only a single summand, namely n itself) is called a *partition* of n. Partitions that differ only by the order of the summands are not considered distinct. From the above example we see that $n = 5$ has seven distinct partitions, or, in symbols, $p(5) = 7$. The series $F(x) = \sum_{n=0}^{\infty} p(n)x^n$, which has as coefficient of x^n precisely $p(n)$, is called a *generating function* of the $p(n)$. It is easy to show (and was already known to Euler) that if we agree to set $p(0) = 1$ (so far $p(0)$ had not been defined), then $F(x) = \prod_{n=1}^{\infty}(1 - x^n)^{-1}$. Here, and in general in work with generating functions, the actual convergence is unimportant. Incidentally, one may show that both representations of $F(x)$ actually do converge for $|x| < 1$. Equating these two different expressions of $F(x)$ and transforming either one or both sides of the equality, one obtains by elementary reasonings such results as

$$p(n) = p(n-1) + p(n-2) - p(n-5) - p(n-7) + \cdots$$
$$+ (-1)^{j+1} p(n - m_j) + \cdots; \tag{4}$$

here $m_j = \frac{1}{2}j(3j \pm 1)$ and the sum obviously breaks off when $3j^2 - j > 2n$. Formula (4) permits one to compute $p(n)$ "by recurrence," if one already knows $p(1)$, $p(2), \ldots, p(r)$, for all $r \leq n - 1$. As a matter of fact, however, $p(n)$ increases very rapidly with n and (4) soon becomes unmanageable. An approximate value of $p(n)$ can easily be obtained by rather simple, combinatorial considerations, with the result that $e^{A\sqrt{n}} < p(n) < e^{B\sqrt{n}}$ holds with some appropriate positive constants $A < B$, but more than that is true. Hardy (1877–1947) and Ramanujan (1887–1920) proved that actually $p(n) \sim (1/(4\sqrt{3}\,n))e^{\pi\sqrt{2/3}\,\sqrt{n}}$. This result required the use of a "Tauberian Theorem," a rather sophisticated reasoning whereby one draws conclusions concerning summands from a knowledge of the behavior of their sum. Finally, the work of Hardy, Ramanujan and Rademacher (1892–1969) led to an *exact* formula for $p(n)$. In order to obtain it, use had to be made of the theory of functions of a complex variable, the theory of modular functions, and many other analytical devices. The theories involved have been materially stimulated by the investigation of the partition function. In 1943 Erdös obtained by elementary (but not easy) reasoning the above-quoted asymptotic formula of Hardy and Ramanujan for $p(n)$, except for the proof that the outside constant is $1/4\sqrt{3}$; the latter was supplied by D. J. Newman in 1951 (see [6]).

8   ELEMENTARY NUMBER THEORY

In the preceding sections we discussed some classical problems of number theory which have sparked the development of advanced branches of mathematics and whose treatment required the tools furnished by these advanced theories. It would be a great mistake, however, to believe that number theory cannot be studied profitably unless, say, complex variables and abstract algebra have been mastered. On the contrary, it might be argued that the study of number theory should precede that of such theories, in agreement with the actual historical sequence of events. Be that as it may, it is a fact that many interesting and challenging problems can be handled by very simple methods.

These problems form the so-called *elementary number theory*. Many questions concerning primes or partitions can be handled by elementary methods. If it happens that a seemingly difficult and deep problem yields to a particularly simple, elementary reasoning, then we experience that striking sensation of elegance, already alluded to in Section 1. Such is, for instance, Euclid's proof that there are infinitely many primes. It also happens that some of the most fundamental concepts of modern algebra (actually, of all of modern mathematics) such as groups, rings, fields, modules, ideals, to name just a few, are obtained by the processes of abstraction and generalization from situations we meet in elementary number theory. Finally, the prerequisites needed for an understanding of this elementary theory are minimal. All this adds up to making elementary number theory an ideal starting point. Therefore some elementary number theory (divisibility, linear and quadratic congruences, number-theoretic functions) will be presented first; this, together with some rudiments of advanced calculus and the theory of functions of a complex variable, and with some algebra should enable us to study in some detail the three problems mentioned earlier in the present chapter. Our aim will be to obtain in each case, as simply as possible, the most important and characteristic results rather than the strongest known formulation of the corresponding theorems. While we shall keep in mind this ultimate aim, it is hoped that many results obtained on the way will prove interesting and rewarding by themselves.

BIBLIOGRAPHY

1. C. L. Baker and F. J. Gruenberger, *The First Six Million Prime Numbers*, West Salem, Wisc.: Microcard Foundation, 1959.
2. J. P. Kulik (Vienna), Manuscript list of primes up to 10^8 (not sufficiently reliable).
3. D. H. Lehmer, *List of Prime Numbers from 1 to* 10,006,721, Washington: Carnegie Institute of Washington, Publication No. **165**, 1914.
4. L. J. Mordell, *Three Lectures on Fermat's Last Theorem*, Cambridge: Cambridge Univ. Press, 1921.
5. O. Neugebauer, (*Die Grundlehren der Mathematischen Wissenschaften in Einzeldarstellungen*, Vol. **13**). Berlin: Springer, 1937.
6. D. J. Newman, The evaluation of the constant in the formula for the number of partitions of *n*. *Amer J. Math.* 73 (1951), 599–601.
7. *Rhind (or Ahmes) Papyrus*, written presumably 1700–1600 B.C., now in the British Museum.
8. S. S. Wagstaff, The irregular primes up to 125,000. *Mathem. of Computation* 32 (1978), 583–591.

Chapter 2

Introductory Remarks and Notations

The theory of numbers is concerned primarily with the properties of the *natural numbers* $1, 2, 3, \ldots$ and, more generally, with those of the *rational integers* $\ldots, -2, -1, 0, 1, 2, 3, \ldots$. Throughout this book, rational integers will be denoted by lowercase italic letters. The set of all rational integers will be denoted by **Z**. In general, sets of numbers will be denoted by boldface capitals.

The sum, difference, and product of two rational integers are again rational integers, that is, they are again elements of the set **Z**. We recall

DEFINITION 1. Whenever an operation defined on the elements of a set **A** is such that it can be performed unrestrictedly and has as a result again an element of **A**, we say that the set **A** is *closed* under that operation.

According to Definition 1, the set **Z** is closed under addition, multiplication, and subtraction. However, there are many other operations that we may want to perform on numbers, such as division, extraction of roots, solving of equations (with, say, coefficients in **Z**), and so on. Unfortunately, **Z** is not closed even under division, because if a and b are rational integers, it is usually impossible to find a third rational integer c such that $a \div b = c$. Attempts to construct sets that permit the unrestricted performance of the desired operations lead to successive generalizations of the concept of "number" and we shall have to consider *rational numbers* (that is, fractions a/b, where a and b are rational integers, $b \neq 0$), *irrational real numbers* (such as $\sqrt{2}$, e, or π), and even *complex numbers* ($z = x + iy$, x, y, real, $i^2 + 1 = 0$). The set of all rational numbers will be denoted by **Q**, that of all real numbers by **R**, and that of all complex numbers by **C**. The reader is assumed to be familiar with these concepts, as well as with the operations of addition, subtraction, multiplication, division, taking powers, extraction of roots, exponentiation, and taking of logarithms.

Current symbols, such as $=$ (equal to), $\neq$ (not equal to), $>$ (greater than), $\geq$ (greater than or equal to), $<$ (less than), $\leq$ (less than or equal to), $a|b$ (a

divides b), $a \nmid b$ (a does not divide b), $\sum_{n=1}^{N} f(n)$ (summation symbol), and so on, will be used without further explanations. As a matter of fact, we already did use them without explanation, comment, or apology in Chapter 1.

Whenever convenient, we shall also make use of a few symbols from set theory and from logic. We list here the most common symbols.

1. Braces { }; these are used to enclose the elements of a given set. For instance, if **P** stands for the set of all prime numbers, we may write **P** = {2, 3, 5, 7, ...}.

2. The symbol $\in$; it is used to show that an element belongs to a given set; thus $b \in$ **B** means that b is an element of the set **B**. Other examples are $12 \in$ **Z**, $7 \in$ **P**. One uses $\notin$ to show that some element does not belong to a given set, for instance $6 \notin$ **P**.

3. The symbols of inclusion, $\supset$ and $\subset$; **A** $\supset$ **B** means that every element of the set **B** is also an element of the set **A**, or in symbols: $\alpha \in$ **B** implies that $\alpha \in$ **A**. By definition, **A** $\subset$ **B** means the same as **B** $\supset$ **A**. For instance, **Z** $\subset$ **Q** $\subset$ **R** $\subset$ **C**. Observe that **A** $\subset$ **B** and **A** $\supset$ **B** imply **A** = **B**.

4. The symbols $\cup$ (union) and $\cap$ (intersection); **A** = **B** $\cup$ **C** (read **A** *equals the union of* **B** *and* **C**) means that the set **A** consists of all (and only) those elements that belong to either **B** or **C**. For instance, if **B** = {3, 5, 6, 7} and **C** = {3, 6, 8, 9}, then **A** = {3, 5, 6, 7, 8, 9,}. **D** = **B** $\cap$ **C** (read **D** *equals the intersection of* **B** *and* **C**) means that the set **D** consists of exactly those elements that belong to both sets **B** *and* **C**. In above example, **D** = {3, 6}.

5. The quantifier $\exists$ (read *there exists*); for instance, to state that there exists a prime between 14 and 18 we may write $\exists p$, $p \in$ **P**, $14 \le p \le 18$.

6. The symbols of implication $\Rightarrow$ (implies), $\Leftarrow$ (is implied by), $\Leftrightarrow$ (implies and is implied by); the latter one means that the statement preceding the symbol $\Leftrightarrow$ is equivalent to the statement following it. Occasionally we use this symbol to define a new concept; in that case the symbols preceding (or following) $\Leftrightarrow$ are defined to mean the same thing as the (already known) symbols following (or preceding) $\Leftrightarrow$. In order to avoid any possible misunderstanding, we shall write "def" under a double arrow used in a definition: $\underset{\text{def}}{\Leftrightarrow}$.

7. The symbol $\ni$ (read *such that*); sometimes $\ni$ is replaced by |, if there is no danger of confusion with the symbol for "is divisible by." This is particularly useful in the definition of sets, characterized by complicated conditions satisfied by their elements.

More symbols will be introduced as needed, often to be used only in a single chapter, or even a single section.

Finally, a few words concerning the proofs. A bewildering variety of methods is used in number theory. Sometimes the same theorem can be proven by several methods (so it seems, according to Professor M. Gerstenhaber ([1], p. 397) that in 1963 there were 152 different proofs for the quadratic reciprocity formula (see Theorem 5.3)); on the other hand, for some important theorems, there exists essentially only one proof—which occasionally (while correct, and hence fully convincing) is not even really satisfactory. As an example consider the following theorem. If $d - 3$ is divisible by 4 and $(-d/n)$ is the symbol of quadratic residuacy defined in Chapter 5, then $d^{-1}\sum_{n=1}^{d-1}n(-d/n)$ is always a negative integer; I wish I could see arithmetically why! Also, as already mentioned, until recently the proof of the Prime Number Theorem (PNT) required the use of complex variables, although the statement of the Theorem has nothing to do with complex quantities.

Everything else being equal, we shall give preference to direct proofs, using concepts germane to the substance of the theorem; however, we shall not be dogmatic about it. If a proof by induction, or an indirect proof (that is, a proof by *reductio ad absurdum*, or by contradiction) is simpler, we shall not hesitate to use it, even if a direct proof is available. It is assumed, of course, that the reader is familiar with all these types of proofs. Similarly, if a proof that uses concepts not occurring in the statement of a theorem is easier or more transparent than one that avoids them, we shall present the more transparent one. For instance, the PNT will be proven by a simple analytic argument (using complex variables), in preference to the (technically) elementary but rather difficult proof which is now available. In the following chapters, in Part II, we shall discuss some topics of elementary number theory; these are of independent interest and some of them will be needed in order to study the Theory of Partitions, the Riemann Zeta function, the PNT, Diophantine equations, and Fermat's Conjecture, which follow in Part III. Occasionally, topics of elementary number theory which are of intrinsic interest are developed beyond the point actually needed in Part III. The reader who is in a hurry and whose curiosity about one of the three main problems treated in Part III cannot wait, may therefore start by reading only the definitions and statements of theorems in Part II (with special emphasis on Chapter 3, on Divisibility, and Chapter 6, on Arithmetic Functions); he may then go to the chapters that interest him most in Part III. Whenever he feels the need for some specific theorem of Part II, he may go back and try to master it.

However, for the reader interested in becoming thoroughly acquainted with the subject matter, it is recommended that he study this elementary Part II carefully. He should not omit working out the problems of each chapter—all of them, if possible—before proceeding to the next chapter.

Part III consists of three units. The first (Chapters 8 and 9) assumes a moderate acquaintance with advanced calculus and the theory of functions of one complex variable. The second unit (Chapters 10, 11, and 12) assumes a certain knowledge of algebra; however, Chapter 12 also depends on some material from Chapter 8. The last unit (Chapters 13 and 14) requires the same moderate knowledge of algebra, and also depends somewhat on the theory of ideals as discussed in Chapter 11. The three units are largely independent of each other and (except as noted here) may be read in any order.

BIBLIOGRAPHY

1. M. Gerstenhaber, The 152-nd proof of the law of quadratic reciprocity, *Am. Math. Monthly* **70** (1963), 397–398.

Part Two

Elementary
Number Theory

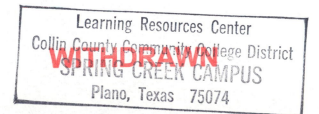

Chapter 3

Divisibility

1 GENERALITIES AND FUNDAMENTAL THEOREM

As already mentioned, we assume that the reader is familiar with the properties of the natural integers and those of the rational integers, as well as with the elementary operations. Given any two rational integers, one can always add, subtract, or multiply them, and again obtain as result an integer. We already observed in Chapter 2 that in general this is no longer the case with the operation of division; hence, the following definition is nontrivial.

DEFINITION 1. An integer a is said to be divisible by an integer $b \neq 0$ if the equation $a = bx$ has a solution with x an integer. If a is divisible by b, we also say that b divides a, or in symbols $b|a$.

We note some immediate consequences of this definition and formulate them in the following:

Theorem 1. *If $a, b, c \in \mathbf{Z}$, then*

(1) $a|0, 1|a, a|a$;
(2) $a|b \Rightarrow a|b \cdot c$;
(3) $a|b, b|c \Rightarrow a|c$;
(4) $a|b \Rightarrow ac|bc$;
(5) $c \neq 0$ and $ac|bc \Rightarrow a|b$;
(6) $a|b_i \ (i = 1, 2, \ldots, r) \Rightarrow a|m_1 b_1 + m_2 b_2 + \cdots + m_r b_r$, for all $m_i \in \mathbf{Z}$;
(7) $a > 0, b > 0, a|b \Rightarrow a \leq b$;
(8) $ab > 0, a|b, b|a \Rightarrow a = b$.

At this point the reader should convince himself that he thoroughly understands the meaning of each symbol used, by translating the hypotheses, as well as each of the eight statements of Theorem 1, into words. The proof of this theorem is left as an exercise for the reader.

In Part I, Definition 1, we have defined the concept of a prime; now our first aim is to prove what is often called the Fundamental Theorem of Arithmetic, stating that the factorization of an integer into primes is essentially unique. However, before we can do this, we must first show that every integer larger than one can be factored into primes at least in *some* way. We state this fact as

Theorem 2. $1 < n \in \mathbf{Z} \Rightarrow n = p_1 p_2 \cdots p_r$.

REMARK 1.   Theorem 2 does *not* assert that the primes $p_1, p_2, \ldots, p_r$ are all distinct.

PROOF (by induction).   The statement is trivially true for $n = 2$, $n = 3$, $n = 5$, and in general for any n that is itself a prime; it also is easily verified for $n = 4 = 2^2$, $n = 6 = 2 \cdot 3$, and so on. Assume that Theorem 2 has already been found to hold for $k = 2, 3, \ldots, n$; then we shall show that the theorem also holds for $k = n + 1$, and hence by induction for all integers. Indeed, either $n + 1$ is a prime p and then $n + 1 = p$, so that the theorem holds with $r = 1$, or else $n + 1 = a \cdot b$, with a and b both less than $n + 1$. Hence, $a \leq n$, $b \leq n$, and by the induction hypothesis $a = p_1 p_2 \cdots p_t$; $b = p_1' \cdots p_s'$, so that $n + 1 = a \cdot b = p_1 \cdots p_t \cdot p_1' \cdots p_s'$, with $p_1, \ldots, p_t, p_1', \ldots, p_s'$ all primes, and the theorem is proven.

The commutativity of multiplication permits us to rearrange the prime factors so that they should be nondecreasing. The product $2 \cdot 5 \cdot 3 \cdot 5 \cdot 2$ ($= 300$), for instance, may be rearranged to read $2 \cdot 2 \cdot 3 \cdot 5 \cdot 5 = 2^2 \cdot 3 \cdot 5^2$. This is called a *standard* or *canonical* factorization of $n = 300$.

DEFINITION 2. $n = p_1^{s_1} p_2^{s_2} \cdots p_r^{s_r}$ is a *standard*, or *canonical* factorization of n if and only if $p_1 < p_2 < \cdots < p_r$ and all exponents are positive integers.
   Now we are able to state

Theorem 3. (Fundamental Theorem of Arithmetic).   *The standard factorization of a natural integer n is unique.*

The proof of Theorem 3 is easy, if we assume for a moment the validity of

Theorem 4. $p \mid a \cdot b, \; p \nmid a \Rightarrow p \mid b$.

Indeed, Theorem 4 may be generalized almost trivially to

Corollary 4.1. $p \mid a_1 a_2 \cdots a_r \Rightarrow \exists i (1 \leq i \leq r) \ni p \mid a_i$. (*In words: if a prime divides a product of integers, then it divides at least one of them.*)

PROOF OF COROLLARY 4.1 (by induction on r).   The statement is trivially true for $r = 1$; for $r = 2$ it is precisely the statement of Theorem 4, which we accept provisionally. We complete the proof by showing that if the corollary holds for $r - 1$ factors, then it also holds for r factors. Indeed, set $a_1 a_2 \cdots a_{r-1} = n$, so that $a_1 a_2 \cdots a_r = n a_r$. Then, by Theorem 4, $p|n a_r \Rightarrow p|a_r$ or $p|n$. In the first alternative, the corollary holds with $i = r$; in the second, $p|n \Rightarrow p|a_1 a_2 \cdots a_{r-1}$ and the corollary holds, by the induction hypothesis, for some i, $1 \le i \le r - 1$.

The proof of Theorem 3, as well as many other proofs to come, becomes neater if we keep in mind the following, which is almost obvious:

REMARK 2.   Every nonempty set of positive integers contains a smallest element.

Indeed, denote the set by **A** and let $a \in$ **A**; such an a exists by the assumption that **A** is nonempty. Next, consider the finite set of positive integers $1, 2, \ldots, a$ and cross out all (finitely many) integers of this set not belonging to **A**. Then some integers will be left standing (a itself, for instance) and the first among those integers that were left is precisely the smallest element of **A**.

It may be mentioned that if in the above considerations the set of positive integers with the *order relation* $\le$ is replaced by some other more complicated sets and order relations, it is not always easy (or even possible) to verify that every subset has a smallest element. A set having this property is called *well ordered*. We are now ready to prove Theorem 3 (always assuming the validity of the as yet unproven Theorem 4).

PROOF OF THEOREM 3 (by contradiction).   If we deny the statement of Theorem 3, then there exists a nonempty set **A** of positive integers, for which the standard factorization is not unique. By Remark 2, there exists a smallest integer in **A**—let it be n. Then, by assumption, $n = p_1 p_2 \cdots p_r = p'_1 p'_2 \cdots p'_s$, where $p_1 \le p_2 \le \cdots \le p_r$ are primes and $p'_1, p'_2, \ldots, p'_s$ are also primes (some, possibly, repeated; we do *not* require the factorization $p'_1 \cdots p'_s$ to be canonical.) From $p_1 | n = p'_1 \cdots p'_s$ and Corollary 4.1 it follows that $p_1 | p'_j$ ($1 \le j \le s$), whence, by the definition of primes, $p_1 = p'_j$. Hence, dividing out this common factor (see Theorem 1[5]) and setting $n_1 = n/p_1$, it follows that $n_1 = p_2 p_3 \cdots p_r = p'_1 p'_2 \cdots p'_{j-1} p'_{j+1} \cdots p'_s$ (the slight modification needed in case $j = 1$ or $j = s$ should be written out in detail by the reader). The factorization $p_2 p_3 \cdots p_r$ is canonical and we may also rearrange the primes p'_k ($k \ne j$) in nondecreasing order, relabeling them, say, as $q_1 \le q_2 \le \cdots \le q_{s-1}$; the unordered set of primes $\{p'_k\}$, $1 \le k \le s$, $k \ne j$ is, of course, identical to the set $\{q_l\}$ ($1 \le l \le s - 1$). We have obtained so far that $n_1 = p_2 p_3 \cdots p_r = q_1 q_2 \cdots q_{s-1}$. These two canonical factorizations of n_1 cannot be the same, because if they were they would lead to a unique factorization for $n = p_1 n_1 = p'_j n_1$, contrary to our assumption. Hence n_1 is an integer with at least two distinct factorizations, $n_1 \in$ **A**. But $n_1 < n$, contrary to the definition of n as *smallest* integer in **A**. This contradiction shows that the set **A** is actually empty, that is, there are *no* positive integers with two distinct canonical factorizations.

In order to complete the proof of Theorem 3, it still remains to prove the key Theorem 4. However, in order to avoid a circular reasoning (*petitio principii*), we *may not* use in this proof of Theorem 4 any results like Corollary 4.1 or Theorem 3, which were obtained precisely by *assuming* Theorem 4. We may, of course, use Theorem 1 and Theorem 2.

In the proof we shall need the concept of *greatest common divisor* and that of a *module*, which we now proceed to define.

DEFINITION 3.   Let $a, b, d \in \mathbf{Z}$; if $d|a$, then d is called a *divisor* of a; if $d|a$ and $d|b$, then d is called a *common divisor* of a and b. The largest positive common divisor of a and b is called their *greatest common divisor* (g.c.d.) and is denoted by $d = (a, b)$. Two integers a and b, such that $(a, b) = d = 1$ holds, are called *relatively prime*, or *coprime*.

Having defined divisors and common divisors, we shall also give here the definition of multiples and common multiples, although these will not be needed immediately.

DEFINITION 4.   Let $a, b, m \in \mathbf{Z}$; if $a|m$, then m is called a *multiple* of a; if $a|m$ and $b|m$, then m is called a *common multiple* of a and b. The smallest positive common multiple of a and b is called their *least common multiple* (l.c.m.) and is denoted by $[a, b]$.

For many purposes it is more convenient to define $d = (a, b)$ as follows:

DEFINITION 3*. $d = (a, b) \underset{\text{def}}{\Leftrightarrow} d \in \mathbf{Z}, d > 0, d|a, d|b$, and $c \in \mathbf{Z}, c|a, c|b \Rightarrow c|d$.

The concepts of g.c.d. and l.c.m. may be generalized as follows:

DEFINITION 3'. d is called the g.c.d. of the integers $a_1, a_2, \ldots, a_r$—in symbols $d = (a_1, a_2, \ldots, a_r)$—provided that

  (i) $d|a_i$ $(i = 1, 2, \ldots, r)$;
  (ii) $d > 0$;
  (iii) $c|a_i$ $(i = 1, 2, \ldots, r) \Rightarrow c|d$.

One may observe that Definition 3' is the direct generalization of Definition 3* rather than of Definition 3.

DEFINITION 4'. m is called the l.c.m. of the integers $a_1, a_2, \ldots, a_r$—in symbols $m = [a_1, a_2, \ldots, a_r]$—provided that

  (i) $a_i|m$ $(i = 1, 2, \ldots, r)$;
  (ii) $m > 0$;
  (iii) $a_i|m_1$ $(i = 1, 2, \ldots, r) \Rightarrow m_1 \geq m$.

Before we proceed further, we have to show first that, given any two rational integers a and b, there exists an integer d with the properties required by Definition 3*, and next, that this integer is the same as $d = (a, b)$ of Definition 3. In order to do that, we shall introduce one more concept, namely that of a *module*. This is not strictly indispensable in the present case, yet the proofs based on this concept are at least as easy as any others available and have the added advantage that they generalize most readily to more difficult situations —including some where even the concept of g.c.d. does not exist (see Section 3). Until we prove the identity of Definitions 3 and 3*, it will be understood that $d = (a, b)$ is the g.c.d. according to Definition 3.

DEFINITION 5.   A *module* **S** is a set of numbers such that for any two elements of **S**, their difference also belongs to **S**.

REMARK 3.   Although the elements of a module need not be integers, in what follows we shall be concerned exclusively with modules[†] consisting of rational integers.

REMARK 4.   The set **S** consisting of zero alone—in symbols **S** = {0}—satisfies the definition of a module. In what follows we shall ignore this trivial module without mentioning it explicitly every time. Hence whenever we speak of a module, we shall assume that it contains at least one element $a \neq 0$.

From the assumptions $0 \neq a \in$ **S** and the defining property of a module it follows that $0 = a - a \in$ **S**; hence

$$a \in \mathbf{S} \Rightarrow -a = 0 - a \in \mathbf{S}.$$

Therefore,

$$a \in \mathbf{S}, \quad b \in \mathbf{S} \Rightarrow -b \in \mathbf{S} \Rightarrow a + b = a - (-b) \in \mathbf{S},$$

and not only the difference but also the sum of any two elements of **S** belongs to **S**. Furthermore, $a \in \mathbf{S} \Rightarrow 2a = a + a \in \mathbf{S}$ and, by induction, $a \in \mathbf{S} \Rightarrow na \in \mathbf{S} \Rightarrow -na \in \mathbf{S}$ for every natural integer n. This suggests the following:

Theorem 5.   *A (nontrivial) module* **S** *of rational integers consists precisely of the multiples of a fixed positive integer d, that is,*

$$\mathbf{S} = \{nd\}.$$

PROOF.   By assumption, $\exists a \ni 0 \neq a \in \mathbf{S}$; also, as seen, $-a \in \mathbf{S}$; but $a \neq 0 \Rightarrow a \neq -a$, and either a or $-a$ is positive. It is only a matter of notation to say that $a > 0$. Hence the set of positive integers belonging to **S** is not empty, and by Remark 2 there exists a smallest positive integer in **S**; let us denote it

[†] The correct plural is of course *moduli*; however, at least in the United States, this Latin plural has acquired a rather pedantic ring. Therefore, we shall follow the common custom and use the plural *modules*. In Great Britain, most authors do write *moduli*.

by d. We already know that with d every integer of the form nd (n any rational integer) also belongs to S, so that $\{nd\} \subset S$. We now want to prove the opposite inclusion $S \subset \{nd\}$; that is, we want to show that if $m \in S$, then there exists a rational integer n, such that $m = nd$. Because $m \in S \Leftrightarrow -m \in S$, it is sufficient to consider only the case $m > 0$ (hence $n > 0$). If $m \neq nd$ for all n, then there exists an integer k such that $kd < m < (k + 1)d$, or $0 < m - kd < d$. The positive integer $r = m - kd$ belongs to S, being the difference of the two elements m and kd of S. However, $r = m - kd < d$, contradicting the definition of d as *smallest* positive integer in S. This contradiction proves that $m = nd$; hence $S \subset \{nd\}$ and $S = \{nd\}$ as asserted.

REMARK 5. It might be worthwhile to remark that in the proof of Theorem 5 we have made use of the "Archimedean Axiom": Given any two positive numbers m (arbitrarily large) and d (arbitrarily small), it is always possible to find a positive integer k such that $m < (k + 1)d$.

Theorem 6. *Let a and b be two given rational integers, and let m and n run independently through the set $\mathbf{Z}$ of rational integers. Then $S = \{am + bn\}$ is a module; actually $S = \{kd\}$, where $d = (a, b)$.*

We recall that here $d = (a, b)$ is the g.c.d. of a and b according to Definition 3.

PROOF OF THEOREM 6. It is trivial to verify that S is a module; it remains to prove the second statement of the Theorem. By Theorem 5, $S = \{kf\}$ for some positive integer f. Now, by its definition, $d|a$ and $d|b$; hence by Theorem 1(6), $d|am + bn$ and every element kf of S is divisible by d. In particular, $d|f$, whence, by Theorem 1(7), $d \leq f$. Selecting now $m = 1$, $n = 0$, it follows that $f|a$; similarly, taking $m = 0$, $n = 1$, it follows that $f|b$. Hence f is a common divisor of a and b, so that, by Definition 3, $f \leq d$. We already obtained the inequality $d \leq f$ so that $d = f$ and Theorem 6 is proven.

Corollary 6.1. $d = (a, b) \Rightarrow \exists m, n \ni ma + nb = d$.

Corollary 6.2. $S = \{am + bn\}, (a, b) = 1 \Rightarrow S = \mathbf{Z}$.

Corollary 6.3. $(a, b) = 1 \Leftrightarrow \exists m, n \ni ma + nb = 1$.

Corollary 6.4. $c|a, c|b \Rightarrow c|am + bn \Rightarrow c|d$.

The ambitious reader may want to prove these Corollaries by himself. It is suggested that he write out the proofs completely, and compare them afterwards with those given here.

PROOFS OF COROLLARIES. The sets $\{am + bn\}$ and $\{kd\}$ are identical by Theorem 6; hence for every rational integer k there exist rational integers m, n such that $ma + nb = kd$. Taking in particular $k = 1$ yields Corollary 6.1. Next, setting $d = 1$ in Theorem 6 reduces the module $\mathbf{S} = \{kd\}$ to $\{k\} = \mathbf{Z}$, proving Corollary 6.2. Corollary 6.3 follows from Corollary 6.1 with $d = 1$. The first implication of Corollary 6.4 follows from Theorem 1(6), the second from Corollary 6.1.

REMARK 6. It follows from Theorem 6 that $d = (a, b)$ could also be defined as the smallest positive element of the module $\mathbf{S} = \{am + bn\}$.

Corollary 6.4 shows that the g.c.d. of a and b, as defined by Definition 3, has the property that every common divisor c of a and b also divides d; from this and Theorem 1(7) it immediately follows that every common divisor c of a and b satisfies $c \le d$. This proves both the meaningfulness of Definition 3* and its identity with Definition 3. After this extended preparation the proof of our key Theorem 4 comes almost as an anticlimax.

PROOF OF THEOREM 4. Assume $p|a \cdot b$, $p \nmid a$. Then $(p, a) = 1$, and by Corollary 6.3, $\exists m, n \ni mp + na = 1$, or $mpb + nab = b$. Now $p|p \cdot b$ trivially, and $p|a \cdot b$ by assumption; hence by Theorem 1(6), $p|mpb + nab = b$, as we wanted to prove. At this point the reader may want to look back and verify that we were not guilty of any circular reasoning. Indeed, using only the definitions of the g.c.d. and of a module, the Archimedean Axiom, and Remark 2 we proved Theorem 5; next, using Theorems 1 and 5 we proved Theorem 6. From Theorem 6 and Theorem 1 the Corollaries to Theorem 6 followed, and Theorem 4 then followed from Corollary 6.3 and Theorem 1. Theorem 4 is easily generalized to any number of factors as Corollary 4.1; finally, using Corollary 4.1, Theorem 2, and Remark 2 we proved Theorem 3.

2 DISCUSSION OF TWO OBJECTIONS

Two objections might have arisen in the reader's mind:

(a) Why was the material not arranged following the outline in the last paragraph?
(b) Why waste so much time and effort in order to prove the "obvious" statement of Theorem 4?

Objection (a) has its merits. Indeed, the proof of Theorem 4 *could* have been arranged differently. For such a rearranged proof, using essentially the same ingredients as the present one, the reader may want to look into the first proof of that theorem in the book by Niven and Zuckerman [6, page 20]. In fact, the reader who has taken the trouble to look into [6] may find it enlightening to also consider the second proof of this theorem, on the same page. This proof uses somewhat different ideas, but is deceptively simple. On the other hand, the proof of the important Theorem 3 is indeed very easy, if one is prepared to accept the plausible—not to say obvious—statement of Theorem 4. This fact was borne out by actually proving Theorem 3 in a few lines, without stopping for a proof of Theorem 4. Still, on second thought, one may want to have a clear conscience and also give a formal proof of the simple statement of Theorem 4; surely the proof of such an "obvious" fact could not be too hard! It may have come as something of a surprise to some readers that it actually took us several definitions, two preparatory theorems (Theorems 5 and 6), and quite a few remarks and corollaries, requiring no less than four pages, to prove Theorem 4. But this already leads us to discuss:

Objection (b). The answer to this objection is that the apparently obvious statement of Theorem 4 is actually far from obvious. In fact, under assumptions only slightly different from those underlying Theorem 4 ($n \in \mathbf{Z}$ being replaced by $n \in \mathbf{A} \supset \mathbf{Z}$, where $\mathbf{A}$ is a set whose structure is only slightly more complicated than that of $\mathbf{Z}$ (both are *rings*), the statement actually becomes false. Hence in order to have a chance to prove Theorem 4, we have to make use of all available information, and exploit thoroughly the advantage of knowing that $n \in \mathbf{Z}$, by using the properties of the rational integers recorded in Theorem 1. This explains the length of the proof of Theorem 4. This consideration should also go a long way toward a justification of the effort spent in proving Theorem 4. Actually, one of the main difficulties in the study of *algebraic integers* (the set called **A** above) is due precisely to the absence of a theorem like our present Theorem 4, and for reasons of just this kind, it has been impossible so far to prove, among others, Fermat's Conjecture. We shall learn more about this topic in Chapters 10–14. (Challenge to the ambitious reader: After becoming familiar with algebraic integers, return to this point and observe that the second proof in [6] seems to work, at least for rings of real algebraic integers, although the statement itself does not hold in general. Where does the proof break down? Hint: Consider the use made of the well ordering of the ordinary integers.) Meanwhile, just to illustrate our point, let us consider here an example of a set of integers where Theorem 3 does *not* hold. This example is due to Hilbert, and while it may be of only limited intrinsic interest, it has the merit of being easily understood. It is hoped that any reader who will have given some thought to this example will be ready to agree that even the most plausible-looking statements (such as Theorem 4, for instance) should be proven formally before they are accepted.

3   AN EXAMPLE OF HILBERT

Let us consider the set $\mathbf{H} = \{h\}$, consisting of precisely those positive rational integers that are of the form $h = 4n + 1$, $0 \le n \in \mathbf{Z}$. When these integers are divided by 4, they leave a remainder equal to one. Such integers are said to be *congruent to one modulo 4*. In symbols this is usually written $h \equiv 1 \pmod 4$; hence $\mathbf{H} = \{ h \in \mathbf{Z} \mid h \equiv 1 \pmod 4, h > 0 \}$.

Writing out the first few elements, we have

$$\mathbf{H} = \{1, 5, 9, 13, 17, 21, 25, 29, 33, 37, 41, 45, 49, 53, 57, \ldots \}.$$

We shall learn more about congruences in Chapter 4; however, we may already observe that the set $\mathbf{H}$ is closed under multiplication. Indeed, $h_1, h_2 \in \mathbf{H} \Rightarrow h_1 = 4n_1 + 1, h_2 = 4n_2 + 1, n_1, n_2 \in \mathbf{Z} \Rightarrow h_1 h_2 = 4(4n_1 n_2 + n_1 + n_2) + 1 = 4n + 1 \in \mathbf{H}$, because $0 < n = 4n_1 n_2 + n_1 + n_2 \in \mathbf{Z}$; hence multiplication is well defined on $\mathbf{H}$. We also observe that some elements h of $\mathbf{H}$ can be represented as products of other elements of $\mathbf{H}$, all smaller than h, while others cannot be so represented. For instance, $25 = 5 \cdot 5$, $5 \in \mathbf{H}$, but 5 cannot be split in this way; nor can 9, because $3 \notin \mathbf{H}$, nor can 21, because $3 \notin \mathbf{H}, 7 \notin \mathbf{H}$. Let us agree to call (*Hilbert*)-*primes* those elements $p > 1$ of $\mathbf{H}$ that cannot be represented as products of elements of $\mathbf{H}$, all smaller than p. Denoting (only in this section) the set of these p by $\mathbf{P}$, we have $\mathbf{P} = \{5, 9, 13, 17, 21, 29, 33, 37, 41, 49, 53, 57, 61, 69, 73, 77, \ldots \}$. The other elements $h > 1$ of $\mathbf{H}$ may be called *composite*. The first few composite elements of $\mathbf{H}$ are $\{25, 45, 65, 81, 85, 105, \ldots \}$. Looking at these numbers, we may check that $25 = 5 \cdot 5$, $45 = 5 \cdot 9$, $65 = 5 \cdot 13, \ldots$, so that we are inclined to believe that Theorem 4 (hence, also Theorem 3) holds for $\mathbf{H}$ just as for $\mathbf{Z}$. Consider, however, $h = 693$. Clearly, $693 = 4 \cdot 173 + 1 \in \mathbf{H}$. Also, $693 = 21 \cdot 33 = 9 \cdot 77$, and we observe that all four factors 33, 21, 9, and 77 not only belong to $\mathbf{H}$, but actually even to $\mathbf{P}$. They are all "primes" and both factorizations are canonical. This shows that Theorem 3 does not hold for $\mathbf{H}$. Then Theorem 4 cannot hold either, because otherwise Theorem 3 (which is an immediate consequence of Theorem 4) would also hold, and it does not.

It is not hard to find the origin of the difficulty: This is due to the fact that we refuse to admit as factors the integers 3, 7, and 11, because these do not belong to $\mathbf{H}$. Indeed if we could "adjoin" them in some way to $\mathbf{H}$ we would obtain $693 = 21 \cdot 33 = 3 \cdot 7 \cdot 3 \cdot 11 = 3 \cdot 3 \cdot 7 \cdot 11 = 9 \cdot 77$, and the essential uniqueness of the factorization would again be saved.

*One final word concerning a way to "adjoin" such "missing elements" to a given set of numbers: One observes that these missing elements, 3, 7, 3, 11, are precisely the g.c.d. (in $\mathbf{Z}$) of the pairs of the factors of 693 in the two factorizations $3 = (21, 9)$, $7 = (21, 77)$, $3 = (33, 9)$, $11 = (33, 77)$. These g.c.d. do not belong to $\mathbf{H}$; but we remember that to any two integers a, b with $d = (a, b)$ there corresponds a module $\mathbf{S} = \{ ma + nb \}$, all of whose elements are multiples of d. Such modules can also be defined over $\mathbf{H}$ (now $m, n \in \mathbf{H}$) instead of over $\mathbf{Z}$, and one still may denote the module corresponding to say,

33 and 9, by (33,9), regardless of whether $d \in \mathbf{H}$ or $d \notin \mathbf{H}$. In fact, if $d = 3 \notin \mathbf{H}$, then the module (33,9), all of whose elements *do* belong to $\mathbf{H}$, succeeds quite effectively to replace in $\mathbf{H}$ in some sense the missing element $d = 3$.

4   TWO FURTHER THEOREMS

Before we change our subject, we shall state and prove two further results that will be needed later.

Theorem 7. $c|ab, (c, a) = 1 \Rightarrow c|b$.

PROOF.   Same as that of Theorem 4.

Theorem 8. $a|n, b|n, m = [a, b] \Rightarrow m|n$.

PROOF.   Consider the module $\mathbf{S} = \{km\}$. If $n \in \mathbf{S}$, then the theorem is proven. Otherwise, by the Archimedean Axiom, $\exists k \ni km < n < (k + 1)m$, or $0 < n - km < m$. If we denote $n - km$ by r, we have $a|n, a|m \Rightarrow a|r$; similarly $b|r$. Hence $r = n - km$ is a positive common multiple of a and b, less than m; but m is by definition the *least* positive common multiple. This contradiction shows that $n \in \mathbf{S}$; hence, indeed $m|n$.

5   SOME RESULTS CONCERNING THE DISTRIBUTION OF PRIME NUMBERS

We already saw in Part I that the problem of the distribution of primes has fascinated the minds of men at least since the Greek antiquity. As already mentioned, some of the deepest methods of analysis and algebra have been brought to bear upon problems arising in this connection—and not always with full success. On the other hand, some of the most important results can be obtained with surprisingly simple reasonings—and were actually known to Euclid. We shall finish this chapter by proving two theorems concerned with the theory of prime numbers and shall come back to this topic in Chapters 8 and 9 for a renewed attack with more powerful weapons.

Theorem 9 (Euclid).   *There exist infinitely many primes.*

The great importance and interest of this theorem could hardly escape the reader; therefore, we pass without further comment to its surprisingly simple proof.

PROOF OF THEOREM 9 (by contradiction).   Let us assume that the set $\mathbf{P} = \{\, p \,\}$ of primes[†] is finite. Let r be the exact number of its elements, and let us list them in increasing order so that $p_1 = 2$, $p_2 = 3$, $p_3 = 5,\ldots$, up to p_r the last (and largest) prime. The set $\mathbf{P}$ of r primes is once more assumed to contain *all* existing prime numbers. Consider now the integer

$$n = p_1 p_2 \cdots p_r + 1 = 2 \cdot 3 \cdot 5 \cdots p_r + 1.$$

We know (see Theorem 2) that n can be factored into primes, $n = q_1 q_2 \cdots q_k$ ($q_j \in \mathbf{P}$, $1 \leq j \leq k$) say, with $k \geq 1$. (Actually $k > 1$, because $k = 1 \Rightarrow n = q_1 > p_r$, contrary to the definition of p_r as largest prime, but this side remark is irrelevant for the proof.) One has $q_1 | n$; also q_1 being a prime, $q_1 \in \mathbf{P}$. Hence, $q_1 = p_j$ for some $j, 1 \leq j \leq r$. Consequently $q_1 | p_1 p_2 \cdots p_r$, and hence by Theorem 1(6), $q_1 | n - p_1 p_2 \cdots p_r = 1$, contrary to the definition of a prime. This contradiction proves that no finite set $\mathbf{P}$ can contain all prime numbers.

Theorem 10.   *There exist arbitrarily large gaps between consecutive primes.*

PROOF.   We denote the product of the first k consecutive integers $1 \cdot 2 \cdot 3 \cdot \cdots \cdot k$ by $k!$ (read k factorial) and observe that $k! + 2$ is divisible by 2, $k! + 3$ is divisible by 3, and in general for $2 \leq q \leq k$, $k! + q$ is divisible by q. Hence, none of these $k - 1$ consecutive integers is a prime; k, however, may be taken arbitrarily large and the theorem is proven.

We can do still better. Indeed, the integers $k! - q(2 \leq q \leq k)$ are also composite, for the same reason as $k! + q$, and clearly $k!$ is composite, too. Concerning the two integers $k! \pm 1$, however, we can say nothing. Either one or both or neither may be a prime. If both are primes, we have an instance of the so-called "twin primes," mentioned in Section 5 of the Introduction. These primes occur very rarely (there might even be only a finite number of them; see the end of this chapter). If exactly one of the integers $k! \pm 1$ is a prime, then there is an unbroken sequence of at least $k - 1$ composite numbers, isolating this prime from the preceding prime as well as from the following prime. Such primes with large gaps before and after them are the "isolated primes" mentioned in Section 5 of the Introduction. If $k! \pm 1$ are both composite, then we have an unbroken sequence of at least $2k + 1$ composite integers (from $k! - k$ to $k! + k$ inclusive), separating consecutive primes $p < k! < p'$.

Taking for instance $k = 6$, $k! = 720$, $k! - 1 = 719$ is a prime, while $k! + 1 = 721 = 7 \cdot 103$ is composite. The prime preceding 719 is 709, and the prime following 719 is 727, leading to the "prime-free" gaps $727 - 719 = 8$ and

---

[†] It should hardly be necessary to recall that $\mathbf{P}$ stands here for the set of all rational primes, as defined in Chapter 1 (Definition 1.1), and should not be confused with the "Hilbert primes" of Section 3.

$719 - 709 = 10$ respectively, both larger than the minimal gap $k - 1 = 6 - 1 = 5$. Similarly, for $k = 7$, $k! - 1 = 7! - 1 = 5039$ is a prime, while $k! + 1 = 7! + 1 = 5041 = 71^2$ is composite, and the primes nearest to 5039 are $5023 < 5039 < 5051$, with gaps of 16 and 12 respectively, both of course larger than $k - 1 = 6$. In the case $k = 3$, however, $k! + 1 = 7$ and $k! - 1 = 5$ are a pair of twin primes. Finally, for $k = 5$, $k! + 1 = 121 = 11^2$ and $k! - 1 = 119 = 7 \cdot 17$ are both composite; the nearest primes are 113 and 127 with a gap of $127 - 113 = 14 > 2k + 1 = 11$.

*Actually, much stronger theorems are known, but the methods needed for the proofs are no longer elementary. Therefore, we shall only mention some results without proofs, and refer the interested reader to the pertinent literature.

Given an arbitrary positive integer g let us count the number of primes less than some fixed integer x and isolated (on both sides) by gaps no less than g; the result is that "almost all" primes are so isolated. This means that the ratio between the number of primes $p \leq x$, with at least one of the two gaps less than g, to the total number of primes $p \leq x$ can be made arbitrarily small if only x has been selected sufficiently large (see [8], p. 164).

Even this statement is only a particular case of a still more general one, where one counts sequences of $r + 1$ consecutive primes $p_i < p_{i+1} < \cdots < p_{i+r}$, so that all gaps between them should be $\geq g$. It can be shown that "almost all" sets of $r + 1$ consecutive primes less than some fixed x satisfy this condition; here "almost all" has the same meaning as in the preceding statement, which clearly is nothing but the particular case $r = 2$ of the present one. Still stronger results are known, but we shall not pursue the matter further (for literature, besides [8] see also [2], [3], [7], [10], and [11]).

Having seen that there are arbitrarily large gaps between consecutive primes and that these occur infinitely often, one may also ask questions in the opposite direction. So for instance one may wonder how often the *smallest* possible difference between consecutive primes may occur. All primes $p > 2$ being odd, this difference is even (except for $p_2 - p_1 = 3 - 2 = 1$); hence $p_{i+1} - p_i = 2$ is the smallest possible difference and the problem may be rephrased as a question concerning the frequency of twin primes less than a given (large) x. This is a famous unsolved problem. As mentioned earlier, it is not even known whether there are infinitely many or only finitely many pairs of twin primes. In 1919, Viggo Brun showed (see [1]), using the "sieve method" he invented essentially for this purpose, that if there are infinitely many twin primes, the sum of their reciprocals converges. The relevance of this result comes from the fact that the series $\Sigma(1/p)$ formed with the reciprocals of all rational primes *diverges*, as we shall prove in Chapter 8 (Corollary 8.2.2.) Hence, in a sense, there cannot be "too many" twin primes—even if the set of twin primes *is* infinite. A heuristic argument of Hardy and Littlewood (see [4];

also [5] (Appendix) and [9]) actually leads to the conclusion that there are about $cx/(\log^2 x)$ pairs of twin primes p, $p + 2$, with $p \leq x$ (and where $c = 2\Pi_{p \geq 3}(1 - (p - 1)^{-2}) = 1.3203236\ldots$). If this is correct, then there are indeed infinitely many twin primes, but while the argument is highly plausible, it has not yet been possible to tighten it into a proof, and the problem is still open.

Accepting provisionally the infinity of twin primes, we have on one hand infinitely often primes that are crowded closely together, and on the other hand isolated primes, strewn among the integers as thinly as we want. This illustrates the irregularity in the distribution of primes. For a study of the regularities of this distribution, we shall have to reconsider the whole problem with analytic tools, and shall do this in Chapters 8 and 9.

PROBLEMS

1. Prove in detail all statements of Theorem 1.

2. Prove that for any two integers a, b with $b > 0$, there exist integers q and r with $0 \leq r < b$, such that $a = bq + r$.

3. (Euclidean algorithm) To find $d = (a, b)$, we may proceed as follows: We apply successively the result of Problem 2 and find a sequence of couples (q_k, r_k), $0 \leq r_k < r_{k-1}$, starting with

$$a = bq_1 + r_1,$$
$$b = r_1 q_2 + r_2,$$
$$r_1 = r_2 q_3 + r_3, \ldots,$$
$$r_{k-1} = r_k q_{k+1} + r_{k+1}, \ldots.$$

   Prove that after a finite number of steps we obtain a remainder $r_m = 0$, and that $r_{m-1} = (a, b) = d$.

4. Prove that if $(a, b) = d$, then $(ka, kb) = kd$.

5. Prove that if $(a, b) = 1$ and $(b, c) = 1$, then $(ac, b) = 1$.

6. Let $m = [a, b]$ denote the least common multiple of a and b. Prove that if $(a, b) = 1$, then $m = a \cdot b$; more generally, if $(a, b) = d$, then $m = a \cdot b \mid d$.

7. Given $a, b \in \mathbf{Z}$ with $(a, b) = d$, suppose that for $m, n \in \mathbf{Z}$ one has $am + bn = f$; does it follow that $f = d$? (See Corollary 6.3; can you generalize it?)

8. Prove that the product of any three consecutive integers is divisible by 6 and if the first integer is even, then the product is divisible by 24.

9. Prove that if $d = (a, b, c)$, then there exist integers m, n, k, such that $ma + nb + kc = d$.

10. Let $a = p_1^{a_1} p_2^{a_2} \cdots p_r^{a_r}$, $b = p_1^{b_1} p_2^{b_2} \cdots p_r^{b_r}$ (p_i are primes that divide either a or b or both, $a_i \geq 0$, $b_i \geq 0$, $i = 1, 2, \ldots, r$) and set

$$c_i = \min(a_i, b_i), \qquad d_i = \max(a_i, b_i).$$

Prove that

$$d = (a, b) = p_1^{c_1} p_2^{c_2} \cdots p_r^{c_r} \quad \text{and} \quad m = [a, b] = p_1^{d_1} p_2^{d_2} \cdots p_r^{d_r}.$$

11. Find the greatest common divisor and the least common multiple of 693 and 144.

12. Prove that the diophantine equation $ax + by = c$ has solutions in intgers x, y if and only if $d|c$, where $d = (a, b)$.

13. Consider the system $\mathbf{H} = \{h \in \mathbf{Z} | h \equiv 1 \pmod 4, h > 0\}$ described in Section 3.

    (a) Find the two smallest positive integers h for which Theorem 3 fails.

    (b) Find $p, a, b \in \mathbf{H}$, $p \nmid a$, $p \nmid b$, but $p | a \cdot b$ (so that Theorem 4 fails; here p means a "Hilbert"–prime as defined in Section 3).

14. Give a formal proof of the fact that $k! \pm q$ is composite for $2 \leq q \leq k$.

15. Find a sequence of at least 20 consecutive composite integers.

BIBLIOGRAPHY

1. V. Brun, La série $(1/5) + (1/7) + (1/11) + (1/13) + \ldots$ est convergente ou finie. *Bull. des Sciences Math.* (2) **43** (1919) 100–104, 124–128.

2. P. Erdös, On the difference of consecutive primes. *Quarterly J. of Math.* (Oxford) **6** (1935) 124–128.

3. P. Erdös, On some applications of Brun's method. *Acta Sci. Math. Szeged* **13** (1949) 57–63.

4. G. H. Hardy and J. E. Littlewood, Some Problems of Partitio Numerorum III. *Acta Math.* **44** (1923) 1–70.

5. G. H. Hardy and E. M. Wright, *An Introduction to the Theory of Numbers*, 3rd ed. Oxford: Clarendon Press, 1954.

6. I. Niven and H. S. Zuckerman, *An Introduction to the Theory of Numbers*, 4th ed. New York: Wiley, 1980.

7. K. Prachar, Ueber ein Resultat von Walfisz. *Monatshefte für Mathem.* **58** (1954) 114–116.

8. K. Prachar, *Primzahlverteilung* (*Die Grundlehren der Math. Wiss. in Einzeldarst.*, Vol. **91**). Berlin: Springer, 1957.

9. D. Shanks, *Solved and Unsolved Problems in Number Theory*, Washington, D.C.: Spartan Books, 1962.

10. W. Sierpinski, Remarques sur la répartition des nombres premiers. *Colloqu. Math.* **1** (1948) 193–194.

11. A. Walfisz, Stark isolierte Primzahlen. *Doklady Akad. Nauk SSSR* **90** (1953) 711–713.

Chapter 4

Congruences

1 CONGRUENCES AS EQUIVALENCE RELATIONS — GENERAL PROPERTIES

We already met with the concept of "congruence" in Section 3.3. In this section we are going to explore it more systematically.

DEFINITION 1.

$$a \equiv b \,(\mathrm{mod}\ m) \underset{\mathrm{def}}{\Leftrightarrow} m|a - b \qquad \text{(read } a \text{ congruent to } b \text{ modulo } m\text{)}.$$

$$a \not\equiv b \,(\mathrm{mod}\ m) \underset{\mathrm{def}}{\Leftrightarrow} m \nmid a - b \qquad \text{(read } a \text{ not congruent—or incongruent—}$$
$$\text{to } b \text{ modulo } m\text{)}.$$

If the "modulus" m is known and there is no danger of confusion, then one may write simply $a \equiv b(m)$ or even simply $a \equiv b$ and similarly $a \not\equiv b(m)$ or simply $a \not\equiv b$.

Theorem 1. *Modulo any integer m, the following congruences are equivalent* (*that is, each implies and is implied by each one of the other three*): $a \equiv b$, $b \equiv a$, $b - a \equiv 0$, $a - b \equiv 0$.

PROOF. Left to the reader.

DEFINITION 2. If x is an integer and $x \equiv b \,(\mathrm{mod}\ m)$, then b is said to be a *residue of x mod m*. If $0 \le b < m$, then b is called a *least positive* residue[†] of x mod m; if $-m/2 < b \le m/2$, then b is called a *least* or *absolutely least* residue of x mod m.

[†] See footnote to Definition 7.

DEFINITION 3.   A set of integers is called a *complete set of residues* (mod m), if no two of them are congruent (mod m) and if every rational integer is congruent to one of them (mod m).

EXAMPLE

$\{7, 8, -5, 10, 18, 40, 48\}$ forms a complete set of residues (mod 7).

Integers that have the same residue with respect to a given modulus stand in a simple relation to each other. While the reader is presumably familiar with the meaning of this sentence, and in general with equivalence relations, for convenience we recall here a few pertinent definitions, followed by some examples.

DEFINITION 4.

(i) Given a set S of elements (not necessarily integers), any set $\mathscr{R} = \{(\alpha, \beta)\}$ of ordered pairs, $\alpha \in S$, $\beta \in S$ is called a (binary) *relation*. If $\alpha \in S$, $\beta \in S$ and $(\alpha, \beta) \in \mathscr{R}$, we say that α is in relation $\mathscr{R}$ to β and write $\alpha \mathscr{R} \beta$.

(ii) A relation $\mathscr{R}$ among the elements of a set S is said to be *reflexive* if $\alpha \in S \Rightarrow \alpha \mathscr{R} \alpha$; $\mathscr{R}$ is said to be *symmetric* if $\alpha, \beta \in S$, $\alpha \mathscr{R} \beta \Rightarrow \beta \mathscr{R} \alpha$; finally, $\mathscr{R}$ is said to be *transitive* if $\alpha, \beta, \gamma \in S$, $\alpha \mathscr{R} \beta$, $\beta \mathscr{R} \gamma \Rightarrow \alpha \mathscr{R} \gamma$.

(iii) A relation $\mathscr{R}$ which is reflexive, symmetric, and transitive is said to be an *equivalence relation*.

(iv) If $\mathscr{R}$ is an equivalence relation and $\alpha \mathscr{R} \beta$, then α is said to be equivalent to β under $\mathscr{R}$.

EXAMPLES

1. Ordinary equality, $a = b$. We check that $a = a$, $a = b \Rightarrow b = a$, and $a = b$, $b = c \Rightarrow a = c$; hence ordinary equality is reflexive, symmetric, and transitive; it is an equivalence relation. Actually, the general concept of an equivalence relation arises by abstracting these key properties of ordinary equality.

2. The relation "less than," $a < b$. Clearly, $a < b$, $b < c \Rightarrow a < c$, and the relation is transitive. However, $a < a$ is false and $a < b$ does not imply $b < a$; hence, this relation is neither reflexive nor symmetric.

3. The relation "divides," $a|b$. By Theorem 3.1(1) and 3.1(3) $a|a$ and $a|b$, $b|c \Rightarrow a|c$; hence, this relation is reflexive and transitive. However, $a|b$ does *not* imply $b|a$; hence this relation is not symmetric.

4. The relation "to be the brother of" is transitive, but not reflexive, nor is it symmetric. (The fact that John is the brother of Mary does not imply that Mary is the brother of John.)

5. The relation "to be the son of" is not reflexive; nor is it symmetric or transitive.
6. The relation "to look exactly alike" is reflexive, symmetric, and transitive; hence it is an equivalence relation. However,
7. The relation "to resemble (somewhat)," while reflexive and symmetric, is not transitive: It is possible that Peter resembles John (somewhat) and John resembles (somewhat) Dick, but that Peter does not appear to resemble Dick at all.

Theorem 2. *For every integer m, the congruence modulo m is an equivalence relation.*

PROOF. By Definition 1 and Theorem 1, modulo any integer m, $a \equiv a$; also $a \equiv b$ implies $b \equiv a$ so that it only remains to check the transitivity: If $a \equiv b$ and $b \equiv c$, then

$$a - b = km, b - c = lm;$$

hence

$$a - c = (a - b) + (b - c) = (k + l)m \Rightarrow m|a - c \Leftrightarrow a \equiv c,$$

and the proof is complete.

Each equivalence relation among the elements of a given set **S** leads to a partition of these elements into "equivalence classes." Formally, we have

DEFINITION 5. Given a set **S** and an equivalence relation $\mathscr{R}$ defined on **S**, all elements equivalent under $\mathscr{R}$ to a given one are said to form[†] an equivalence class.

EXAMPLE

In a box there are 24 marbles, of which 10 are red, 5 green, and 9 yellow. The relation "to have the same color" is easily seen to be an equivalence relation (see Example 6 above). Under it the 24 marbles are partitioned into three equivalence classes: one consisting of 10 red marbles, another of 5 green marbles, and the last of 9 yellow marbles.

DEFINITION 6. The equivalence classes induced by the congruence modulo m are called *residue classes* modulo m.

Theorem 3. *The sets $0 \leq r \leq m - 1$ and $-m/2 < r \leq m/2$ form complete sets of residues.*

PROOF. Left as an exercise for the reader.

[†] See Problem 2, page 61.

DEFINITION 7.   The set $0 \leq r \leq m - 1$ is called a *complete set of least positive*[†] *residues*; the set $-m/2 < r \leq m/2$ is called a *complete set of least residues*.

Theorem 4.   *Let a, b, c, d, r, s be integers and m and n be positive integers. If congruences (1) to (5) are understood modulo m, then:*

(1) $a \equiv b \Rightarrow ca \equiv cb$;

(2) $a \equiv b, c \equiv d \Rightarrow a + c \equiv b + d$;

(3) $a \equiv b, c \equiv d \Rightarrow ar + cs \equiv br + ds$;

(4) $a \equiv b, c \equiv d \Rightarrow ac \equiv bd$;

(5) $a \equiv b \Rightarrow a^n \equiv b^n$,

(6) $a \equiv b \pmod{mn} \Rightarrow a \equiv b \pmod{m}$;

(7) $a \equiv b \pmod{m} \Leftrightarrow a \equiv b \pmod{-m}$.

All these statements are immediate consequences of Definition 1; their proofs are left to the reader.

Theorem 5.   *If $p(x)$ is a polynomial with integer coefficients and $a \equiv b \pmod{m}$, then $p(a) \equiv p(b) \pmod{m}$.*

PROOF.   Follows from Theorem 4(5) and 4(3).

So far, the properties of congruences appeared to be almost identical with the corresponding properties of ordinary equality. This observation illustrates well the point that to some extent the properties of an equivalence relation are due to the fact that *it is an equivalence relation*. Ordinary equality and congruence (mod m), both being equivalence relations, of course share all those properties, due precisely to the fact that they *are* equivalence relations. It might, therefore, be appropriate to point out at least one difference between the two relations.

If $ca = cb$ and $c \neq 0$, then we may cancel the common factor c and infer that $a = b$; however, if $c \not\equiv 0 \pmod{m}$ and $ca \equiv cb \pmod{m}$, we *cannot*, in general, conclude that $a \equiv b \pmod{m}$. In other words, the converse of Theorem 4(1) is false. Consider, for instance, the following:

EXAMPLE

$a = 21, b = 16, c = 12, m = 10$. Then $c \equiv 2 \not\equiv 0 \pmod{10}$, $ca = 252 \equiv 2 \pmod{10}$, $cb = 192 \equiv 2 \pmod{10}$, so that $ca - cb = 252 - 192 = 60 \equiv 0 \pmod{10}$; but $a = 21 \equiv 1 \pmod{10}$, $b = 16 \equiv 6 \pmod{10}$ and clearly, $1 \not\equiv 6 \pmod{10}$.

[†] This is the customary definition; it might be better to call these *non-negative*, rather than *positive* residues, and to reserve the name of complete set of least positive residue for the set $1 \leq r \leq m$. Occasionally, the distinction is relevant.

In view of this situation, it is comforting to know that a common factor may be canceled in a congruence, provided it is coprime to the modulus.

Indeed, if $ca \equiv cb \pmod{m}$, then $m|c(a - b)$; hence if also $(c, m) = 1$, then it follows from Theorem 3.7 that $m|a - b \Leftrightarrow a \equiv b \pmod{m}$, as asserted. More generally, the following cancellation rule holds.

Theorem 6. $ca \equiv cb \pmod{m} \Rightarrow a \equiv b \pmod{m/(m, c)}$.

REMARK 1. One observes that the condition $c \not\equiv 0 \pmod{m}$ *does not* appear as an assumption in Theorem 6. Indeed, if $c \equiv 0 \pmod{m}$, $(m, c) = |m|$, and the last congruence reduces to $a \equiv b \pmod{\pm 1}$ (see Theorem 4(7)), which holds trivially for all integers a and b.

PROOF OF THEOREM 6. Let $(c, m) = d$; then $c = dc_1$, $m = dm_1$ with $(c_1, m_1) = 1$. By Definition 1, $ca \equiv cb \pmod{m} \Leftrightarrow ca - cb = mk, k \in \mathbf{Z}$, or $c(a - b) = mk$, so that $dc_1(a - b) = dm_1k \Rightarrow c_1(a - b) = m_1k \Rightarrow m_1|c_1(a - b) \Rightarrow m_1|a - b$ by $(c_1, m_1) = 1$ and Theorem 3(7); hence, $a \equiv b \pmod{m_1}$ and the theorem is proven, because $m_1 = m/d = m/(m, c)$.

By setting $d = (m, c) = 1$, we obtain as an immediate corollary from Theorem 6,

Corollary 6.1. $(c, m) = 1, ca \equiv cb \pmod{m} \Rightarrow a \equiv b \pmod{m}$,

a result we had already found directly.

In the study of congruences with a composite modulus, the following theorem is useful.

Theorem 7. $a \equiv b \pmod{m_1}, a \equiv b \pmod{m_2} \Rightarrow a \equiv b \pmod{[m_1, m_2]}$.

PROOF. The assumptions state that $a - b$ is a common multiple of m_1 and of m_2; hence the result follows from Theorem 3.8.

Corollary 7.1. $(m_1, m_2) = 1, a \equiv b \pmod{m_1}, a \equiv b \pmod{m_2} \Rightarrow a \equiv b \pmod{m_1 m_2}$.

The easy proof is left to the reader.

2 OPERATIONS WITH RESIDUE CLASSES

Let us consider the set $\{0, 1, 2, \ldots, m - 1\}$ of least positive residues modulo m. By Theorem 3, each of these integers belongs to exactly one residue class. All congruences being understood modulo m, let **A** be a residue class to which belongs the least positive residue r_1; then $\mathbf{A} = \{a|a \equiv r_1\}$. Similarly, let $\mathbf{B} = \{b|b \equiv r_2\}$. By Theorem 4, if $a \in \mathbf{A}$ and $b \in \mathbf{B}$, then $a + b \equiv r_1 + r_2$ and

$a \cdot b \equiv r_1 \cdot r_2$. If r_3 and r_4 are least positive residues such that $r_1 + r_2 \equiv r_3$ and $r_1 \cdot r_2 \equiv r_4$, then for *every* element a of **A** and *every* element b of **B** one has $a + b \equiv r_3$ and $a \cdot b \equiv r_4$, respectively; it is important to observe here that r_3 and r_4 are independent of a and b and depend only on r_1 and r_2. Moreover, if we define the residue classes $\mathbf{C} = \{c \mid c \equiv r_3\}$ and $\mathbf{D} = \{d \mid d \equiv r_4\}$ then $a + b \equiv c$ and $a \cdot b \equiv d$ hold, regardless of the particular choice of elements within their residue class. This shows that the residue class of a sum or of a product does not depend at all on the summands and factors themselves, but only on their respective residue classes.

This fact permits us to define *addition* and *multiplication* of *residue classes*. We define $\mathbf{A} + \mathbf{B} = \mathbf{C}$ to mean that any element of the residue class **A** added to any element of the residue class **B** equals an element of the residue class **C**, and previous considerations have shown that this operation is indeed well defined. The equality $\mathbf{A} \cdot \mathbf{B} = \mathbf{D}$ has to be interpreted in a similar way. The results of these reasonings may be formalized in

Theorem 8. *The operations of addition and multiplication of residue classes are well-defined; the set of residue classes is closed under both operations.*

The last statement means, of course, that the sum and the product of any two residue classes is again a residue class. The least positive residue of a residue class may be considered (and is sometimes called) the "representative" of that class, and $\mathbf{A} = \{a \mid a \equiv r\}$ is conveniently represented by the symbol (r). Practically, in order to add or multiply residue classes, it is sufficient to add or multiply their representatives, always dropping multiples of the modulus so as to stay within the range $0 \leq r \leq m - 1$. This is called addition or multiplication modulo m. For instance, to the modulus 5, we have $(2) + (4) = (1)$ and $(2)(3) = (1)$. In this way we may construct addition and multiplication tables for these residue classes. The tables for $m = 6$ and $m = 5$ below should be self-explanatory.

A comparison of these tables shows that the addition tables for $m = 5$ and $m = 6$ are very similar—Increasing integers succeed each other on successive parallels to one of the diagonals on which we find $m - 1$. Of more interest are the multiplication tables. From that for $m = 6$, we read off such results as $(2)(2) = (4)$, $(3)(1) = (3)$; more generally, $(a)(1) = (1)(a) = (a)$. These relations look very much like the corresponding ones among ordinary integers and they do not surprise us. The result $(4)(5) = (2)$ is also easily interpreted, because $4 \cdot 5 = 20 \equiv 2 \pmod{6}$. However, we also find $(3)(2) = (0)$. This is of course correct, because $3 \cdot 2 = 6 \equiv 0 \pmod{6}$, but we have to note the fact that we obtain "zero" as product of two factors, both different from zero. We observe that this situation does not occur in the multiplication table for $m = 5$, where the products of nonvanishing factors (inside the heavy frame) are all different from zero. The reason for this difference is that 5 is a prime, while 6 is

Table 4.1. $m = 6$

(A) Addition

	0	1	2	3	4	5
0	0	1	2	3	4	5
1	1	2	3	4	5	0
2	2	3	4	5	0	1
3	3	4	5	0	1	2
4	4	5	0	1	2	3
5	5	0	1	2	3	4

(B) Multiplication

	0	1	2	3	4	5
0	0	0	0	0	0	0
1	0	1	2	3	4	5
2	0	2	4	0	2	4
3	0	3	0	3	0	3
4	0	4	2	0	4	2
5	0	5	4	3	2	1

Table 4.2. $m = 5$

(A) Addition

	0	1	2	3	4
0	0	1	2	3	4
1	1	2	3	4	0
2	2	3	4	0	1
3	3	4	0	1	2
4	4	0	1	2	3

(B) Multiplication

	0	1	2	3	4
0	0	0	0	0	0
1	0	1	2	3	4
2	0	2	4	1	3
3	0	3	1	4	2
4	0	4	3	2	1

not. Technically, the set of residue classes modulo a prime form a *field*, while those modulo a composite integer form a *commutative ring with divisors of zero*. The simplest way to avoid these "divisors of zero" is to restrict our attention to residue classes that are relatively prime to the modulus; this motivates the following definitions.

DEFINITION 8.   A residue class $\mathbf{A} = \{a|a \equiv r \pmod{m}\}$ is called a *prime* residue class, if $(r, m) = 1$.

DEFINITION 9. A *complete set of reduced (or prime) residues* is a set $\mathbf{S} = \{r_i\}$ satisfying the following conditions:

(i) $i \neq j \Rightarrow r_i \not\equiv r_j \pmod{m}$;

(ii) $r \in \mathbf{S} \Rightarrow (r, m) = 1$;

(iii) $(a, m) = 1 \Rightarrow \exists r \in \mathbf{S} \ni a \equiv r \pmod{m}$.

If, in addition, $0 < r \leq m$, then $\mathbf{S}$ is called a *reduced set of least positive residues*; if $-m/2 < r \leq m/2$, then $\mathbf{S}$ is a *reduced set of least residues*.

In words, a complete set of reduced residues consists of a set of mutually incongruent integers, all coprime to the modulus, and such that every integer coprime to the modulus is congruent to (exactly) one of them.

Theorem 9. *Let[†] $\mathbf{R}$ be the set of prime residue classes* mod m. *If $\mathbf{A}, \mathbf{B} \in \mathbf{R}$, then the equation $\mathbf{AX} = \mathbf{B}$ has exactly one solution $\mathbf{X} \in \mathbf{R}$.*

Corollary 9.1. *Every prime residue class $\mathbf{A}$ has exactly one "inverse" class $\mathbf{B} = \mathbf{A}^{-1}$, such that $\mathbf{A} \cdot \mathbf{A}^{-1} = \mathbf{I}$, where $\mathbf{I} = \{n | n \equiv 1 \pmod{m}\}$.*

The statement of Theorem 9 may be reformulated as follows: If $(a, m) = (b, m) = 1$, then the congruence $ax \equiv b \pmod{m}$ has exactly one solution modulo m; if $x = c$ is that solution, then $(c, m) = 1$.

PROOF. Let $\mathbf{S} = \{r_1, r_2, \ldots, r_k\}$ be a set of least positive reduced residues mod m, and let $\mathbf{A} = (a)$, $a \in \mathbf{S}$. Next, consider the set $ar_1, ar_2, \ldots, ar_k$. Clearly, $(ar_j, m) = 1$, because $(a, m) = (r_j, m) = 1$. Also if $i \neq j$, $ar_i \not\equiv ar_j \pmod{m}$, because otherwise $m | a(r_i - r_j)$ and $(m, a) = 1 \Rightarrow m | r_i - r_j$, which is impossible for $0 < r_i, r_j < m$. Hence, the k integers $ar_j (1 \leq j \leq k)$ form a complete set of reduced residues mod m and each of the elements of this set is congruent to exactly one element of $\mathbf{S}$. In particular, if $\mathbf{B} = (b)$, $b \in \mathbf{S}$, there is exactly one element ar_j satisfying $ar_j \equiv b \pmod{m}$. Hence if $\mathbf{C} = \{c | c \equiv r_j \pmod{m}\}$, $ax \equiv b \pmod{m}$ holds if and only if $x \in \mathbf{C}$. The proof is completed by the remark that $c \in \mathbf{C} \Rightarrow (c, m) = 1$.

From Theorem 9 and Corollary 9.1, together with obvious remarks concerning closure and associativity, follows:

Corollary 9.2. *The set $\mathbf{R}$ of prime residue classes* mod m *form a group under residue class multiplication; the identity element of the group is the class $\mathbf{I} = \{n | n \equiv 1 \pmod{m}\}$.*

[†] There is no danger that this notation will lead to confusion with the set of reals, as this set does not occur at all in the present chapter.

3 THEOREMS OF FERMAT, EULER, AND WILSON

DEFINITION 10. The number of positive integers r not exceeding m and coprime with m is denoted by $\phi(m)$. In symbols

$$\phi(m) = \sum_{\substack{(m, r)=1 \\ 0 < r \leq m}} 1.$$

EXAMPLES

$$\phi(1) = 1, \quad \phi(2) = 1, \quad \phi(3) = 2, \quad \phi(4) = 2, \quad \phi(5) = 4.$$

$$\phi(p) = p - 1, \quad \phi(p^r) = p^r - p^{r-1} = p^r\left(1 - \frac{1}{p}\right).$$

This function $\phi(m)$ is usually called Euler's ϕ-function.

REMARK 2†. A complete set of reduced residues consists of exactly $\phi(m)$ integers.

Theorem 10 (Euler). $(a, m) = 1 \Rightarrow a^{\phi(m)} \equiv 1 \pmod{m}$.

FIRST PROOF. Let $r_1, r_2, \ldots, r_k$ be a reduced set of residues mod m. If $(a, m) = 1$, then as seen in the proof of Theorem 9, $ar_1, ar_2, \ldots, ar_k$ also form a reduced set of residues; hence each element of the first set is congruent (mod m) to one and only one element of the second set. Multiplying all these congruences termwise (see Theorem 4) we obtain $r_1 r_2 \cdots r_k \equiv a^k r_1 r_2 \cdots r_k$ (mod m). But $(r_1 r_2 \cdots r_k, m) = 1$; hence by Corollary 6.1 we may cancel the common factor $r_1 \cdots r_k$ and obtain $1 \equiv a^k \pmod{m}$. As already observed (Remark 2), $k = \phi(m)$, and the theorem is proven.

SECOND PROOF. The order of the multiplicative group **R** of prime residues is $k = \phi(m)$. Also, $(a, m) = 1 \Rightarrow A = (a) \in \mathbf{R}$; hence by Lagrange's Theorem, $\mathbf{A}^k = \mathbf{I}$, or $a^k = a^{\phi(m)} \equiv 1 \pmod{m}$.

The brevity and simplicity of this argument illustrates again the advantage of the modern algebraic approach.

Theorem 11 (Fermat). $p \nmid a \Rightarrow a^{p-1} \equiv 1 \pmod{p}$.

PROOF. Take $m = p$ in Theorem 10 and use $\phi(p) = p - 1$ (see preceding examples).

† See Problem 18, page 63.

EXAMPLES

Modulo 7 one has $5^1 = 5$, $5^2 = 25 \equiv 4$, $5^3 = 125 \equiv 6$, $5^4 = 625 \equiv 2$, $5^5 = 3125 \equiv 3$, $5^6 = 15625 \equiv 1$. On the other hand, modulo 6 one finds $5^1 = 5$, $5^2 = 25 \equiv 1$, and indeed $\phi(6) = 2$ (a complete set of reduced residues mod 6 is $\{1, 5\}$).

Theorem 12 (Wilson). $(n - 1)! + 1 \equiv 0 \pmod{n} \Leftrightarrow n = p$.

In other words, said congruence holds if n is a prime and is false if n is a composite number. Hence the verification of this congruence may be considered as a test (not necessarily a very practical one) of primality for n.

PROOF.

(i) Theorem 12 holds trivially for $n = 2$ and $n = 3$; hence let $n = p \geq 5$ and select an r satisfying $1 \leq r \leq p - 1$; then $(r, p) = 1$, and by Corollary 9.1 there exists exactly one m such that $mr \equiv 1 \pmod{p}$ and $1 \leq m \leq p - 1$. In general, $m \neq r$. Indeed if $m = r$, $m^2 - 1 \equiv 0 \pmod{p}$ so that $p \mid (m + 1)(m - 1)$, possible only if $m = 1$ or $m = p - 1$. In particular all $p - 3$ integers $2 \leq r \leq p - 2$ fall into pairs (observe that $p - 3$ is even), such that the corresponding products are congruent to $1 \pmod{p}$. Hence $2 \cdot 3 \cdot 4 \cdot \cdots \cdot (p - 2) \equiv 1 \pmod{p}$ so that $(p - 1)! \equiv (p - 1) \equiv -1 \pmod{p}$, and the implication $\Leftarrow$ of the theorem is proven.

(ii) The other implication is rather trivial. Indeed, in case n is not a prime, $n = a \cdot b$, $1 < a < n$, so that $a \mid (n - 1)!$, $a \nmid ((n - 1)! + 1)$ and, *a fortiori*,

$$n \nmid ((n - 1)! + 1) \Leftrightarrow (n - 1)! + 1 \not\equiv 0 \pmod{n},$$

and the proof is complete.

* The preceding proof essentially follows Gauss [2] and is quite satisfactory. Theorem 12 was actually known long before Wilson, presumably already by Leibniz. It was first published by Waring (1734–1798), who ascribes it to Wilson (1741–1793). It also may be of interest to indicate a different proof—not because it is superior to Gauss' proof, but because it is of an entirely different character, and illustrates the use that can be made of analysis in number-theoretic problems. This second proof is due to M. A. Stern (see [6], p. 391). The first published proof of Theorem 12 is due to Lagrange (see Problem 21).

SECOND PROOF OF THEOREM 12.   The well-known Maclaurin series

$$-\log(1 - x) = \log\frac{1}{1 - x} = x + \frac{x^2}{2} + \frac{x^3}{3} + \cdots$$

shows that

$$\exp\left(x + \frac{x^2}{2} + \frac{x^3}{3} + \cdots\right) = \frac{1}{1-x} = 1 + x + x^2 + \cdots.$$

This may be written as

$$e^x e^{x^2/2} e^{x^3/3} \cdots = 1 + x + x^2 + x^3 + \cdots.$$

The left side may be expanded in a power series as follows:

$$\left(1 + \frac{x}{1!} + \frac{x^2}{2!} + \cdots\right)\left(1 + \frac{x^2/2}{1!} + \frac{(x^2/2)^2}{2!} + \cdots\right)\cdots$$

$$\left(1 + \frac{x^p/p}{1!} + \frac{(x^p/p)^2}{2!} + \cdots\right)$$

$$= 1 + \frac{x}{1!} + x^2\left(\frac{1}{2!} + \frac{1}{2}\right) + x^3\left(\frac{1}{3!} + \frac{1}{1!}\frac{1/2}{1!} + \frac{1/3}{1!}\right) + \cdots$$

$$+ x^p\left(\frac{1}{p!} + \frac{1}{(p-2)!}\frac{1/2}{1!} + \cdots + \frac{1/p}{1!}\right) + \cdots.$$

For prime p the coefficient of x^p is seen to be of the form $1/p! + r/s + 1/p$, where r/s is the sum of a finite number of rational fractions that do not contain the factor p in their denominators. Hence if $(r, s) = 1$, $p \nmid s$. The coefficient of x^p (as of all powers of x) is unity, so that $1/p! + r/s + 1/p = 1$, whence $1 - r/s = 1/p! + 1/p = (1 + (p-1)!)/p!$ or $\alpha \underset{\text{def}}{=} (p-1)!(s-r)/s = (1 + (p-1)!)/p$. The right term here shows that there are only two alternatives: Either $p|((p-1)! + 1)$, and then α is an integer; or else it is a rational fraction with denominator p. However, as seen, $p \nmid s$, so that $\alpha = (p-1)!(s-r)/s$ rules out this second alternative, and $p|((p-1)! + 1)$ as we wanted to prove. The other (trivial) implication may be proven as before.

4 LINEAR CONGRUENCES

The problem of solving single equations or systems of equations in one or several unknowns is familiar from elementary algebra. Similar problems may be asked concerning congruences. If $f(x)$ is a polynomial with integral coefficients and m is an integer, one may ask for those values of x for which $f(x) \equiv 0 \pmod{m}$. This problem is of course equivalent to that of solving the Diophantine equation $f(x) = ym$ in integers x and y. The theory developed so far permits us to handle some of these problems, especially in case $f(x)$ is a linear polynomial.

Theorem 13. *If* $(a, m) = 1$, *then*

$$ax \equiv b \,(\mathrm{mod}\ m) \tag{1}$$

has a unique solution mod m.

PROOF. By Theorem 10, $y = a^{\phi(m)-1}$ is a solution of $ay \equiv 1$; hence, also of $aby \equiv b$. Setting $x = by$, $x = ba^{\phi(m)-1}$ satisfies (1) and the existence of a solution is proven. If x_1 and x_2 are two solutions, then $ax_1 \equiv b$, $ax_2 \equiv b \Rightarrow$ $a(x_1 - x_2) \equiv 0 \Leftrightarrow m|a(x_2 - x_1)$, and using $(a, m) = 1$ and Theorem 3.7, $m|x_2 - x_1 \Leftrightarrow x_1 \equiv x_2$. Hence if x_1 and x_2 are both least positive residues mod m, $x_1 = x_2$, thus proving the uniqueness (mod m) of the solution.

Corollary 13.1. *If* $(a, m) = 1$, *then* x *is a solution of* (1) *if and only if it is of the form* $x = ba^{\phi(m)-1} + km$, $k \in \mathbf{Z}$.

PROOF. By the proof of Theorem 13 we know that $x_0 = ba^{\phi(m)-1}$ is a solution and that all other solutions of (1) differ from x_0 by an integral multiple of m. It is therefore sufficient to verify that for every $k \in \mathbf{Z}$, x satisfies (1):

$$ax = ba^{\phi(m)} + kma \equiv ba^{\phi(m)} \equiv b \,(\mathrm{mod}\ m)$$

by Theorem 10.

Having settled the case $(a, m) = 1$, we now consider the general case $(a, m) = d$.

Theorem 14. *If* $(a, m) = d$, *then* (1) *has no solution if* $d \nmid b$, *and has a unique solution* mod m/d *if* $d|b$.

PROOF. Congruence (1) is equivalent to the Diophantine equation

$$ax = b + km; \tag{2}$$

if $x = c \in \mathbf{Z}$ is a solution, then $b = ac - km$, and by Theorem 3.1, $d|ac - km = b$; hence if $d \nmid b$, no such solution can exist. If $d|b$, set $a = da_1$, $b = db_1$, $m = dm_1$; then $(a_1, m_1) = 1$, and after an obvious simplification (2) becomes $a_1 x = b_1 + km_1$, which is equivalent to $a_1 x \equiv b_1 \,(\mathrm{mod}\ m_1)$, $(a_1, m_1) = 1$. By Theorem 13 we know that this congruence has a unique solution modulo $m_1 = m/d$, and the proof of Theorem 14 is complete.

We have already repeatedly observed that the problem of solving (1) is identical to that of solving the Diophantine equation $ax + my = b$. It is often convenient to have the results just obtained reformulated directly in terms of Diophantine equations; this, of course, involves hardly any new work, but we shall modify the notation slightly to make it more symmetrical.

Theorem 15 (Bachet, ca. 1612). *Consider the Diophantine equation*

$$ax + by = c \qquad (3)$$

with $(a, b) = d$. *If* $d \nmid c$, *then* (3) *has no solutions; if* $d | c$, *then* (3) *has infinitely many solutions. If* x_0, y_0 *is a solution of* (3), *then all other solutions are given by*

$$x = x_0 + n(b/d)$$
$$y = y_0 - n(a/d), \qquad (4)$$

with n running through all rational integers.

PROOF OF THEOREM 15.   All statements are immediate consequences of Theorem 14, except the assertion that (4) represents all solutions of (3).

Let $a = a_1 d$, $b = b_1 d$, $c = c_1 d$; then, after cancellation of the common factor d, equation (3) becomes

$$a_1 x + b_1 y = c_1, \qquad (a_1, b_1) = 1. \qquad (5)$$

All solutions x of (5) satisfy $a_1 x \equiv c_1 \pmod{b_1}$, $(a_1, b_1) = 1$. Theorem 14 guarantees that all solutions of this congruence (hence also all solutions of (5)) are of the form $x = x_0 + nb_1$, and by the symmetry (in x and y) of (5) it also follows that $y = y_0 + ma_1$. Substituting these in (5), $a_1 x_0 + b_1 y_0 + a_1 b_1(m + n) = c_1$; we now select x_0 and y_0 as solutions of (5), obtaining $a_1 b_1(m + n) = 0$, so that $m = -n$ and the proof of (4) is complete.

REMARK 3.   A particular solution x_0, y_0 of (5), hence also of (3), may be found as follows: By Theorem 3 there are integers m and n such that $a_1 m + b_1 n = 1$; hence $a_1(mc_1) + b_1(nc_1) = c_1$ and we may take $x_0 = mc_1$, $y_0 = nc_1$. A practical way to find m and n follows from the Euclidean algorithm (see Chapter 3, Problem 3).

Systems of k congruences in k unknowns, all taken to the same modulus m, can be reduced to k independent congruences involving only one unknown. Once this is done, we may use Theorem 14. It is sufficient to illustrate the situation for $k = 2$. Consider, therefore, the following system, all congruences being understood modulo m:

$$a_1 x + b_1 y \equiv c_1,$$
$$a_2 x + b_2 y \equiv c_2. \qquad (6)$$

Let[†]

$$D = \begin{vmatrix} a_1 b_1 \\ a_2 b_2 \end{vmatrix}, \qquad D_1 = \begin{vmatrix} c_1 b_1 \\ c_2 b_2 \end{vmatrix}, \qquad D_2 = \begin{vmatrix} a_1 c_1 \\ a_2 c_2 \end{vmatrix}.$$

[†] For convenience we here denote these determinants, which are rational integers, by capitals, rather than lower case letters.

Then, as in the proof of Cramer's rule, one shows that the system (6) implies

$$Dx \equiv D_1,$$
$$Dy \equiv D_2. \tag{6'}$$

If $(D, m) = 1$, then Theorem 13 is applicable, and (6') has solutions x_0, y_0, which are unique modulo m. By direct substitution in (6) one easily verifies (making use once more of $(D, m) = 1$) that these solutions of (6') actually satisfy (6), so that (6) and (6') are equivalent systems. The case $(D, m) > 1$ can be handled similarly, using Theorem 14, but we shall not pursue this matter further.

5   THE CHINESE REMAINDER THEOREM

As we just saw, systems of linear congruences involving *several unknowns*, but all to the *same modulus*, reduce quite trivially to ordinary congruences in a single unknown. The situation is rather different if we have to deal with a system of linear congruences in a *single unknown*, each taken to a *different modulus*. This is a famous problem; the main result, our Theorem 16, was known to Sun Tsu (see [7] and Example 1 on page 49), and possibly even to Chinese mathematicians before the Christian era, whence the name under which it is generally known. It might be of some interest to observe that thousands of miles away from Sun Tsu, but almost exactly at the same time (ca. 100 A.D.) Nicomachus (Neo-Pythagorean born in Gerasa (Palestine)) stated and solved *exactly* the same problem of Example 1.

Theorem 16 (Chinese Remainder Theorem).   *Consider the system of k linear congruences* $a_i x \equiv b_i \pmod{m_i}$, *$(i = 1, 2, \ldots, k)$, where $(m_i, m_j) = 1$ for all $i \neq j$ and $(a_i, m_i) = 1$ for $1 \leq i \leq k$; let $m = m_1 m_2 \cdots m_k$. This system has a unique solution modulo m; one solution x_0 is given by (7) and all others are of the form $x = x_0 + nm$, where n is any rational integer.*

REMARK 4.   If we try to solve the congruence $ax \equiv b \pmod{m}$, where $m = p_1^{s_1} p_2^{s_2} \cdots p_k^{s_k}$, then it follows from Corollary 7.1 that the problem is equivalent to that of solving the k congruences $ax \equiv b \pmod{m_i}$ with $m_i = p_i^{s_i}$. This shows that the problem considered by Theorem 16, as well as the condition $i \neq j \Rightarrow (m_i, m_j) = 1$ (which may seem strange, and at first sight, unduly restrictive), are actually both quite natural.

PROOF OF THEOREM 16.   By $(a_i, m_i) = 1$ and Theorem 13, we know that each congruence has a solution of the form $x \equiv c_i \pmod{m_i}$. Set $t_i = m/m_i$ and let y_i be a solution of $t_i y \equiv 1 \pmod{m_i}$. We know, again by Theorem 13, that each of these congruences has (mod m_i) a unique solution y_i, because also $(t_i, m_i) = 1$. Now consider

$$x_0 = y_1 t_1 c_1 + y_2 t_2 c_2 + \cdots + y_k t_k c_k. \tag{7}$$

Clearly, $t_j \equiv 0 \pmod{m_i}$ for every $i \neq j$; hence, $x_0 \equiv y_i t_i c_i \equiv c_i \pmod{m_i}$, because $y_i t_i \equiv 1 \pmod{m_i}$, and x_0 is a solution of the given system. With x_0 also $x_0 + nm$ is a solution, because $x_0 + nm \equiv x_0 \pmod{m} \Rightarrow x_0 + nm \equiv x_0 \pmod{m_i}$ by Theorem 4(6). It only remains to verify that we obtain all solutions in this way.

Let x_1 be another solution of the system; then $x_1 - x_0 \equiv 0 \pmod{m_i}$ for every $i (1 \leq i \leq k)$. Hence by Corollary 7.1, $x_1 - x_0 \equiv 0 \pmod{m}$, so that $x_1 = x_0 + nm$, and the proof is complete.

EXAMPLES

1. Find the least positive integer that upon division by 3, leaves a remainder of 2, upon division by 5 leaves a remainder of 3, and upon division by 7 leaves a remainder of 2 (Sun-Tsu, first century A.D.).

   We have to solve the system

$$x \equiv 2 \pmod{3},$$
$$x \equiv 3 \pmod{5},$$
$$x \equiv 2 \pmod{7}.$$

   Following the steps of the proof of Theorem 16, set $m = 3 \cdot 5 \cdot 7 = 105$, $t_1 = 35$, $t_2 = 21$, $t_3 = 15$. Then y_1, y_2, y_3 are determined by the congruences $35 y_1 \equiv 1 \pmod{3}$, $21 y_2 \equiv 1 \pmod{5}$, and $15 y_3 \equiv 1 \pmod{7}$; these simplify to $2 y_1 \equiv 1 \pmod{3}$, $y_2 \equiv 1 \pmod{5}$, and $y_3 \equiv 1 \pmod{7}$, with the obvious solution $y_1 = 2$, $y_2 = 1$, $y_3 = 1$. Hence by (7),

$$x = 2 \cdot 35 \cdot 2 + 1 \cdot 21 \cdot 3 + 1 \cdot 15 \cdot 2 = 233 \equiv 23 \pmod{105}.$$

   The smallest positive solution is $x = 23$, and all solutions (not asked for by Sun-Tsu) are given by $x = 23 + 105n$.

2. Find a positive integer such that, when divided by 3, 4, 5, and 6, it should leave the remainders 2, 3, 4, and 5, respectively (Brahmegupta, 7th century A.D.).

   Here the system is

$$x \equiv 2 \pmod{3},$$
$$x \equiv 3 \pmod{4},$$
$$x \equiv 4 \pmod{5},$$
$$x \equiv 5 \pmod{6}.$$

Theorem 16 is not directly applicable, because the moduli[†] are not two-by-two coprime, but we may use Theorem 16 to solve the system formed by the first three congruences. Proceeding as in Example 1, set $m = 3 \cdot 4 \cdot 5 = 60$, $t_1 = 20$, $t_2 = 15$, $t_3 = 12$; then $20y_1 \equiv 1$ (mod 3), $15y_2 \equiv 1$ (mod 4), $12y_3 \equiv 1$ (mod 5), so that $y_1 = 2$, $y_2 = 3$, $y_3 = 3$, and

$$x_0 = 2 \cdot 20 \cdot 2 + 3 \cdot 15 \cdot 3 + 3 \cdot 12 \cdot 4 = 359 \equiv -1 \ (\text{mod } 60).$$

Hence $x = 60n - 1$ is the general solution of the system formed by the first three congruences. Now, also considering the last congruence, $x \equiv 5$ (mod 6), we see that it is automatically satisfied by all solutions of the subsystem of the first three congruences so that $x = 60n - 1$ is the general solution of the whole system, and in particular, $x = 59$ is the smallest positive solution.

One also observes that if the last condition would have been, say, $x \equiv 2$ (mod 6) (or, generally, $x \equiv a \not\equiv 5$ (mod 6)), then the system would have had no solution at all.

REMARK 5. Suppose that in some of the congruences $(a_i, m_i) = d_i > 1$; then (see Theorem 14) if $d_i \nmid b_i$, the ith congruence (and, a fortiori, the whole system) has no solution. If $d_i | b_i$, then the factor d_i may be cancelled in the ith congruence, which will become $a_i'x \equiv b_i'$ (mod m_i'), with $a_i' = a_i/d_i$, $m_i' = m_i/d_i$, $(a_i', m_i') = 1$, and the new system equivalent to the original one satisfies the conditions of coprimality of the coefficients a_i' and the moduli m_i'.

We shall not discuss systematically the situations that may arise if the other coprimality conditions of Theorem 16 are violated, that is if $(m_i, m_j) = d_{ij} > 1$. The kinds of difficulties that may arise were partly illustrated in the discussion of Brahmegupta's example. The interested reader is referred to LeVeque's book [3], where this problem is treated, and to Dickson's History [1], Vol. 2, p. 58, where further references are given.

6   ON PRIMITIVE ROOTS

This section has been inserted mainly for its intrinsic interest; its results will be used mainly in the proof of Theorem 26, which itself is not really indispensable for what follows.

DEFINITION 11. If $x^r \equiv c$ (mod m) has a solution with integral x, then c is said to be an *rth power residue* (mod m).

By Theorem 10, $(a, m) = 1 \Rightarrow a^{\phi(m)} \equiv 1$ (mod m); hence there exists a set **K** of positive integers k such that $a^k \equiv 1$ (mod m). By Remark 3.2, **K** contains a

---

[†] For the plural of modulus in the present sense, we shall use "moduli"; see footnote p. 23.

smallest positive integer h. It is easy to see that $h|\phi(m)$. Indeed otherwise, $\phi(m) = hn + r$, with $1 \leq r < h$; but this would imply that

$$1 \equiv a^{\phi(m)} = a^{hn+r} = a^{hn}a^r = (a^h)^n \cdot a^r \equiv 1 \cdot a^r \equiv a^r \,(\mathrm{mod}\ m),$$

contrary to the definition of h as *smallest* positive exponent for which $a^h \equiv 1$ (mod m).

DEFINITION 12. If h is the smallest positive integer such that $a^h \equiv 1$ (mod m), then a is said to *belong to the exponent h* modulo m.

REMARK 6. If $a^h \equiv 1$ (mod m), $0 < h \in \mathbf{Z}$, then $(a, m) = 1$; indeed, the congruence means that $\exists l \in \mathbf{Z} \ni a^h = 1 + lm$, or $1 = a^h - lm$ and $d = (a, m)$ divides $a^h - lm$; hence $d|1 \Rightarrow d = 1$.

REMARK 7. If $a^n \equiv 1$ (mod m), then $n = kh$.

PROOF. $h \nmid n \Rightarrow n = lh + r\ (1 \leq r < h) \Rightarrow 1 \equiv a^n = a^{lh+r} = (a^h)^l \cdot a^r \equiv a^r$ (mod m), contrary to the definition of h.

REMARK 8. $h \leq \phi(m)$; this follows trivially (see Theorem 3.1(7)) from $h|\phi(m)$ already established.

DEFINITION 13. If g belongs to the exponent $\phi(m)$ modulo m, then g is called a *primitive root* mod m.

Theorem 17. *If g is a primitive root mod m, then $g, g^2, \ldots, g^{\phi(m)}$ are all incongruent modulo m and form a complete set of reduced residues.*

PROOF. The number of elements in the set is exactly $\phi(m)$ (see Remark 2), and satisfies $(g^r)^{\phi(m)} = (g^{\phi(m)})^r \equiv 1$ (mod m); hence by Remark 6 all are coprime to m; therefore, all we have to show is that they are pairwise incongruent mod m. Let us assume the contrary—that is, let $1 \leq r < s \leq \phi(m)$ and assume that $g^r \equiv g^s$ (mod m); then $g^{s-r} \equiv 1$ (mod m) with $1 \leq s - r \leq \phi(m) - 1$, contrary to the definition of g as a primitive root.

Corollary 17.1. *Necessary and sufficient conditions for g to be a primitive root mod m are that $(g, m) = 1$ and that $g, g^2, \ldots, g^{\phi(m)}$ be all incongruent mod m.*

PROOF. Left to the reader.

Corollary 17.2. *If g is a primitive root mod m, then $g^r \equiv 1$ (mod m) $\Rightarrow$ $\phi(m)|r$.*

PROOF. Remark 7 with $h = \phi(m)$.

Corollary 17.3. *If g is a primitive root mod p, then $g, g^2, \ldots, g^{p-1} \equiv 1$ (mod p) are all incongruent mod p and $g^r \equiv 1$ (mod p) $\Rightarrow (p - 1)|r$.*

PROOF. Corollaries 17.1 and 17.2 for $m = p$.

DEFINITION 14.   If $(a, m) = 1$ and m admits the primitive root g, then the exponent k for which $a \equiv g^k \pmod{m}$ is called the *index of a* (mod m).

REMARK 9.   The index of an integer is uniquely defined if we also require that $0 \le k < \phi(m)$. (Why?).

REMARK 10.   The indices behave very much like logarithms, and there exists a calculus of indices of considerable interest, but it will not be presented here; the interested reader may wish to consult [3], Chapter 4.

Theorem 18.   *If b belongs to the exponent h* (mod m) *and* $(k, h) = d$, *then* b^k *belongs* mod m *to the exponent* h/d.

PROOF (all congruences are mod m).   Let b^k belong to the exponent n; then $(b^k)^n \equiv 1$. By Remark 7,

$$h \mid kn \implies \frac{h}{d} \left| \frac{k}{d} \cdot n \implies \frac{h}{d} \right| n,$$

because $(h/d, k/d) = 1$. Conversely, $(b^k)^{h/d} = b^{kh/d} = (b^h)^{k/d} \equiv 1$; hence, still by Remark 7, $n \mid (h/d)$, so that $n = (h/d)$ (see Theorem 3.1(8)).

Corollary 18.1.   *If g is a primitive root* mod m, *then* g^k *is also a primitive root if and only* $(k, \phi(m)) = 1$.

PROOF.   Take $h = \phi(m)$ in Theorem 18.

Theorem 19.   *If m has any primitive roots, then it has exactly* $\phi(\phi(m))$ *distinct ones.*

PROOF.   Let g be a primitive root mod m. By Theorem 17, $g, g^2, \ldots, g^{\phi(m)}$ form a complete set of reduced residues and any primitive root has to occur among them. By Theorem 18, the exponent to which g^k belongs is $n = \phi(m)/(k, \phi(m))$. Now, g^k is a primitive root precisely when $n = \phi(m)$; for that it is necessary and sufficient to have $(k, \phi(m)) = 1$. However, the number of integers k satisfying $1 \le k \le \phi(m)$ and prime to $\phi(m)$ is, by definition, $\phi(\phi(m))$.

At this point the reader may wonder about the significance of the clause "If m has any primitive roots ... ." Do not all integers have *some* primitive root? The answer turns out to be NO! In fact, it is the exception rather than the rule that an integer should have a primitive root, and the following theorem holds:

Theorem 20.   *An integer m has primitive roots if and only if* $m = 2, 4, p^r, 2p^r$ (*p an odd prime*).

We shall not prove this theorem completely but shall prove the "only if" part, and settle the cases $m = 2^c$ and $m = p$. (The proofs omitted here, i.e., the existence of primitive roots if $m = 2p^r$ and $m = p^r$ ($r \ge 2$) are the object of

Problems 30–33 at the end of this chapter; they also may be found in [4] and [5].) In these proofs we shall have to make use of some properties of the Euler ϕ-function, which could easily be presented now; however, we shall investigate the ϕ-function systematically in Chapter 6 and prefer here to simply anticipate the needed results, which are

(A) $(m_1, m_2) = 1 \Rightarrow \phi(m_1)\phi(m_2) = \phi(m_1 m_2)$.

(B) $\phi(m) = m \prod_{p|m} \left(1 - \dfrac{1}{p}\right)$.

(C) (Corollary of B). If m is divisible by an odd prime, then $\phi(m)$ is even (in symbols $m \neq 2^b \Rightarrow 2|\phi(m)$).

(D) $\sum_{d|m} \phi(d) = m$.

Theorem 21. $(m_1, m_2) = 1$, $a^{h_1} \equiv 1 \pmod{m_1}$, $a^{h_2} \equiv 1 \pmod{m_2} \Rightarrow a^{[h_1, h_2]} \equiv 1 \pmod{m_1 m_2}$.

PROOF.   Let $d = (h_1, h_2)$ so that $h_1 = dk_1$, $h_2 = dk_2$, $(k_1, k_2) = 1$, and set $h = [h_1, h_2] = h_1 h_2/d$. Then $(a^d)^{h_1/d} \equiv 1 \pmod{m_1}$, $(a^d)^{h_2/d} \equiv 1 \pmod{m_2}$; hence $a^h = (a^d)^{(h_1/d)(h_2/d)} \equiv 1^{h_2/d} \equiv 1 \pmod{m_1}$, and also $a^h = (a^d)^{(h_2/d)(h_1/d)} \equiv 1^{h_1/d} \equiv 1 \pmod{m_2}$, so that $a^h \equiv 1 \pmod{m_1 m_2}$, by Theorem 7.

Corollary 21.1. If $(m_1, m_2) = 1$ and if a belongs to $h_1 \pmod{m_1}$ and to $h_2 \pmod{m_2}$, then setting $[h_1, h_2] = h$, $a^h \equiv 1 \pmod{m_1 m_2}$.

Theorem 22. If b is odd and $c \geq 3$, then $b^{2^{c-2}} \equiv 1 \pmod{2^c}$.

PROOF.   For $c = 3$, Theorem 22 reads $b^2 \equiv 1 \pmod 8$ which is true (observe that $1^2 \equiv 3^2 \equiv 5^2 \equiv 7^2 \equiv 1 \pmod 8$). For $c > 3$ we use induction on c. If the Theorem holds for some c, then $b^{2^{c-2}} = 1 + k \cdot 2^c$ and, squaring both sides,

$$\left(b^{2^{c-2}}\right)^2 = b^{2^{c-1}} = 1 + 2k \cdot 2^c + k^2 \cdot 2^{2c} \equiv 1 \pmod{2^{c+1}}$$

so that Theorem 22 holds for $c + 1$, and hence for all $c \geq 3$.

Corollary 22.1. If $m = 2^c$, $c \geq 3$, then m can have no primitive roots.

PROOF.   If $(b, m) = 1$, then b is odd; also, $\phi(m) = 2^{c-1}$. Hence, $b^{(\phi(m))/2} = b^{2^{c-2}} \equiv 1 \pmod m$ by Theorem 22, and all integers b belong to exponents $\leq (\phi(m))/2 < \phi(m)$.

Theorem 23. If m is divisible by two distinct odd primes, it can have no primitive root.

PROOF.   Let $m = p^s m_1$, $(p, m_1) = 1$, $m_1 \neq 2^c$. Then, if $0 < b \in \mathbf{Z}$, $(b, m) = 1$, one has $b^{\phi(p^s)} \equiv 1 \pmod{p^s}$, $b^{\phi(m_1)} \equiv 1 \pmod{m_1}$ by Theorem 10, and by

Theorem 21, $b^h \equiv 1 \pmod{m}$, where $h = [\phi(p^s), \phi(m_1)]$. By Property (C), $d = (\phi(p^s), \phi(m_1))$ is even; hence $d \geq 2$, so that using Property (A),

$$h = \frac{\phi(p^s)\phi(m_1)}{d} = \frac{\phi(m)}{d} \leq \frac{\phi(m)}{2}.$$

Consequently, every b coprime to m belongs to an exponent $\leq (\phi(m))/2$, and m has no primitive root.

We can now settle the case $m = 2^c$ and prove the "only if" part of Theorem 20: $m = 2$ has the primitive root $g = 1$, $m = 4$ has the primitive root $g = 3$, and on account of Corollary 22.1, $m = 2^c$ has no primitive root for $c \geq 3$. If m contains two odd primes, it can have no primitive root, by Theorem 23. Consider now the case $m = 2^c p^s$. Let $c \geq 3$, and take any b such that $(b, m) = 1$. Then $b^{\phi(p^s)} \equiv 1 \pmod{p^s}$, by Theorem 10; $b^{2^{c-2}} \equiv 1 \pmod{2^c}$, by Theorem 22, and using Theorem 21, $b^h \equiv 1 \pmod{m}$, where $h = [\phi(p^s), 2^{c-2}]$. By Property (C), $d = (\phi(p^s), 2^{c-2}) = 2d_1 \geq 2$, and clearly, $\phi(2^c) = 2^{c-1}$; hence,

$$h = \frac{\phi(p^s) \cdot 2^{c-2}}{2d_1} = \frac{\phi(p^s) \cdot 2^{c-1}}{4d_1} = \frac{\phi(p^s)\phi(2^c)}{4d_1} = \frac{\phi(m)}{4d_1} \leq \frac{\phi(m)}{4},$$

so that $m = 2^c p^s$ can have no primitive root for $c \geq 3$. If $c = 2$, $m = 4p^s$, and proceeding as before, for any $(b, m) = 1$ one obtains successively that $b^2 \equiv 1 \pmod{4}$ and $b^h \equiv 1 \pmod{m}$, with $h = [\phi(p^s), 2] = \phi(p^s) = \frac{1}{2}\phi(m)$; hence again, m can have no primitive root. As stated by Theorem 20, in all cases where the existence of a primitive root has not been ruled out, it actually exists, but we shall prove this statement here only in the special case $m = p$.

The integers $1, 2, \ldots, p - 1$ form a complete set of reduced residues mod p. Each of them belongs to some exponent $h \leq \phi(p) = p - 1$ and we know (see Remarks following Definition 11) that actually $h | (p - 1)$. If we select an arbitrary divisor h of $p - 1$, there might or might not exist positive integers $b, 1 \leq b \leq p - 1$, belonging to the exponent h. But if any exist, then there exist exactly $\phi(h)$ of them. Indeed, if a belongs to the exponent h, then $a, a^2, \ldots, a^h \equiv 1$ are h incongruent solutions of the congruence $x^h \equiv 1 \pmod{p}$. These are actually all its solutions, because in strong analogy with algebraic equations, the number of solutions of a congruence modulo a prime cannot exceed its degree.[†]

Hence all reduced residues belonging to the exponent h are found among these a^k ($1 \leq k \leq h$). As seen in the proof of Theorem 19, the power a^k belongs to the exponent $h/(h, k)$ and this reduces to h if and only if $(h, k) = 1$; the number of such exponents $k \leq h$ is precisely $\phi(h)$. Given an arbitrary integer h, we shall denote by $N(h)$ the number of positive integers

---

[†] This fact, stated and proven as Theorem 24, could easily be established here; however, in order not to interrupt the present proof, it seems preferable to anticipate this result.

$b \leq p - 1$ belonging to the exponent h, and we just proved that there are only two alternatives: Either $N(h) = 0$, or else $N(h) = \phi(h)$, the last one being possible only if $h|(p-1)$. Consequently, $\Sigma_{h|p-1} N(h) \leq \Sigma_{h|p-1}\phi(h)$. However, each integer b belongs to some exponent $h \leq p - 1$; hence, $\Sigma_{h|p-1} N(h) = \Sigma_{h \leq p-1} N(h) = p - 1$. Also, by property D, $\Sigma_{h|p-1}\phi(h) = p - 1$. Putting these results together, $p - 1 = \Sigma_{h|p-1} N(h) \leq \Sigma_{h|p-1}\phi(h) = p - 1$. This shows that $\leq$ is actually an equality. However, this is not possible unless $N(h) = \phi(h)$ for every $h|p-1$. In particular, for $h = p - 1$, we obtain the result (slightly stronger than what we want to prove) that there are $N(p - 1) = \phi(p - 1)$ integers $b \leq p - 1$ belonging to the exponent $p - 1 = \phi(p)$, i.e., that are primitive roots. We have actually proven

Corollary 20.1. *Each prime p has exactly $\phi(p - 1)$ primitive roots incongruent (mod p).*

7 *CONGRUENCES OF HIGHER DEGREES

Let

$$f(x) = a_0 x^n + a_1 x^{n-1} + \cdots + a_n \tag{8}$$

be a polynomial of degree n with integer coefficients. The congruence $f(x) \equiv 0$ (mod m) is said to be of degree n if $a_0 \not\equiv 0$ (mod m); more generally, one has:

Definition 15. *If $a_0 \equiv a_1 \equiv \cdots \equiv a_{n-j-1} \equiv 0$ (mod m), $a_{n-j} \not\equiv 0$ (mod m), then*

$$a_0 x^n + a_1 x^{n-1} + \cdots + a_{n-j-1} x^{j+1} + a_{n-j} x^j + \cdots$$

$$+ a_{n-1} x + a_n \equiv 0 \ (\text{mod } m)$$

is said to be a congruence of degree j.

REMARK 11. If $f(x)$ is defined by (8), then it follows from Taylor's Theorem that

$$f(x) = f(a) + (x - a)\frac{f'(a)}{1!} + \cdots + (x - a)^n \frac{f^{(n)}(a)}{n!}$$

$$= f(a) + (x - a)g(x).$$

Here $g(x)$ is a polynomial with integral coefficients because $(f^{(m)}(a))/m!$ is an integer for every m and integer a (the reader is invited to give a proof of this fact in Problem 37); hence if $f(a) \equiv 0$ (mod m), then

$$f(x) = f(a) + (x - a)g(x) \equiv (x - a)g(x) \ (\text{mod } m).$$

The converse, that $f(x) \equiv (x - a)g(x)$ (mod m) implies $f(a) \equiv 0$ (mod m), is trivial. By analogy with the definition of a multiple root of an equation,

Remark 11 suggests the following:

DEFINITION 16. The integer a is said to be a *k-fold solution of the congruence* $f(x) \equiv 0 \pmod{m}$ if $f(x) \equiv (x - a)^k g(x) \pmod{m}$, with k a positive integer and $g(x)$ a polynomial with integral coefficients and $g(a) \not\equiv 0 \pmod{m}$. In particular, if $k \geq 2$, a is said to be a *multiple solution*.

While an algebraic equation of degree n always has exactly n complex roots (if we count their multiplicities properly), the situation is more complicated in the case of congruences mod m. We may have either no solution at all or infinitely many, because if x_0 is a solution, so is $x = x_0 + km$. It is clearly of interest to consider as distinct only solutions that are not congruent mod m. Even with this restriction, congruences may have no solution, or one single solution mod m, or a number of solutions less than the degree or equal to it or higher than the degree. For illustration, consider the following:

EXAMPLES

$x^2 + 1 \equiv 0 \pmod{7}$ has no solution;
$x^2 - 3 \equiv 0 \pmod{6}$ has only one solution, $x \equiv 3 \pmod{6}$;
$x^2 + 1 \equiv 0 \pmod{17}$ has two solutions, $x \equiv 4$ and $x \equiv 13 \pmod{17}$;
$x^2 - 1 \equiv 0 \pmod{8}$ has four solutions (namely
$\quad x \equiv 1, 3, 5,$ or $7 \pmod{8}$), as already seen in the proof of Theorem 22.

As usual it is easier to bring some order into this apparent chaos when the modulus m is a prime power p^k, and we conclude this section with two theorems concerning prime and prime power moduli.

Theorem 24 (Lagrange). *If*

$$f(x) \equiv 0 \pmod{p} \tag{9}$$

is a congruence of degree $n \geq 1$, then it has at most n solutions.

Before proving this theorem, let us make a few remarks.

REMARK 12. In general, let $f(x) \equiv x^r g(x) \pmod{p}$, with $g(0) \not\equiv 0 \pmod{p}$, $r \geq 0$; then $g(x)$ can be selected so that its degree should not exceed $p - 2$. Indeed, for integral $x \not\equiv 0 \pmod{p}$, $x^{p-1} \equiv 1$, $x^p \equiv x$, $x^{k(p-1)+s} \equiv x^s \pmod{p}$ with $0 \leq s \leq p - 2$.

REMARK 13. If all coefficients of $f(x)$ are divisible by p, then it is clear that every integer is a solution of (9). In such a case we say that $f(x)$ *vanishes identically* mod p.

REMARK 14. One has to be careful in the interpretation of a congruence like $f(x) \equiv g(x) \pmod{p}$, where $f(x)$ is given by (8) and $g(x) = \sum_{j=0}^{m} b_j x^{m-j}$. It may stand for the ordinary congruence $h(x) \equiv 0 \pmod{p}$, where $h(x) = f(x)$

$- g(x)$, but usually it stands for the identical congruence, i.e., for the statement that $f(a) \equiv g(a) \pmod{p}$ holds for every integer a. By Theorem 24 that means that $a_j \equiv b_j \pmod{p}$ for $0 \leq j \leq \max(m, n)$. This is also, of course, the correct interpretation of the congruence mod m in the preceding Remark 12 and of the second congruence in Definition 16.

PROOF OF THEOREM 24.   For $n = 1$, this is precisely Theorem 13 with $m = p$; for $n > 1$ we use induction on n. Assume, contrary to the assertion, that Theorem 24 holds up to a certain degree $n - 1$, but that there exists a congruence $a_0 x^n + a_1 x^{n-1} + \cdots + a_n \equiv 0 \pmod{p}$ of degree n with $n + 1$ solutions $u_1, u_2, \cdots, u_{n+1}$, all incongruent mod p. Set $g(x) = f(x) - a_0(x - u_1) \cdots (x - u_n)$. Then the degree of $g(x)$ is at most $n - 1$. If $g(x)$ vanishes identically mod p (that is, if all coefficients of $g(x)$ are divisible by p) then $f(x) \equiv a_0(x - u_1) \cdots (x - u_n) \pmod{p}$ so that

$$0 \equiv f(u_{n+1}) \equiv a_0(u_{n+1} - u_1) \cdots (u_{n+1} - u_n) \pmod{p};$$

this, however, is possible only if $a_0 \equiv 0 \pmod{p}$ (here we use the assumption $u_i \not\equiv u_j \pmod{p}$ for $i \neq j$, and Corollary 3.4.1) so that $f(x)$ vanishes identically, contrary to the assumption $n > 1$. Consequently, $g(x) \equiv 0 \pmod{p}$ is a congruence of degree n_1, with $1 \leq n_1 \leq n - 1$; hence by the induction assumption, it can have at most $n - 1$ solutions incongruent mod p. However, $u_1, u_2, \ldots, u_n$ are clearly solutions and are also, by assumption, incongruent $\pmod{p}$; this contradiction shows that (9) could not have had $n + 1$ incongruent solutions and finishes the proof.

It is interesting to observe[†] that if (9) has exactly $m(\leq n)$ solutions $\pmod{p}$, then in general,

$$f(x) \equiv 0 \pmod{p^r} \tag{10}$$

also has exactly m distinct solutions; this now means, of course, that (10) has m solutions incongruent mod p^r.

To state the result with precision, we recall the definition and an important property of the discriminant[‡] D of the polynomial (1) with integral coefficients.

DEFINITION 17.   If $x_1, x_2, \ldots, x_n$ are the (not necessarily integral, not even necessarily real) roots of $f(x) = 0$, so that $f(x) = a_0(x - x_1) \cdots (x - x_n)$, then the product $D = a_0^{2n-2} \prod_{i<k}(x_i - x_k)^2$ is called the *discriminant* of $f(x)$; alternatively, D may be defined by

$$D = (-1)^{n(n-1)/2} a_0^{n-2} \prod_{i=1}^{n} f'(x_i),$$

where $f'(x)$ stands for the derivative of $f(x)$.

[†] This remark is of considerable importance in the study of p-adic valuations, and in connection with a famous theorem known as Hensel's Lemma.

[‡] To conform to a generally accepted custom, the discriminant will be denoted by a capital D, although here it is a rational integer.

REMARK 15.   The identity of the two definitions is a non-trivial theorem (see, e.g., [8] Section 28). The fact that D is an integer is not obvious from either definition, but becomes so using still another representation of D, namely[†] $D = (-1)^{n(n-1)/2} a_0^{-1} R(f, f')$, where $R(f, f')$ stands for the resultant of $f(x)$ and $f'(x)$; this is a certain[‡] determinant, having as entries the (integral) coefficients of $f(x)$ and $f'(x)$ and divisible by a_0.

As an illustration, let us consider the polynomial

$$f(x) = 3(x - 1)(x + 2)\left(x - \tfrac{2}{3}\right) = 3x^3 + x^2 - 8x + 4,$$

with $n = 3$ and $f'(x) = 9x^2 + 2x - 8$. According to the first definition,

$$D = 3^4(1 + 2)^2\left(1 - \tfrac{2}{3}\right)^2\left(-2 - \tfrac{2}{3}\right)^2 = 3^2 \cdot 1^2 \cdot 8^2 = 576.$$

According to the second definition,

$$D = (-1)^{(3 \cdot 2)/2} \cdot 3^{3-2} f'(1) f'(-2) f'\left(\tfrac{2}{3}\right) = -3 \cdot 3 \cdot 24 \cdot (-8/3) = 576.$$

Finally,

$$R(f, f') = \begin{vmatrix} 3 & 1 & -8 & 4 & 0 \\ 0 & 3 & 1 & -8 & 4 \\ 9 & 2 & -8 & 0 & 0 \\ 0 & 9 & 2 & -8 & 0 \\ 0 & 0 & 9 & 2 & -8 \end{vmatrix}$$

$$= 3 \cdot 4 \cdot 2 \begin{vmatrix} 1 & 1 & -8 & 2 & 0 \\ 0 & 3 & 1 & -4 & 1 \\ 3 & 2 & -8 & 0 & 0 \\ 0 & 9 & 2 & -4 & 0 \\ 0 & 0 & 9 & 1 & -2 \end{vmatrix} = -3 \cdot 24^2,$$

so that $D = (-1)^{(3 \cdot 2)/2} \cdot 3^{-1}(-3 \cdot 24^2) = 576.$

[†]Observe erroneous omission of the factor $(-1)^{n(n-1)/2}$ in [8].
[‡]In general the resultant of $f(x) = \sum_{j=0}^{n} a_j x^{n-j}$ and $g(x) = \sum_{j=0}^{m} b_j x^{m-j}$ is

$$R(f, g) = \left. \begin{vmatrix} a_0 & a_1 & \cdots & a_n & 0 & 0 & \cdots & 0 \\ 0 & a_0 & \cdots & a_{n-1} & a_n & 0 & \cdots & 0 \\ \vdots & \vdots & & & & & & \\ 0 & 0 & \cdots & a_0 & a_1 & & \cdots & a_n \\ b_0 & b_1 & \cdots & b_m & 0 & \cdots & 0 \\ 0 & b_0 & \cdots & b_{m-1} & b_m & \cdots & 0 \\ \vdots & \vdots & & & & & \\ 0 & 0 & \cdots & b_0 & b_1 & \cdots & b_m \end{vmatrix} \right\} \begin{matrix} \\ m \text{ rows} \\ \\ \\ \\ n \text{ rows} \\ \\ \end{matrix}$$

Let us make still another observation: If $r \geq 2$ and $x = a$ is a solution of the congruence (10), then it is, *a fortiori*, a solution of the congruence

$$f(x) \equiv 0 \pmod{p^{r-1}}. \tag{10'}$$

The converse is not necessarily true; if $x = b$ is a solution of (10′) with $0 \leq b < p^{r-1}$, it may or may not also be a solution of (10). We may try, however, to find a solution of (10) of the form

$$a = b + kp^{r-1}, 0 \leq k < p, 0 \leq b < p^{r-1}. \tag{11}$$

Indeed, as observed, if a is a solution of (10), then it is also a solution of (10′); hence so also is every b satisfying $b \equiv a \pmod{p^{r-1}}$, and in particular, the unique $b < p^{r-1}$, congruent to $a \pmod{p^{r-1}}$. Consequently, every solution a of (10) can be written in the form (11); such a solution a of (10) is said to *correspond* to the solution b of (10′). Now the key theorem may be stated as follows.

Theorem 25. *For $r \geq 2$, to every solution $b \pmod{p^{r-1}}$ of (10′) there corresponds exactly one solution $\pmod{p^r}$ of (10), provided that $p \nmid f'(b)$; if $p \mid f'(b)$ but $p^r \nmid f(b)$, then there are no solutions of (10) corresponding to b; finally if $p \mid f'(b)$ and $p^r \mid f(b)$, then there are exactly p solutions (distinct $\mathrm{mod}\ p^r$) of (10), corresponding to the solution b of (10′).*

PROOF. By Taylor's theorem,

$$f(x) = f(c) + (f'(c)/1!)(x - c) + \cdots + (f^{(n)}(c)/n!)(x - c)^n,$$

and for integral c as already seen, the coefficients $f^{(j)}(c)/j!$ are all integers. In particular, for $a = b + kp^{r-1}$,

$$f(b + kp^{r-1}) = f(b) + f'(b) \cdot kp^{r-1} + \sum_{j=2}^{n} \frac{f^{(j)}(b)}{j!} k^j p^{j(r-1)}.$$

If $r \geq 2, j \geq 2$, then $j(r - 1) \geq r$, so that the last sum is divisible by p^r; hence, if $a = b + kp^{r-1}$ is a solution of (10), so that $f(a) \equiv 0 \pmod{p^r}$, we obtain $f(b) + f'(b)kp^{r-1} \equiv 0 \pmod{p^r}$. However, $p^{r-1} \mid f(b)$, because b is a solution of (10′); hence dividing by p^{r-1}, the last congruence may be written as

$$kf'(b) \equiv -f(b)p^{1-r} \pmod{p}. \tag{12}$$

If $p \nmid f'(b)$, then it follows by Theorem 13 that the congruence (12) has a unique solution $k \pmod{p}$. If $p \mid f'(b)$ but $p \nmid f(b)p^{1-r}$, then (12) cannot be satisfied by any k, while if $p \mid f'(b)$, $p \mid f(b)p^{1-r}$ (i.e., if $p^r \mid f(b)$), then (12) is trivially satisfied for every value of $k \pmod{p}$ and the Theorem is proven.

From Theorem 25 easily follows the result we are actually aiming at, namely:

Corollary 25.1. *If D is the discriminant of the polynomial (8) and if $p \nmid D$, then all congruences (10) have the same number $m(\leq n)$ of solutions, which is in particular the number of solutions of (9).*

In the proof of the Corollary we shall use the following lemmas, which are close analogs of a well-known theorem in the algebra of polynomials.

Lemma 1.   *A solution b of* (9) *is a multiple solution if and only if* $p|f'(b)$.

PROOF.   By Taylor's theorem,

$$f(x) = f(b) + (x - b)f'(b) + \cdots + (x - b)^n f^{(n)}(b)/n!.$$

As already observed in the proof of Theorem 25, the coefficients $f^{(k)}(b)/k!$ are integers if b is an integer; hence

$$f(x) = f(b) + (x - b)f'(b) + (x - b)^2 g(x).$$

By assumption, $p|f(b)$; if $p|f'(b)$ also holds, then $f(x) \equiv (x - b)^2 g(x)$ (mod p), and in view of Definition 16, b is a multiple solution of (9). Conversely, if $x \equiv b$ (mod p) is a multiple solution of (9), then by Definition 16, $f(x) \equiv (x - b)^2 g_1(x)$ (mod p); but $p|f(b)$, so that by Taylor's theorem $f(x) \equiv (x - b)f'(b) + (x - b)^2 g(x)$ (mod p) also holds. Consequently, $(x - b)f'(b) \equiv (x - b)^2(g_1(x) - g(x))$ (mod p) or $f'(b) \equiv (x - b)h(x)$ (mod p), where $h(x) = g_1(x) - g(x)$ is a polynomial with integral coefficients (the reader should supply a justification for the cancellation of the factor $x - b$). The left side being independent of x, we may substitute any integer for x on the right side without changing its value (mod p). Setting in particular, $x = b$ we obtain $f'(b) \equiv 0$ (mod p), and Lemma 1 is proven.

Lemma 2.   *If congruence* (9) *has a multiple solution, then* $p|D$.

PROOF.   If b is a multiple solution of (9), then, by Definition 16, $f(x) \equiv (x - b)^2 g(x)$ (mod p). By Remark 14, the coefficients $a_j(0 \leq j \leq n)$ of $f(x)$ are congruent mod p to the coefficients c_j of $h(x) = (x - b)^2 g(x) = \sum_{j=0}^m c_j x^{m-j}(m \geq n)$. Consequently, if D_h is the discriminant of $h(x)$, then $D \equiv D_h$ (mod p), because both discriminants are polynomials in the respective coefficients. However $D_h = 0$, because the equation $h(x) = 0$ has the multiple root $x = b$; hence $D \equiv 0$ (mod p).

PROOF OF COROLLARY 25.1.   Let b_i ($i = 1, 2, \ldots, m$) be one of the m solutions of (9) incongruent mod p, i.e., of (10) with $r = 1$. Then $f(b_i) \equiv 0$ (mod p), and from $p \nmid D$ and Lemma 2 it follows that (9) has no multiple solutions; hence by Lemma 1, $p \nmid f'(b_i)$. By Theorem 25 it follows that to each b_i corresponds (mod p^2) exactly one solution of (10) with $r = 2$. This proves the Corollary for $r = 2$. Let us assume then that we already know that (10′) has exactly m solutions (mod p^{r-1}), $r \geq 2$. If b_i is one of them, then $p^{r-1}|f(b_i)$; hence, a fortiori, $p|f(b_i)$. Consequently, using $p \nmid D$ and the two Lemmas, it follows that $p \nmid f'(b_i)$. It now follows again from Theorem 25 that to each b_i corresponds (mod p^r) exactly one solution of (10); hence, as (10′) has m solutions (mod p^{r-1}), (10) also has exactly m solutions (mod p^r), and the Corollary is proven.

Theorem 26.   *Let p be a prime and n an arbitrary positive integer. Let a be an integer not divisible by p, and set $d = (n, p - 1)$. Then the congruence*

$$x^n \equiv a \;(\text{mod } p) \tag{13}$$

has d solutions modulo p, if $a^{(p-1)/d} \equiv 1$ (mod p); otherwise, (13) has no solutions.

PROOF. $a \equiv x^n \Rightarrow a^{(p-1)/d} \equiv x^{n(p-1)/d} \equiv (x^{p-1})^{n/d} \equiv 1$ (mod p) by Theorem 11; hence no solution can exist if $a^{(p-1)/d} \not\equiv 1$ (mod p). On the other hand, if $a^{(p-1)/d} \equiv 1$ (mod p), let g be a primitive root mod p. Then by Theorem 17, $a \equiv g^j$ (mod p) for some integer j, and any solution x of the given congruence can be written as $x \equiv g^y$; moreover, to solutions x incongruent mod p correspond exponents y incongruent mod $p - 1$ (the reader should prove this formally). Substituting in (13), the congruence becomes $g^{yn} \equiv a \equiv g^j$ (mod p) $\Leftrightarrow ny \equiv j$ (mod $p - 1$). By Theorem 14, the last congruence has no solution if d does not divide j, and has exactly one solution mod($p - 1)/d$) if $d \mid j$. To decide between these two alternatives we observe that

$$\left(g^j\right)^{(p-1)/d} \equiv a^{(p-1)/d} \equiv 1 \;(\text{mod } p)$$

by assumption; hence by Remark 7 applied to g with $(g, p) = 1$ and $h = p - 1$, one has $p - 1 \mid (j(p-1)/d)$, so that j/d is an integer and $d \mid j$. Consequently, $ny \equiv j$ (mod $p - 1$) has exactly one solution y_0 (mod($p - 1)/d$) and the d solutions $y = y_0 + k(p-1)/d (k = 0, 1, \ldots, d - 1)$, incongruent mod $p - 1$. To each corresponds a solution x incongruent mod p, and the Theorem is proven.

We shall not pursue further the general theory of higher congruences, but pass on immediately to study in some detail a most important particular case, that of quadratic congruences. The reader interested in the general theory is advised to consult such books as the excellent *Introduction to Number Theory*, by Nagell [5].

PROBLEMS

1. Prove Theorem 1 in detail.
2. Prove that Definition 5 is meaningful; that is, show that each element of the set **A** belongs to one and only one equivalence class, so that $\mathscr{R}$ indeed induces a partition of the elements of **A**.

3. (i) Consider the set whose elements are the ordered couples of integers, the second being different from zero. Two couples (a, b), (c, d) are said to be "similar" if $ad = bc$. Prove that this "similarity" is an equivalence relation.

(ii) Define "addition" among the equivalence classes of couples by $(a, b) + (c, d) = (ad + bc, bd)$ and "multiplication" by $(a, b) \times (c, d) = (ac, bd)$. Show that

$$(a, b) + (c, d) = (e, f) \Leftrightarrow \frac{a}{b} + \frac{c}{d} = \frac{e}{f}$$

and

$$(a, b) \times (c, d) = (g, h) \Leftrightarrow \frac{a}{b} \cdot \frac{c}{d} = \frac{g}{h}.$$

Show further that if we establish a correspondence $(a, b) \Leftrightarrow (r/s)$, with r and s uniquely determined by $(r/s) = (a/b)$, $(r, s) = 1$, then (r/s) is independent of the couple (a, b) selected and depends only on its equivalence class. These results are expressed succinctly by stating that the set of equivalence classes of ordered couples is *isomorphic* to the set of rational fractions in reduced form. What is the subset of equivalence classes of couples (a, b) corresponding to the rational integers? To zero? To one?

4. Prove Theorem 3 in detail.

5. Prove Theorem 4 in detail.

6. Write out in complete detail the proof of Theorem 5.

7. Find a complete set of residues mod 7, all of which are divisible by 10. Can you solve the same problem if 7 is replaced by 6? Why?

8. Prove Corollary 7.1.

9. Prove that for every integer n, $n(n + 1)(2n + 1)/6$ is also an integer.

10. Prove that $(a + b)^p \equiv a^p + b^p \pmod{p}$; more generally, that

$$(a + b + \cdots + z)^p \equiv a^p + b^p + \cdots + z^p \pmod{p}.$$

11. Prove that if n is an integer, then $3n^2 - 1$ is never the square of an integer. (Hint: Use congruences mod 3.)

12. Prove that for every odd n, $n^2 \equiv 1 \pmod{8}$.

13. Prove that for every integer n, $n^3 \equiv -1, 0,$ or $+1 \pmod{9}$.

14. Prove that the following congruences hold for every integer n:
(a) $2^{2n} - 1 \equiv 0 \pmod{3}$;
(b) $2^{3n} - 1 \equiv 0 \pmod{7}$;
(c) $2^{4n} - 1 \equiv 0 \pmod{15}$.

15. Prove that there exist infinitely many primes $p \equiv 3 \pmod 4$.

16. Prove that if $n = 2^m + 1$ is prime, then m must be a power of 2 (n is then called a *Fermat prime*).

17. Prove that if $n = 2^m - 1$ is prime, m is also a prime (n is then called a *Mersenne prime*).

18. Prove Remark 2 in detail.

19. Find the smallest positive solution of the congruence $240^{37} \equiv x \pmod 7$.

20. Find the greatest common divisor $(p!, (p-1)! - 1)$, where p is a prime.

21. (Lagrange) Let p be an odd prime. Define the coefficients $a_1, a_2, \ldots, a_{p-1}$ by the identity

$$(x-1)(x-2)\cdots(x-p+1) = x^{p-1} - a_1 x^{p-2} + \cdots + a_{p-1}$$

or

$$x(x-1)\cdots(x-p+1) = x^p - a_1 x^{p-1} + \cdots + a_{p-1} x.$$

Here replace x by $x - 1$, getting

$$(x-1)(x-2)\cdots(x-p)$$
$$= (x-p)[(x-1)(x-2)\cdots(x-p+1)]$$
$$= (x-p)\left[x^{p-1} - a_1 x^{p-2} + \cdots + a_{p-1}\right]$$
$$= (x-1)^p - a_1(x-1)^{p-1} + \cdots + a_{p-1}(x-1).$$

Equate coefficients of equal powers in the last two expressions, obtain equations for the a_i $(i = 1, 2, \ldots, p-1)$ and infer that

$$p \mid a_i \ (i = 1, 2, \ldots, p-2), \quad a_{p-1} \equiv -1 \pmod p.$$

Obtain from this a third proof of Wilson's Theorem 12. Next, observing that $x^{p-1} - a_1 x^{p-2} + \cdots + a_{p-1} \equiv x^{p-1} - 1 \pmod p$, find a new proof of Fermat's Theorem 11.

22. Find the general solution of the Diophantine equation $10x - 49y = 3$.

23. Find the general solution of the congruence $3^x - 2^x \equiv 5 \pmod 7$. What can you say about the congruence $3^x - 2^x \equiv 4 \pmod 7$?

24. Let $d = (a, b)$. Show how the Euclidean algorithm (Chapter 3, Problem 3) leads to a method of finding integers m and n such that $am + bn = d$. (Hint: Observe that $d = r_{m-1} = r_{m-3} - r_{m-2}q_{m-1}$, and that one may continue to replace r_k by $r_{k-2} - r_{k-1}q_k$ all the way back to a and b.) *Application*: $a = 12$, $b = 22$.

25. Solve the system of simultaneous congruences:

$$3x + 7y - 10 \equiv 0 \pmod{14},$$
$$11x - 8y + 8 \equiv 0 \pmod{14}.$$

26. Let g be a primitive root mod p. Then $p - g$ is also a primitive root mod p if $p \equiv 1 \pmod{4}$, and belongs to the exponent $(p - 1)/2$ if $p \equiv 3 \pmod{4}$.

27. Justify the example: $\phi(p^r) = p^r - p^{r-1}$ on page 43.

28. Prove Corollary 17.1.

29. Prove that if $m = p^r$, then g is a primitive root if and only if $g, g^2, \ldots, g^{p^r - p^{r-1}}$ are all incongruent mod m (Hint: Use Problem 27.)

30. Prove that for every prime p there exists a primitive root g such that $g^{p-1} \not\equiv 1 \pmod{p^2}$. (Hint: With g, $g_1 = g + p$ is also a primitive root mod p, and if $g^{p-1} \equiv 1 \pmod{p^2}$, then $g_1^{p-1} \not\equiv 1 \pmod{p^2}$.)

31. Let g be a primitive root mod p, with $g^{p-1} \not\equiv 1 \pmod{p^2}$; then $g^{(p-1)p^{m-2}} \not\equiv 1 \pmod{p^m}$ for every integer $m \geq 2$. (Hint: By Problem 30, the statement is true for $m = 2$ and some g; now use induction by assuming the statement to be true with $m - 1$ instead of m. By Theorem 10, $g^{\phi(p^{m-2})} = g^{p^{m-3}(p-1)} \equiv 1 \pmod{p^{m-2}}$, so that by the induction hypothesis, $g^{p^{m-3}(p-1)} = 1 + qp^{m-2}$, $p \nmid q$. The proof by induction may be concluded by elevating both sides to the power p.)

32. If g is a primitive root mod p and $g^{p-1} \not\equiv 1 \pmod{p^2}$, then g is a primitive root mod p^m. (Hint: From $(g, p^m) = 1$ it follows (see Definition 12 and the remarks preceding it) that there exists a positive integer v to which g belongs mod p^m, so that $g^v \equiv 1 \pmod{p^m}$ and $v \mid p^{m-1}(p - 1)(= \phi(p^m))$. But $g^v \equiv 1 \pmod{p}$ also holds; hence $p - 1 \mid v$, so that $v = (p - 1)p^a$ with $0 \leq a \leq m - 1$ and $g^{(p-1)p^a} \equiv 1 \pmod{p^m}$. Assuming $a < m - 1$, one gets a contradiction with Problem 31; hence $v = \phi(p^m)$.)

33. If $n = 2p^m$, then n has a primitive root. (Hint: If g is an odd primitive root mod p^m, then it is also a primitive root mod $2p^m$; if g is an even primitive root mod p^m, then $g + p^m$ is an odd primitive root mod p^m, hence also mod $2p^n$.)

34. Prove that $x^p - x \equiv 0 \pmod{p}$ has all rational integers as solutions.

35. Justify Remark 9.

36. Let $f(x) \equiv 0 \pmod{p}$ be a congruence of degree n. Prove that it has exactly n solutions mod p if and only if $f(x)$ is a factor of $x^p - x \pmod{p}$, that is, if $x^p - x + pg(x) = f(x)h(x)$ for some polynomials $g(x), h(x)$ with integral coefficients.

37. Let $F(x)$ be a polynomial with integral coefficients and let a and n be integers, $n \geq 0$; prove that $F^{(n)}(a)/n!$ is an integer.

38. Show that if $h(x)$ is a polynomial of degree at most $p - 2$ and $(x - b)h(x) \equiv 0 \pmod{p}$ holds identically, then one may cancel the factor $x - b$. (This fact has been used in the proof of Lemma 1, p. 60.)

BIBLIOGRAPHY

1. L. E. Dickson, *History of the Theory of Numbers*. New York: Chelsea, 1952.

2. C. F. Gauss, *Disquisitiones Arith.* (Translated from the second Latin edition, 1870, by A. Clark). New Haven: Yale Univ. Press, 1966, see Art. 24, p. 77.

3. W. J. LeVeque, *Topics in Number Theory*, Vol. 1. Reading, Mass.: Addison Wesley, 1956.

4. N. H. McCoy, *The Theory of Numbers*. New York: Macmillan, 1965.

5. T. Nagell, *Introduction to Number Theory*. New York: Wiley, 1951.

6. M. A. Stern, *Lehrbuch der Algebraischen Analysis*. Leipzig: Winter'sche Verlagsbuchhandlung, 1860.

7. Sun Tsu, *Suan-ching* (arithmetic), edited by Y. Mikaini, *Abhandlungen-Geschichte der Mathematischen Wissenschaften* **30** (1912), 32.

8. B. L. Van der Waerden, *Modern Algebra*, Vol. 1. New York: Frederick Ungar, 1953.

Chapter 5

Quadratic Residues

1 INTRODUCTION

Consider the general congruence of second degree modulo[†] an odd prime p:

$$ax^2 + bx + c \equiv 0, p \nmid a. \tag{1}$$

On account of $p \nmid a$, (1) is equivalent to

$$4a(ax^2 + bx + c) = (2ax + b)^2 - (b^2 - 4ac) \equiv 0.$$

$p \nmid a$ also implies (Theorem 4.13) that $2ax + b \equiv x_1$ has a unique solution mod p; hence setting $2ax + b \equiv x_1$ and $b^2 - 4ac \equiv a_1$, (1) is seen to be equivalent to the system

$$x_1^2 \equiv a_1, \tag{2}$$

$$2ax \equiv x_1 - b. \tag{2'}$$

Here (2') is a linear congruence and can be handled according to Section 4.4 (see in particular Theorem 4.13); hence the general problem of quadratic congruences reduces to that of solving (2). Concerning (2), the following fundamental theorem holds:

Theorem 1 (Legendre).

(1) $p \nmid a \Rightarrow x^2 \equiv a$ has $\begin{cases} 2 \text{ solutions if } a^{(p-1)/2} \equiv 1, \\ 0 \text{ solutions otherwise.} \end{cases}$

(2) $a^{(p-1)/2} \not\equiv 1 \Rightarrow a^{(p-1)/2} \equiv -1.$

FIRST PROOF. The first assertion follows by taking $n = 2$ in Theorem 4.26. For the second, observe that by Theorem 4.11 one has for every $a \not\equiv 0$ that

---

[†] In this section all congruences without indication of a modulus are understood to be mod p, with p an *odd* prime.

$0 \equiv a^{p-1} - 1 = (a^{(p-1)/2} - 1)(a^{(p-1)/2} + 1)$; hence $a^{(p-1)/2} \not\equiv 1 \Rightarrow a^{(p-1)/2} \equiv -1$, thus completing the proof.

We present also a second proof of Theorem 1, independent of Theorem 4.26 and of the concept of primitive roots.

SECOND PROOF.   Assertion (2) follows as before from the remark $a \not\equiv 0 \Rightarrow$ either $a^{(p-1)/2} \equiv -1$, or $a^{(p-1)/2} \equiv 1$; it only remains to prove (1). If $a^{(p-1)/2} \equiv -1$, then $x^2 \equiv a$ can have no solution. Indeed if $x \equiv c$ were a solution, then

$$a^{(p-1)/2} \equiv (c^2)^{(p-1)/2} \equiv c^{p-1} \equiv 1$$

(by Theorem 4.11), contrary to the assumption $a^{(p-1)/2} \equiv -1$. This proves the last assertion of (1). Next, if $a^{(p-1)/2} \equiv 1$, then by Theorem 4.24, a is one of the at most $(p-1)/2$ solutions (incongruent mod p) of $x^{(p-1)/2} \equiv 1$. A set of $(p-1)/2$ solutions, incongruent mod p, is furnished by 1^2, $2^2, \ldots, ((p-1)/2)^2$. These are indeed solutions, because for $1 \leq n \leq (p-1)/2$, $(n^2)^{(p-1)/2} = n^{p-1} \equiv 1$ by Theorem 4.11; they are incongruent because, if $1 \leq n < m \leq (p-1)/2$, then $0 < m - n < m + n \leq p - 1 < p$, so that $m^2 - n^2 = (m-n)(m+n) \not\equiv 0$. Consequently, a is necessarily of the form $a \equiv n^2$ $(1 \leq n \leq (p-1)/2)$.

Now let $a \equiv n^2$ be any of these $(p-1)/2$ residues; then $x^2 \equiv a$ has the two incongruent solutions $x \equiv n$ and $x \equiv p - n$ (indeed, $n \equiv p - n$ would imply $p|n$, impossible for $1 < n < (p-1)/2$), and this also finishes the proof of the first assertion of (1).

2   THE LEGENDRE SYMBOL AND THE LAW OF QUADRATIC RECIPROCITY

In view of Theorem 1, it seems appropriate to introduce

DEFINITION 1.   Let $(r, m) = 1$; then r is said to be a *quadratic residue modulo m* if there exists some integer x, such that $x^2 \equiv r$ (mod m). We call n a *quadratic nonresidue modulo m* if the congruence $x^2 \equiv n$ (mod m) has no solutions.

Corollary 1.1.   *There exist exactly* $(p-1)/2$ *quadratic residues and* $(p-1)/2$ *quadratic nonresidues modulo p.*

PROOF.   Left to the reader.

DEFINITION 2.   For $p \nmid a$ we define the *Legendre Symbol* (a/p) of quadratic residuacy as follows:

$(a/p) = +1$ if a is a quadratic residue and $(a/p) = -1$ if a is a quadratic nonresidue modulo p.

EXAMPLE

Modulo 7, one has $1^2 \equiv 6^2 \equiv 1$; $2^2 \equiv 5^2 \equiv 4$; $3^2 \equiv 4^2 \equiv 2$; hence, 1, 2, and 4 are quadratic residues and 3, 5, 6 are quadratic nonresidues mod 7.

Using Legendre's symbol, $(1/7) = (2/7) = (4/7) = 1$ and $(3/7) = (5/7) = (6/7) = -1$.

Theorem 2. *Let p be an odd prime, $p \nmid a \cdot b$; then*

(1) $a \equiv b \pmod{p} \Rightarrow (a/p) = (b/p)$;
(2) $(a^2/p) = 1$;
(3) $(1/p) = 1$;
(4) $(a/p)(b/p) = (ab/p)$;
(5) $(a/p) \equiv a^{(p-1)/2} \pmod{p}$;
(6) $(-1/p) = (-1)^{(p-1)/2} = \begin{cases} 1 & \text{if } p \equiv 1 \ (\mathrm{mod}\,4), \\ -1 & \text{if } p \equiv 3 \ (\mathrm{mod}\,4). \end{cases}$

PROOFS.   Assertions (1) and (2) should be evident, (3) is the particular case $a = 1$ of (2), (5) is a rephrasing of Theorem 1, and (6) is the particular case $a = -1$ of (5). Proof of (4):

$$(a/p)(b/p) \equiv a^{(p-1)/2}b^{(p-1)/2} = (ab)^{(p-1)/2}$$
$$\equiv (ab/p) \pmod{p} \ (\text{we used (5) twice}).$$

However $(a/p)(b/p) = \pm 1$, $(ab/p) = \pm 1$; hence, $(a/p)(b/p) \equiv (ab/p)$ $(\mathrm{mod}\ p) \Rightarrow (a/p)(b/p) = (ab/p)$ because $p > 2$.

The Theory of quadratic residues is dominated by the famous Law of Quadratic Reciprocity of Gauss and our next aim is its proof.

Theorem 3 (Quadratic Reciprocity Law of Legendre-Gauss).   *If p and q are distinct odd primes, then*

$$\left(\frac{p}{q}\right)\left(\frac{q}{p}\right) = (-1)^{[(p-1)/2][(q-1)/2]}$$

REMARK 1.   Theorem 3 may be rephrased as follows: The symbols (p/q) and (q/p) are equal, except when p and q are both congruent to 3 (mod 4), in which case $(p/q) = -(q/p)$.

Theorem 3 had already been stated by Euler and an incomplete proof of it was given by Legendre. Gauss gave eight proofs of it. Many other mathematicians have given proofs (Bachmann counts 45, Gerstenhaber claimed (in 1963) that there were 152, the last one his own). The shortest proof known today is probably due to Frobenius (See [1]; for recent versions of it, see [2] pp. 69–71, and [3]). We shall present a version of one of Gauss' own proofs, due essentially to his student Eisenstein. The reader may be interested to know that in Gauss' opinion the three greatest mathematicians were Archimedes, Newton, and Eisenstein. Most people today would replace Eisenstein's name by that of Gauss himself, but nevertheless, Eisenstein's contribution to mathematics is highly impressive, especially if one keeps in mind that he died in 1852 at the age of only 29. Who knows whether we would not have agreed with Gauss' opinion, if Eisenstein had lived as long as Gauss himself! As do most other proofs (including Frobenius'), ours will make essential use of *Gauss' Lemma*, which we state as

Theorem 4 (Gauss's Lemma). *Let p be an odd prime, $p \nmid a$. Let $\mathbf{S} = \{r\}$ consist of the least positive residues of the set $\{ma\}$, $m = 1, 2, \ldots, (p-1)/2$. Denote by n the number of integers $r \in \mathbf{S}, r > p/2$; then $(a/p) = (-1)^n$.*

PROOF.   We make two observations: First, all elements of $\mathbf{S}$ are incongruent (mod p). Indeed if $m_1 \neq m_2$, then $m_1 a \equiv m_2 a \Rightarrow p|(m_1 - m_2)a \Rightarrow p|m_1 - m_2$, which is impossible because $0 < m_1, m_2 \leq (p-1)/2$. Next, denoting by $s_1, s_2, \ldots, s_n$ the values of $r > p/2$ and by $t_1, t_2, \ldots, t_k$ the values of $r < p/2$, also $p - s_i \neq t_j$ (indeed $p - s_i \equiv t_j \Rightarrow s_i + t_j \equiv 0 \Rightarrow p|m_i a + m_j a \Leftrightarrow p|m_i + m_j$, which is impossible because $2 \leq m_i + m_j \leq p - 1$). Consequently, $p - s_1, p - s_2, \ldots, p - s_n; t_1, \ldots, t_k$ are a set of $n + k = (p-1)/2$ integers, all belonging to the interval from 1 to $(p-1)/2$ and incongruent to each other mod p, hence they are precisely the integers $1, 2, \ldots, (p-1)/2$ in some order, and their product $(p - s_1)(p - s_2) \cdots (p - s_n)t_1 t_2 \cdots t_k$ equals $((p-1)/2)!$ However the product multiplied out equals $(-1)^n s_1 s_2 \cdots s_n t_1 \cdots t_k$ plus a multiple of p, or taking congruences mod p and replacing s_i and t_j by their values $m_i a$ and $m_j a$, $(-1)^n a^{(p-1)/2}((p-1)/2)! \equiv ((p-1)/2)!$ From $(p, ((p-1)/2)!) = 1$ follows, by Corollary 4.6.1, that $a^{(p-1)/2} \equiv (-1)^n$, and further, by Theorem 2(5), that $(-1)^n \equiv (a/p)$, whence using $p > 2$ that $(a/p) = (-1)^n$, as in the proof of Theorem 2(4).

In the proof of Theorem 3, and often afterwards, we shall use

DEFINITION 3.   The *greatest integer function*, in symbols $y = [x]$, is defined as the largest rational integer not in excess of x.

EXAMPLES

$[3] = 3$, $[-1] = -1$; in general, for every integer n, $[n] = n$; $[5/2] = 2$, $[e] = 2$, $[\pi] = 3$, $[-\pi] = -4$; in general, for nonintegral $x > 0$, $[-x] = -[x] - 1$.

Lemma 1 (Eisenstein). *Let m, n be odd coprime integers, $m \neq 1$, $n \neq 1$. Set $m' = \frac{1}{2}(m - 1)$, $n' = \frac{1}{2}(n - 1)$; then*

$$\sum_{r=1}^{m'} \left[\frac{nr}{m} \right] + \sum_{r=1}^{n'} \left[\frac{mr}{n} \right] = m'n'.$$

PROOF. The result is geometrically evident, if we observe that the sum on the left counts the points with integral coordinates (so-called *lattice points*) *inside* the rectangle of sides $m/2$ and $n/2$; indeed, there are no lattice points on the diagonal. (The coordinates (s, t) of a point of the diagonal satisfy $(s/t) = (m/n)$ with $s < m$, $t < n$, and m/n is in reduced form, because $(m, n) = 1$; hence, s and t cannot both be integers.) In the lower triangle, the vertical of abscissa r meets the diagonal at the ordinate rn/m; hence there are exactly $[rn/m]$ points with integral ordinates on it and if we let $r = 1, 2, \ldots, m'$, $m' = (m - 1)/2$, we obtain altogether $\sum_{r=1}^{m'}[rn/m]$ lattice points inside the lower triangle. Similarly, there are $\sum_{r=1}^{n'}[rm/n]$ lattice points in the upper triangle, while the total number of lattice points inside the rectangle is clearly $m' \cdot n'$, and the Lemma is proven.

Lemma 2. *If $p \nmid a$, $p' = (p - 1)/2$, and n is defined as in Theorem 4, then*

$$\sum_{m=1}^{p'} \left[\frac{ma}{p} \right] + \frac{1}{8}(a - 1)(p^2 - 1) \equiv n \; (\text{mod} \, 2).$$

PROOF. As in Gauss' Lemma (Theorem 4), consider the set $\mathbf{S} = \{r\}$ of least positive residues $ma \equiv r$ (mod p), where $m = 1, 2, \ldots, p'$, $p' = (p - 1)/2$.

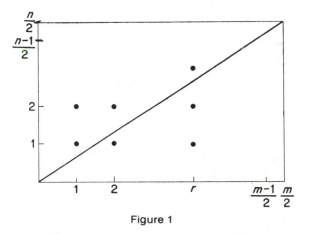

Figure 1

Given m, the corresponding r is the difference $ma - p[ma/p]$. Summing over all m, $a\sum_{m=1}^{p'} m - p\sum_{m=1}^{p'}[ma/p] = \sum r$, or

$$a \sum_{m=1}^{p'} m = p \sum_{m=1}^{p'} \left[\frac{ma}{p} \right] + \sum r. \tag{3}$$

Again, as in Gauss' Lemma, we distinguish between the r's less than $p/2$, denoted by $t_1,\ldots,t_k$, and those larger than $p/2$, denoted by $s_1, s_2,\ldots,s_n$. Then $\sum r = \sum_{j=1}^{n} s_j + \sum_{j=1}^{k} t_j$. We also recall that the set $\{ p - s_j, t_j \}$ is precisely the set of all integers from 1 to p' in some order; hence

$$\sum_{j=1}^{n} (p - s_j) + \sum_{j=1}^{k} t_k = np - \sum_{j=1}^{n} s_j + \sum_{j=1}^{k} t_j$$

$$= \sum_{m=1}^{p'} m = \frac{[(p-1)/2][(p+1)/2]}{2} = \frac{p^2 - 1}{8}. \tag{4}$$

(3) and (4) may be rewritten as

$$a\frac{p^2 - 1}{8} = p \sum_{m=1}^{p'} \left[\frac{ma}{p} \right] + \sum_{j=1}^{n} s_j + \sum_{j=1}^{k} t_j$$

and

$$-\frac{p^2 - 1}{8} = -np + \sum_{j=1}^{n} s_j - \sum_{j=1}^{k} t_j,$$

respectively. Adding, we obtain

$$(a - 1)\frac{p^2 - 1}{8} = p\left\{ \sum_{m=1}^{p'} \left[\frac{ma}{p} \right] - n \right\} + 2 \sum_{j=1}^{n} s_j.$$

Modulo 2, the right side is congruent to

$$p\left\{ \sum_{m=1}^{p'} \left[\frac{ma}{p} \right] - n \right\} \equiv \sum_{m=1}^{p'} \left[\frac{ma}{p} \right] - n \equiv n - \sum_{m=1}^{p'} \left[\frac{ma}{p} \right]$$

because $p \equiv 1 \pmod 2$ and $2u \equiv 0 \pmod 2 \Leftrightarrow u \equiv -u \pmod 2$, and the Lemma is proven.

PROOF OF THEOREM 3. By Gauss' Lemma, we know how to define integers n and m, so that $(q/p) = (-1)^n$, $(p/q) = (-1)^m$; whence $(p/q)(q/p) = (-1)^{n+m}$. By Lemma 2,

$$n \equiv \sum_{r=1}^{p'} \left[\frac{rq}{p} \right] + \frac{(q-1)(p^2-1)}{8} \equiv \sum_{r=1}^{p'} \left[\frac{rq}{p} \right] \pmod 2,$$

because p and q are odd, so that

$$(q - 1)(p^2 - 1) = (q - 1)(p + 1)(p - 1) \equiv 0 \ (\text{mod} \ 16).$$

In the same way $m \equiv \sum_{r=1}^{q'}[rp/q]$ (mod 2), where $q' = (q - 1)/2$. Hence $m + n \equiv \sum_{r=1}^{p'}[rq/p] + \sum_{r=1}^{q'}[rp/q]$ (mod 2). By Lemma 1, the second member equals $p'q'$, so that $(p/q)(q/p) = (-1)^{m+n} = (-1)^{p' \cdot q'}$ and the Theorem is proven.

Theorem 3 is valid only when the "numerator" and "denominator" in Legendre's symbol are both odd primes. One complements it by indicating the corresponding results for $(-1/p)$ and $(2/p)$. These are often called the "complementary laws."

Theorem 5 (Complementary Laws of Quadratic Reciprocity). *If p is an odd prime, then $(-1/p) = (-1)^{(p-1)/2}$ and $(2/p) = (-1)^{(p^2-1)/8}$.*

PROOF.   The first result is a restatement of Theorem 2(6). To prove the second, set $a = 2$ in Lemma 2:

$$\sum_{m=1}^{p'} \left[\frac{2m}{p} \right] + \frac{2 - 1}{8}(p^2 - 1) \equiv n \ (\text{mod} \ 2).$$

In the first member, $2m/p \leq 2p'/p = (p - 1)/p < 1$; hence $[2m/p] = 0$ and $n \equiv (p^2 - 1)/8$ (mod 2). The result now follows from Theorem 4.

Corollary 5.1. -1 *is a quadratic residue of the primes $p \equiv 1$ (mod 4) and a quadratic nonresidue of the primes $p \equiv 3$ (mod 4).*

Corollary 5.2.   2 *is a quadratic residue of the primes $p \equiv \pm 1$ (mod 8) and a quadratic nonresidue of the primes $p \equiv \pm 3$ (mod 8).*

PROOF.   Left to the reader.

So far, the Legendre Symbol (a/p) has been defined only for $p \nmid a$. It often is convenient to have a meaning attached to (a/p) for all a. Therefore, Definition 2 is complemented by

DEFINITION 2′.   If $p|a$, then $(a/p) = 0$.

To illustrate the advantages of Definition 2′, we observe for instance how simply one can now state, in full generality and without any need for case distinctions, that the number of solutions of the congruence $x^2 \equiv a$ (mod p) is $1 + (a/p)$. The cases $(a/p) = \pm 1$ are settled by Theorem 1, while for $p|a$ there is only one solution, namely $x \equiv 0$ (mod p). To illustrate the usefulness of the Reciprocity Law let us compute the Legendre Symbol $\left(\dfrac{213}{499}\right)$.

$\left(\dfrac{213}{499}\right) = \left(\dfrac{3}{499}\right)\left(\dfrac{71}{499}\right)$ by Theorem 2(4); $\left(\dfrac{3}{499}\right) = -\left(\dfrac{499}{3}\right)$ by the Reciproc-

ity Law, because $499 \equiv 3 \pmod{4}$; $-\left(\dfrac{499}{3}\right) = -\left(\dfrac{1}{3}\right) = -1$ by Theorem 2(1)

and 2(3). Similarly, $\left(\dfrac{71}{499}\right) = -\left(\dfrac{499}{71}\right) = -\left(\dfrac{2}{71}\right) = -(-1)^{(71^2-1)/8} = -1$ by

Theorems 3, 2(1), and 5 respectively; consequently, $\left(\dfrac{213}{499}\right) = +1$.

3   ⋆THE JACOBI AND KRONECKER SYMBOLS

DEFINITION 4.   Let a and b be coprime integers, with $b = p_1 p_2 \cdots p_r$ odd (the prime factors need not be distinct); then the Jacobi Symbol (a/b) is defined by $(a/b) = \prod_{i=1}^{r}(a/p_i)$, where (a/p_i) are Legendre Symbols.

Paralleling Definition 2' we might also add;

DEFINITION 4'.   If $(a, b) > 1$, then the Jacobi Symbol (a/b) is defined and $(a/b) = 0$.

REMARK 2.   The Legendre symbol (a/p_i) occurs in the product as often as the highest power of p_i that divides b.

REMARK 3.   If a is a quadratic residue mod b, then the congruence $x^2 \equiv a \pmod{b}$ has solutions $b \mid c^2 - a$ for appropriate c and, a fortiori, $p_i \mid c^2 - a$; consequently, $x^2 \equiv a \pmod{p_i}$ has solutions for $p_i \mid b$, $(a/p_i) = 1$, and hence $(a/b) = 1$. Therefore, if a is a quadratic residue mod b, then $(a/b) = 1$, as in the case of the Legendre symbol, but the converse is not true any more. It is possible to have $(a/p_i) = -1$ for an even number (say, $2h$) of prime factors p_i, so that $x^2 \equiv a \pmod{p_i}$ has no solutions; hence, a fortiori, $x^2 \equiv a \pmod{b}$ has none, while $(a/b) = \prod_i(a/p_i) = (-1)^{2h} = +1$.

From the definition of Legendre and Jacobi Symbols immediately follows:

Theorem 6.   Let b, b_i be odd, with $(a, b) = (a_i, b) = (a, b_i) = 1$ $(i = 1, 2)$; then

(1) $a_1 \equiv a_2 \pmod{b} \Rightarrow \left(\dfrac{a_1}{b}\right) = \left(\dfrac{a_2}{b}\right);$

(2) $\left(\dfrac{a}{b_1}\right)\left(\dfrac{a}{b_2}\right) = \left(\dfrac{a}{b_1 b_2}\right);$

(3) $\left(\dfrac{a_1}{b}\right)\left(\dfrac{a_2}{b}\right) = \left(\dfrac{a_1 a_2}{b}\right);$

(4) $\left(\dfrac{a}{b^2}\right) = \left(\dfrac{a^2}{b}\right) = 1;$

(5) $a = a_1 m^2,\ b = b_1 n^2 \Rightarrow \left(\dfrac{a}{b}\right) = \left(\dfrac{a}{b_1}\right).$

PROOF.   The proofs are left to the reader.

The Law of Reciprocity and the Complementary Laws also generalize immediately to Jacobi Symbols, as follows:

Theorem 7.   *If a, b are odd, $(a, b) = 1$, then*

(1) $\left(\dfrac{a}{b}\right)\left(\dfrac{b}{a}\right) = (-1)^{a' \cdot b'}$ with $a' = \tfrac{1}{2}(a - 1),\ b' = \tfrac{1}{2}(b - 1);$

(2) $\left(\dfrac{-1}{b}\right) = (-1)^{(b-1)/2};$

(3) $\left(\dfrac{2}{b}\right) = (-1)^{(b^2-1)/8}.$

PROOFS.   Let a_1 and a_2 be odd integers. Then one easily checks the congruences $(a_1 - 1)(a_2 - 1) \equiv 0 \pmod 4 \Rightarrow a_1 a_2 - a_1 - a_2 + 1 \equiv 0 \pmod 4 \Rightarrow a_1 a_2 - 1 \equiv (a_1 - 1) + (a_2 - 1) \pmod 4 \Rightarrow (a_1 a_2 - 1)/2 \equiv (a_1 - 1)/2 + (a_2 - 1)/2 \pmod 2$, and by induction on the number r of factors,

$$\frac{a_1 a_2 \cdots a_r - 1}{2} \equiv \sum_{j=1}^{r} \frac{a_j - 1}{2} \pmod 2. \tag{5}$$

To prove Theorem 7(1), let $a = p_1 p_2 \cdots p_r,\ b = p_1' \cdots p_s'$. By Definition 4, $(a/b) = \prod_{1 \le j \le s}(a/p_j')$ so that, also using Theorem 2(4)

$$\left(\frac{a}{b}\right) = \prod_{1 \le j \le s}\left(\frac{a}{p_j'}\right) = \prod_{1 \le j \le s}\prod_{1 \le i \le r}\left(\frac{p_i}{p_j'}\right).$$

By Theorem 3,

$$\left(\frac{p_i}{p_j'}\right) = \left(\frac{p_j'}{p_i}\right)(-1)^{[(p_i-1)/2][(p_j'-1)/2]},$$

so that

$$\prod_{1 \le i \le r}\left(\frac{p_i}{p_j'}\right) = (-1)^{[(p_j'-1)/2]\Sigma_{i=1}^{r}(p_i-1)/2} \prod_{1 \le i \le r}\left(\frac{p_j'}{p_i}\right)$$

and

$$\left(\frac{a}{b}\right) = (-1)^{\Sigma_{j=1}^{s}(p_j'-1)/2\,\Sigma_{i=1}^{r}(p_i-1)/2} \prod_{1 \le j \le s}\prod_{1 \le i \le r}\left(\frac{p_j'}{p_i}\right).$$

Having a finite number of factors, we may rearrange the last product and obtain

$$\prod_{1\leq i\leq r}\prod_{1\leq j\leq s}\left(\frac{p'_j}{p_i}\right) = \prod_{1\leq i\leq r}\left(\frac{b}{p_i}\right) = \left(\frac{b}{a}\right),$$

by Theorem 2(4) and Definition 4, respectively. Also, by (5),

$$\sum_{j=1}^{s}\frac{p'_j - 1}{2} \equiv \frac{p'_1 \cdots p'_s - 1}{2} \equiv \frac{b-1}{2} = b' \ (\mathrm{mod}\,2);$$

similarly, $\Sigma_{i=1}^{r}(p_i - 1)/2 \equiv a'$ (mod 2), so that $(a/b) = (-1)^{a'\cdot b'}(b/a)$, as claimed. Again,

$$\left(\frac{-1}{b}\right) = \prod_{1\leq j\leq s}\left(\frac{-1}{p'_j}\right) = \prod_{1\leq j\leq s}(-1)^{(p'_j-1)/2} = (-1)^{\Sigma_{j-1}(p'_j-1)/2} = (-1)^{b'}$$

(use being made successively of Definition 4, Theorem 5, and congruence (5)), which proves Theorem 7(2).

Theorem 7(3) is similarly proven, using instead of (5) the congruence

$$\frac{(a_1 a_2 \cdots a_r)^2 - 1}{8} \equiv \sum_{i=1}^{r}\frac{a_i^2 - 1}{8} \ (\mathrm{mod}\,2), \qquad (6)$$

obtained as follows: $a_1 \equiv a_2 \equiv 1 \ (\mathrm{mod}\,2) \Rightarrow (a_1^2 - 1)(a_2^2 - 1) \equiv 0 \ (\mathrm{mod}\,16) \Rightarrow a_1^2 a_2^2 - 1 \equiv (a_1^2 - 1) + (a_2^2 - 1) \ (\mathrm{mod}\,16) \Rightarrow (a_1^2 a_2^2 - 1)/8 \equiv (a_1^2 - 1)/8 + a_2^2 - 1)/8 \ (\mathrm{mod}\,2) \Rightarrow (6)$ by induction on r. Using (6),

$$\left(\frac{2}{b}\right) = \prod_{1\leq j\leq s}\left(\frac{2}{p'_j}\right) = \prod_{1\leq j\leq s}(-1)^{(p'^2_j-1)/8} = (-1)^{\Sigma_{j-1}(p'^2_j-1)/8}$$

$$= (-1)^{((p'_1\cdots p'_s)^2-1)/8} = (-1)^{(b^2-1)/8},$$

and Theorem 7(3) is proven.

While Definition 4 of the Jacobi Symbol is more general than that of the Legendre Symbol, it is still restricted by the conditions $(a, b) = 1$ and b odd. We already dispensed with the first restriction, simply by stipulating (see Definition 4′) that $(a, b) > 1 \Rightarrow (a/b) = (b/a) = 0$. It is somewhat less trivial to discard the restriction $2 \nmid b$. This is partially done by:

DEFINITION 5.   The *Kronecker Symbol* (a/b) is defined as follows:

(i) $(a/b) = 0$ if $(a, b) > 1$.

(ii) (a/b) is identical with the Jacobi Symbol if

$$b \equiv 1 \ (\mathrm{mod}\,2) \quad \text{and} \quad (a, b) = 1.$$

(iii) If $a \equiv 0$ or $1 \pmod 4$ and $b = 2^c b_1$, $b_1 \equiv 1 \pmod 2$, then $(a/b) = (a/2)^c (a/b_1)$; here (a/b_1) is a Jacobi symbol and

$$\left(\frac{a}{2}\right) = \left(\frac{a}{-2}\right) = \begin{cases} 0 & \text{if } a \text{ is even,} \\ +1 & \text{if } a \equiv 1 \pmod 8, \\ -1 & \text{if } a \equiv 5 \pmod 8. \end{cases}$$

REMARK 4.   By Theorem 7(3),

$$\left(\frac{2}{a}\right) = (-1)^{(a^2-1)/8} = \begin{cases} 1 & \text{if } a \equiv 1 \pmod 8, \\ -1 & \text{if } a \equiv 5 \pmod 8; \end{cases}$$

hence for $a \equiv 1 \pmod 4$, one has $(a/2) = (2/a)$; this also holds for $a \equiv 0 \pmod 4$ if we use Definition 4'.

We finish this chapter by listing some properties of the Kronecker Symbol not covered by those of the Jacobi Symbol.

Theorem 8.   *Let* m, n, a, a_1, a_2, k *be positive integers* $a \equiv 0$ *or* $1 \pmod 4$; *then*

(1) $a_1 \equiv a_2 \pmod 8 \Rightarrow \left(\dfrac{a_1}{2}\right) = \left(\dfrac{a_2}{2}\right)$;

(2) $\left(\dfrac{a_1}{2}\right)\left(\dfrac{a_2}{2}\right) = \left(\dfrac{a_1 a_2}{2}\right)$;

(3) $\left(\dfrac{a}{2}\right) = \left(\dfrac{2}{a}\right)$;

(4) $\left(\dfrac{a}{2}\right)^k = \left(\dfrac{a}{2^k}\right)$;

(5) $\left(\dfrac{a}{m}\right)\left(\dfrac{a}{n}\right) = \left(\dfrac{a}{mn}\right)$;

(6) $m \equiv n \pmod a \Rightarrow \left(\dfrac{a}{m}\right) = \left(\dfrac{a}{n}\right)$.

All proofs, except that of (6), are very simple; they are left as an exercise for the reader.

One of the important advantages of the Jacobi and Kronecker Symbols is that they permit a rapid calculation of Legendre Symbols. The computation of $(213/499)$, already performed using exclusively the theory of the Legendre Symbol, can now proceed as follows:

$$\left(\frac{213}{499}\right) = \left(\frac{499}{213}\right) = \left(\frac{73}{213}\right) = \left(\frac{213}{73}\right) = \left(\frac{67}{73}\right) = \left(\frac{73}{67}\right) = \left(\frac{6}{67}\right) = \left(\frac{2}{67}\right)\left(\frac{3}{67}\right)$$

$$= -\left(\frac{3}{67}\right) = \left(\frac{67}{3}\right) = \left(\frac{1}{3}\right) = +1.$$

There are more steps than in the previous method, but each step can be performed mentally with ease. In practice of course, one would combine both methods so as to obtain the greatest simplification of the numerical work.

PROBLEMS

1. Prove in detail the equivalence of congruence (1) with the system (2), (2′) of Section 1.

2. Prove Corollary 1.1.

3. Find a complete set of quadratic residues and of quadratic nonresidues modulo 11 and modulo 13.

4. Find $\left(\frac{3}{5}\right)$ and $\left(\frac{5}{11}\right)$ by Gauss' Lemma, that is, by determining in each case the integer n for which $(a/p) = (-1)^n$.

5. Prove Corollaries 5.1 and 5.2.

6. Compute $(101/131)$, $(100/131)$, and $(99/131)$ using only the theory of the Legendre Symbol. Compute the same symbols considered as Jacobi Symbols.

7. Prove that the product of two quadratic residues or two quadratic nonresidues (mod p) is a quadratic residue, while the product of a quadratic residue and a quadratic nonresidue is a quadratic nonresidue mod p.

8. Prove Theorem 6.

9. Prove Theorem 8 in detail; special care is needed for (6). (Hint: Use the reciprocity formula to replace (a/m) and (a/n) by (m/a) and (n/a), respectively, then use Theorem 6(1) and Theorem 8(1).)

10. Find all primes for which 3 is a quadratic residue and all primes for which 3 is a quadratic nonresidue. Solve the same problem for -5.

11. Prove that for every $p > 3$ the sum of the quadratic residues is divisible by p.

12. Show that $p \nmid a \Rightarrow \sum_{m=0}^{p-1}((am+b)/p) = 0$. (Hint: First show that $\sum_{m=0}^{p-1}(m/p) = 0$.)

★13. Let $S(a) = \sum_{m=1}^{p-1}((m^2+a)/p)$. Show that $S(a)$ depends only on (a/p), and in particular show that $p|a \Rightarrow S(a) = p - 1$. (Hint: Show that for every $k \not\equiv 0 \pmod p$, $S(a) = (k^2/p)S(a) = S(ak^2)$, and use Problem 7.)

★14. Prove that $\sum_{a=0}^{p-1} S(a) = 0$. (Hint: Invert the order of summations and use Problem 12.)

*15. Prove

(i) $\left(\dfrac{a}{p}\right) = -1 \Rightarrow S(a) = 0.$

(ii) $\left(\dfrac{a}{p}\right) = +1 \Rightarrow S(a) = -2.$

(iii) $p \nmid a \Rightarrow S(a) + \left(\dfrac{a}{p}\right) = -1.$

(Hints: For (i), observe that by Problem 13

$$S(a) = S(an^2) = \sum_{m=0}^{p-1} \left(\frac{m^2 + an^2}{p}\right).$$

For every $n \not\equiv 0 \pmod{p}$, $p \nmid a$ implies that

$$S(a) = \left(\frac{a}{p}\right) \sum_{m=1}^{p-1} \left(\frac{m^2 a + a^2 n^2}{p}\right).$$

Sum over $n(1 \le n \le p-1)$, and conclude that if $(a/p) = -1$, then $S(a) = -S(a)$.

For (ii) split the sum in Problem 14 into 3 parts according to the values of (a/p), and use Problem 13 and Part (i).

Part (iii) follows immediately from (i) and (ii).)

*16. Let $f(x) = ax^2 + bx + c$, set $d = b^2 - 4ac$, and define

$$T(f) = \sum_{m=0}^{p-1} \left(\frac{f(m)}{p}\right).$$

Prove the following:

(i) If $p \nmid ad$, then $T(f) = -(a/p)$.

(ii) If $p|a$ but $p \nmid d$, or $p \nmid a$ but $p|d$, then $T(f) = (p-1)(a/p)$.

(iii) If $p|a$ and $p|d$ then $T(f) = p(c/p)$.

(Hints: The case $p|a$, $p \nmid d$ of Part (ii) follows from Problem 12. If $p \nmid a$, then

$$\left(\frac{a}{p}\right) T(f) = \left(\frac{4a}{p}\right) T(f) = \sum_{m=0}^{p-1} \left(\frac{(2am+b)^2 - d}{p}\right) = \sum_{m=0}^{p-1} \left(\frac{k^2 - d}{p}\right)$$

where $k = 2am + b$, or $T(f) = \left(\dfrac{a}{p}\right) \displaystyle\sum_{m=0}^{p-1} \left(\frac{m^2 - d}{p}\right)$;

if $p|d$, the result of (ii) follows immediately; if $p \nmid d$, use Problem 15 (iii) to obtain (i). Case (iii) is trivial.)

*17. Prove:

   (i) If b is kept fixed, the Jacobi Symbol (a/b) is a periodic function of a; find its least period.

   (ii) If a is kept fixed, the Jacobi Symbol (a/b) is a periodic function of b; find its least period.

*18. (i) Let $a \equiv 1 \pmod 4$ and $b = 2^k b_1$, $2 \nmid b_1$. Using the definition of the Jacobi Symbol (a/b_1) and of the Kronecker Symbols $(a/2)$ and $(a/b) = (a/2)^k (a/b_1)$, find the relation between (a/b) and (b/a).

   (ii) Same question in case $a \equiv 0 \pmod 4$ (Caution: If $a = a_1 \cdot 2^k$, $k > 0$ and $b \equiv 3 \pmod 4$, then (a/b) is defined, but (b/a) is not.)

19. Determine the number of solutions of the following two congruences:

   (a) $x^2 \equiv 231 \pmod{997}$,

   (b) $x^2 \equiv 997 \pmod{231}$. (Hint: 997 is a prime.)

BIBLIOGRAPHY

1. G. Frobenius, Ueber das quadratische Reziprozitätsgesetz. *Sitzungsber. der königl. Preuss. Akad. der Wiss.*, Berlin (1914), 335–349 and 484–488.
2. W. J. LeVeque, *Topics in Number Theory*, Vol. 1. Reading, Mass.: Addison-Wesley, 1956.
3. D. Shanks, *Solved and Unsolved Problems in Number Theory*. Washington, D.C.: Spartan Books, 1962.

Chapter 6

Arithmetical Functions

1 INTRODUCTION

Let us denote by $\mathbf{Z}^+$ the set of natural integers; clearly $\mathbf{Z}^+ \subset \mathbf{Z}$. Functions whose domain is $\mathbf{Z}$ or $\mathbf{Z}^+$ are usually called arithmetical functions (sometimes also number-theoretic functions), regardless of their range. It is, of course, easy to fabricate such functions out of any functions defined over $\mathbf{Q}$, $\mathbf{R}$, or $\mathbf{C}$, simply by considering their restrictions to $\mathbf{Z}$ or $\mathbf{Z}^+$. But this rarely leads to interesting results. So for instance if we restrict the function $y = x^2$ to $x \in \mathbf{Z}$, we simply obtain the sequence of squares; if we restrict $y = \sin \pi x$ to $x \in \mathbf{Z}$, we obtain $y = 0$ for all $x \in \mathbf{Z}$, and so on. It is much more interesting to consider functions that have $\mathbf{Z}$ or $\mathbf{Z}^+$ as their *natural domain*, which means that we cannot give them a simple, sensible interpretation unless the independent variable is an integer. So for instance it makes sense to speak about the number of divisors of an integer m, but no simple meaning can be attached to the number of divisors of π, or of e, or of i. We already met with some arithmetical functions. One of them, the Legendre-Jacobi-Kronecker symbol, has been discussed in Chapter 5; another is the number of divisors of an integer; still another one is Euler's ϕ-function. In the present chapter we shall study these and also a few other arithmetical functions. In addition, we shall discuss two functions that are not, properly speaking, arithmetic functions, being defined over the reals; their connection with arithmetical functions is so close, though, that this seems the logical place to study them. One, the function $y = [x]$ which we already met, has $\mathbf{Z}$ at least as its range; the other, $y = x - [x] - \frac{1}{2}$, not even that.

2 THE FUNCTION $[x]$

The function $[x]$ has already been defined verbally (see Definition 5.3); it stands for the greatest integer not in excess of the real number x. If we think of

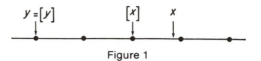

Figure 1

the integers as being represented by equidistant dots on the real line, then, if x is one of the dots, $[x] = x$; otherwise, $[x]$ stands for the dot next to the left of x. For convenience we define once more the function $y = [x]$.

DEFINITION 1. $[x] = n \ni x - 1 < n \leq x, n \in \mathbf{Z}$.

The reader is invited to convince himself that Definition 5.3 and the present one are equivalent. The statements of the following theorem follow almost immediately from the definition.

Theorem 1.   *Let x be a real number; then*

(1) $[x + n] = [x] + n$ *for any integer n*;

(2) $\left[\dfrac{[x]}{n}\right] = \left[\dfrac{x}{n}\right]$ *for any natural integer n*;

(3) $\displaystyle\sum_{1 \leq n \leq x} 1 = [x]$;

(4) $0 \leq x - [x] < 1$;

(5) $|x - [x] - \tfrac{1}{2}| \leq \tfrac{1}{2}$;

(6) $[x] + [y] \leq [x + y] \leq [x] + [y] + 1$;

(7) $x \notin \mathbf{Z} \Rightarrow [-x] = -[x] - 1$;

(8) $[2x] - 2[x] = \begin{cases} 1 & \text{if } [2x] \text{ is odd}, \\ 0 & \text{if } [2x] \text{ is even}. \end{cases}$

PROOF.   To prove the last assertion, let $x = n + \alpha$, $0 \leq \alpha < 1$; then $[x] = n$. If $0 \leq \alpha < \tfrac{1}{2}$, then $[2x] = [2n + 2\alpha] = 2n$ is even and $[2x] - 2[x] = 2n - 2n = 0$; if $\tfrac{1}{2} \leq \alpha < 1$, then $[2x] = [2n + 2\alpha] = 2n + 1$ is odd and $[2x] - 2[x] = 2n + 1 - 2n = 1$. The proof of the other statements is left to the reader.

Theorem 2. $n! = \displaystyle\prod_{p \leq n} p^{e_p}$, *where* $e_p = \displaystyle\sum_{m \geq 1} [n/p^m]$.

REMARK 1.   If $p^m > n$ (i.e., for $m > \log n / \log p$), then $[n/p^m] = 0$; therefore, the sum in e_p contains exactly $[\log n / \log p]$ nonvanishing terms.

PROOF OF THEOREM 2.   Consider all integers $m \leq n$, written in natural order. Every pth integer is divisible by p; if we divide p out of them, we obtain the factor p exactly $[n/p]$ times. At the same time, integers that contained p precisely to the first power are no more divisible by p, while those that

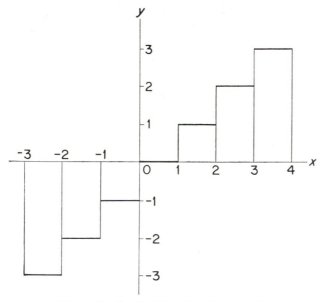

Figure 2. Graph of the function $y = [x]$

contained p to the kth power $(k > 1)$ now contain it only to the $(k - 1)$th power. In particular, every p^2th integer had been divisible by p^2 and is now divisible by p (at least!). We factor out p from each of these integers, obtaining another $[n/p^2]$ times the factor p; all integers $m \leq n$ that had contained p to the first or second power do not contain the factor p any more, while every p^3th integer, which originally was divisible by p^3, now is still divisible by at least the first power of p. In general, let k_a be the exact power of p that divides a given integer a and set $v_a(= v_a(m, p)) = \max(0, k_a - m)$; then, after m operations of the kind described, the factor p has been obtained $[n/p] + [n/p^2] \cdots + [n/p^m]$ times and $n! = p^{\{[n/p]+ \cdots +[n/p^m]\}} \cdot c$, where $c = b_1 \cdot b_2 \cdots b_n$, and the power of p that divides b_a is v_a. In particular, if $m = [\log n/\log p]$, then $v_a = 0$ for all b_a; hence $p \nmid c$ and $e_p = \sum_{m \geq 1}[n/p^m]$, as asserted.

3 THE FUNCTION $y = ((x))$

DEFINITION 2. $((x)) = x - [x] - \frac{1}{2}$ if $x \notin \mathbf{Z}$, $((n)) = 0$ for $n \in \mathbf{Z}$.

From Theorem 1 we know that $|((x))| \leq \frac{1}{2}$. Actually, $-\frac{1}{2} < ((x)) < \frac{1}{2}$, because $((x)) = 0$ for integral x. We observe that $((x))$ is a periodic function of period one, which for $0 < x < 1$ reduces to the straight line segment $((x)) = x - \frac{1}{2}$. This function is sometimes called the "sawtooth function."

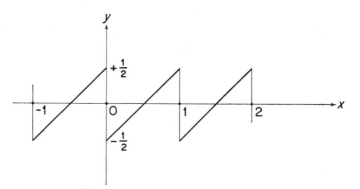

Figure 3. Graph of the function $y = ((x))$

Consider the first-degree polynomial $B_1(x) = x - \frac{1}{2}$. Then $((x)) = B_1(x - [x])$. The following theorem is an immediate consequence of preceding remarks.

THEOREM 3.

(1) *The function $y = ((x))$ is periodic, of period one, and piecewise linear.*
(2) *Let $B_1(x) = x - \frac{1}{2}$; then for $x \notin \mathbf{Z}$, $((x)) = B_1(x - [x])$;*
(3) $-\frac{1}{2} < ((x)) < \frac{1}{2}$;
(4) $((-x)) = -((x))$;
(5) $\int_0^1 ((x))\, dx = 0$;
(6) $\left| \int_\alpha^\beta ((x))\, dx \right| \leq 1/8$ *for all real α, β;*
(7) $y(x) = \int_1^x ((x))\, dx$ *satisfies $|y(x)| \leq 1/8$;*
(8) $h, k \in \mathbf{Z}^+ \Rightarrow \sum_{m=1}^{k-1} \left(\left(\frac{mh}{k}\right)\right) = 0$.

PROOF. The proofs for statements (1) to (7) are very easy and are left to the reader. For statement (8), observe that setting $m = k - n$, one has

$$\sum_{m=1}^{k-1} \left(\left(\frac{mh}{k}\right)\right) = \sum_{n=1}^{k-1} \left(\left(\frac{(k-n)h}{k}\right)\right) = \sum_{m=1}^{k-1} \left(\left(\frac{(k-m)h}{k}\right)\right) = \sum_{m=1}^{k-1} \left(\left(\frac{-mh}{k}\right)\right)$$

$$= \sum_{m=1}^{k-1} \left\{-\left(\left(\frac{mh}{k}\right)\right)\right\} = -\sum_{m=1}^{k-1} \left(\left(\frac{mh}{k}\right)\right)$$

so that $2 \sum_{m=1}^{k-1}((mh/k)) = 0$ and the statement is proven: The first, second, and last equality are trivially justified; the third holds by Theorem 3(1), and the fourth by Theorem 3(4).

4 THE EULER FUNCTION $\phi(n)$

The Euler function $\phi(n)$ has already been defined (see Definition 4.10) as the number of integers not exceeding $n(\in \mathbf{Z}^+)$ and coprime to n, and some of its important properties have already been stated. Most of them follow quite easily from the property denoted by (A) in Chapter 4, Section 6 (see p. 53), which $\phi(n)$ shares with many other arithmetic functions; this property, called *multiplicativity*, is described as follows.

DEFINITION 3. A function $f(n)$ defined on $\mathbf{Z}^+$ is said to be *multiplicative* if $(m, n) = 1 \Rightarrow f(m)f(n) = f(mn)$; $f(n)$ is said to be *totally multiplicative* if $f(m)f(n) = f(mn)$ for all $m, n \in \mathbf{Z}^+$ (without the restriction $(m, n) = 1$).

We shall ignore the trivial case of $f(n) = 0$ for all $n \in \mathbf{Z}^+$. We also note that if $f(n)$ is not equal to zero for all integers n, then $f(n) = f(1 \cdot n) = f(1) \cdot f(n)$ (with $f(n) \neq 0$ for some n) shows that $f(1) = 1$.

Theorem 4. $\phi(n)$ *is multiplicative.*

This statement is precisely Property (A), anticipated in Chapter 4. In the proof we shall use two lemmas.

Lemma 1. *Let* $(m_1, m_2) = 1$; *let* h_1 *run through a complete set of residues* mod m_1, *and let* h_2 *run through a complete set of residues* mod m_2. *Then* $h = h_2 m_1 + h_1 m_2$ *runs through a complete set of residues* mod $m_1 m_2$.

PROOF. h_1 takes m_1 values, h_2 takes m_2 values, all integral; the set $\mathbf{H} = \{h\}$ consists, therefore, of $m_1 m_2$ integers. Hence we shall have proven Lemma 1 if we show that these $m_1 m_2$ integers are all incongruent mod $m_1 m_2$. To do this, assume that $h = h_2 m_1 + h_1 m_2$ and $h' = h_2' m_1 + h_1' m_2$ are two integers of $\mathbf{H}$, congruent mod $m_1 m_2$, and that h_1, h_2 are not both the same as h_1', h_2'. Then $(h_2 - h_2')m_1 + (h_1 - h_1')m_2 \equiv 0 \pmod{m_1 m_2} \Rightarrow (h_2 - h_2')m_1 + (h_1 - h_1')m_2 \equiv 0 \pmod{m_1} \Rightarrow (h_1 - h_1')m_2 \equiv 0 \pmod{m_1} \Rightarrow h_1 - h_1' \equiv 0 \pmod{m_1}$, because $(m_1, m_2) = 1$. However, h_1 and h_1' belong to the same complete set of residues mod m_1; hence they can be congruent $(\mathrm{mod}\ m_1)$ only if they are identical and $h_1 = h_1'$. In the same way we also obtain $h_2 = h_2'$, contrary to our assumption that[†] $(h_1, h_2) \neq (h_1', h_2')$, and the Lemma is proven.

[†] It should cause no confusion that here (h_1, h_2) stands for the ordered set and not for the greatest common divisor.

Lemma 2. *Let* $(m_1, m_2) = 1$, *and let* h_1 *run through a complete set of reduced residues* mod m_1 *and* h_2 *through a complete set of reduced residues* mod m_2; *then* $h = m_1 h_2 + m_2 h_1$ *runs through a complete set of reduced residues* mod $m_1 m_2$.

PROOF. Consider the set $\mathbf{H} = \{h\}$ discussed in Lemma 1. We already know that all its elements are incongruent mod $m_1 m_2$. We now have to show that $(h, m_1 m_2) = 1$ if and only if $(h_1, m_1) = 1$ and $(h_2, m_2) = 1$. On account of $(m_1, m_2) = 1$ the condition $(m_1 h_2 + m_2 h_1, m_1 m_2) = 1$ is equivalent to the two simultaneous conditions $(m_1 h_2 + m_2 h_1, m_1) = 1$ and $(m_1 h_2 + m_2 h_1, m_2) = 1$, which simplify to $(m_2 h_1, m_1) = 1$ and $(m_1 h_2, m_2) = 1$. Once more using $(m_1, m_2) = 1$, these conditions reduce to $(h_1, m_1) = 1$ and $(h_2, m_2) = 1$; hence

$$(h, m_1 m_2) = 1 \Leftrightarrow \begin{cases} (h_1, m_1) = 1 \\ (h_2, m_2) = 1, \end{cases}$$

and Lemma 2 is proven.

PROOF OF THEOREM 4. By the definition of Euler's ϕ-function, there are $\phi(m_1)$ distinct values of h_1 with $1 \le h_1 \le m_1$, $(h_1, m_1) = 1$, $\phi(m_2)$ distinct values of h_2 with $1 \le h_2 \le m_2$, $(h_2, m_2) = 1$, and $\phi(m_1 m_2)$ distinct values of h with $1 \le h \le m_1 m_2$, $(h, m_1 m_2) = 1$. By Lemma 2, each of these $\phi(m_1 m_2)$ values of h is obtained exactly once by setting $h = m_1 h_2 + m_2 h_1$, with h_1 running through a complete set of reduced residues modulo m_1 and h_2 through a complete set of reduced residues modulo m_2. Consequently, there are $\phi(m_1)\phi(m_2)$ distinct values of h, $\phi(m_1 m_2) = \phi(m_1)\phi(m_2)$, and the Theorem is proven.

Theorem 5. $\phi(p^r) = p^r - p^{r-1}$.

PROOF. If $n \le p^r$, then $(n, p^r) = 1$ except for $n = kp$, where k may take the values $1, 2, \ldots, p^{r-1}$ (see Chapter 4, Problem 27).

Theorem 6. $\phi(m) = m\prod_{p|m}(1 - (1/p))$.

This is Property (B) anticipated in Chapter 4.

PROOF. Let $m = p_1^{s_1} p_2^{s_2} \cdots p_r^{s_r}$. Then by Theorem 4, $\phi(m) = \prod_{p_i|m}\phi(p_i^{s_i})$; by Theorem 5, $\phi(p_i^{s_i}) = p_i^{s_i} - p_i^{s_i-1}$, so that $\phi(m) = \prod_{p_i|m} p_i^{s_i}(1 - (1/p_i)) = m\prod_{p|m}(1 - (1/p))$.

Corollary 6.1. *If* $m \ne 1, 2$ *then* $2|\phi(m)$.

This of course implies Property (C) of Chapter 4.

PROOF.   By Theorem 6,

$$m = p_1^{s_1} p_2^{s_2} \cdots p_r^{s_r} \Rightarrow \phi(m)$$

$$= p_1^{s_1-1} p_2^{s_2-1} \cdots p_r^{s_r-1} (p_1 - 1)(p_2 - 1) \cdots (p_r - 1)$$

and, by assumption, either some odd $p_j | m$ so that $2 \nmid (p_j - 1) \Rightarrow 2 | \phi(m)$ or $m = 2^s$, $s \geq 2$, so that $\phi(m) = 2^{s-1}$ and again $2 | \phi(m)$.

Theorem 7. $\sum_{d|m} \phi(d) = m$.

This is Property (D) anticipated in Chapter 4.

PROOF.   Let $m = p_1^{s_1} p_2^{s_2} \cdots p_r^{s_r}$ and consider the product $P = \prod_{p_i | m}(1 + \phi(p_i) + \cdots + \phi(p_i^{s_i}))$. Multiplying it out and observing that $(p_i^t, p_j^s) = 1$ if $p_i \neq p_j$, we obtain by Theorem 4 that P consists of a sum of terms of the form

$$\phi(p_1^{t_1}) \phi(p_2^{t_2}) \cdots \phi(p_r^{t_r}) = \phi(d),$$

where $d = p_1^{t_1} p_2^{t_2} \cdots p_r^{t_r} (0 \leq t_i \leq s_i)$ runs precisely through all the divisors of m; furthermore, Theorem 3.3 guarantees that each divisor is obtained exactly once. Hence $P = \sum_{d|m} \phi(d)$. However by Theorem 5,

$$1 + \phi(p) + \cdots + \phi(p^s) = 1 + (p - 1) + (p^2 - p) + \cdots$$

$$+ (p^{s-1} - p^{s-2}) + (p^s - p^{s-1}) = p^s;$$

hence $P = \prod_{p_i | m} p_i^{s_i} = m$. Writing that the two expressions of P are equal we obtain the Theorem. A different, neater proof of Theorem 7 will be given in Section 5.

One should observe that the proof of the equality $P = \sum_{d|m} \phi(d)$ uses only the fact that $\phi(n)$ is multiplicative; hence, we have incidentally proven the following corollary.

Corollary 7.1.   Let $f(n)$ be multiplicative and let $n = p_1^{s_1} p_2^{s_2} \cdots p_r^{s_r}$ be the canonical factorization of $n > 1$; then $\sum_{d|n} f(d) = \prod_{p|n}(1 + f(p) + \cdots + f(p^s))$. If $n = 1$, then $\sum_{d|n} f(d) = 1$.

The last statement follows from the fact that, as already observed, $f(1) = 1$ because $f(n)$ is multiplicative.

5   THE MÖBIUS FUNCTION $\mu(n)$

The Möbius Function $\mu(n)$ is defined on the set $\mathbf{Z}^+$ of natural integers $n = p_1^{s_1} p_2^{s_2} \cdots p_r^{s_r}$ ($s_j \geq 1$) as follows.

DEFINITION 4.

$$\mu(n) = \begin{cases} 0 & \text{if } \sum_{i=1}^{r} s_i > r \left(\text{i.e., if } \exists j, 1 \leq j \leq r \ni p_j^2 | n \right), \\ (-1)^r & \text{if } \sum_{i=1}^{r} s_i = r \ (\text{i.e., if } s_1 = s_2 = \cdots = s_r = 1). \end{cases}$$

$M(x) = \sum_{n \leq x} \mu(n)$ is called the *sum function* of $\mu(n)$.

In particular, for $r = 0$, we obtain $\mu(1) = 1$; for $r = 1$, $\mu(p) = -1$ for every prime p; also

$$\mu(p_1 p_2) = 1, \qquad \mu(p^2) = \mu(p^3) = \cdots = 0.$$

REMARK 2. $M(x)$ is defined over **R**, not only over $\mathbf{Z}^+$; by the general convention on empty sums, $M(x) = 0$ if $x < 1$.

DEFINITION 5.   An integer not divisible by the square of any integer $n \neq 1$ is called *squarefree*.

Theorem 8. $\mu(n)$ *is multiplicative.*

PROOF.   Let $(m, n) = 1$; hence $p^2 | mn$ only if either $p^2 | m$ or $p^2 | n$, and in either case $\mu(m)\mu(n) = \mu(mn) = 0$. If m and n are both squarefree, then $m = p_1 p_2 \cdots p_r$, $n = q_1 q_2 \cdots q_s$, $mn = p_1 p_2 \cdots p_r q_1 q_2 \cdots q_s$, the primes p_i and q_j being all distinct. Then $\mu(m) = (-1)^r$, $\mu(n) = (-1)^s$, and $\mu(mn) = (-1)^{r+s}$, so that the Theorem holds.

Theorem 9.

$$\sum_{d|n} \mu(d) = \begin{cases} 1 & \text{if } n = 1; \\ 0 & \text{otherwise}. \end{cases}$$

FIRST PROOF.   By Theorem 8, $\mu(n)$ is multiplicative; hence remembering that for any $p, 1 + \mu(p) + \mu(p^2) + \cdots = 1 + \mu(p) = 0$, the result follows from Corollary 7.1.

SECOND PROOF.   Let $n = p_1^{s_1} \cdots p_r^{s_r}$, $n_1 = p_1 p_2 \cdots p_r$. Then $\mu(d) = 0$ unless $d | n_1$. Hence $\sum_{d|n} \mu(d) = \sum_{d|n_1} \mu(d)$. The divisors of n_1 are 1; $p_1, p_2, \ldots, p_r$; $p_1 p_2, \ldots, p_{r-1} p_r; \ldots;$ $p_1 p_2 \cdots p_r = n$. In general there are $\binom{r}{k} = (r!/k!(r-k)!)$ distinct divisors $d_s^{(k)}\left(s = 1, 2, \ldots, \binom{r}{k}\right)$ of n_1 containing exactly $k(\leq r)$ prime factors; the corresponding $\mu(d_s^{(k)})$ equals $(-1)^k$ and $\sum_{s=1}^{\binom{r}{k}} \mu(d_s^{(k)}) = (-1)^k \binom{r}{k}$. Hence $\sum_{d|n} \mu(d) = \sum_{k=0}^{r} (-1)^k \binom{r}{k} = (1-1)^r = 0$, provided that $r > 0$. In case $r = 0$, $n = 1$, and $\sum_{d|1} \mu(d) = \mu(1) = 1$. If in this proof we replace everywhere $\mu(d)$ by $|\mu(d)|$, we obtain

Corollary 9.1. *Let r be the number of distinct primes that divide n; then* $\sum_{d|n}|\mu(d)| = 2^r$.

Still another proof of Theorem 9 may be based on the function $\zeta(s) = \sum_{n=1}^{\infty} n^{-s}$.

Let $s > 1$; then one observes (for a proof see Chapter 7) that the series which defines $\zeta(s)$ converges and equals $\prod_p (1 - p^{-s})^{-1}$, the product being taken over all primes. Hence $\zeta(s)^{-1} = \prod_p (1 - p^{-s}) = \sum_{n=1}^{\infty} a_n n^{-s}$. It may easily be checked that only squarefree numbers actually occur in the last sum; hence, $a_n = 0$ if $p^2 | n$; also $a_n = +1$ or -1, depending on whether n is a product of an even or an odd number of primes. This identifies a_n as the Möbius function $\mu(n)$ and $\zeta(s)^{-1} = \sum_{n=1}^{\infty} \mu(n) n^{-s}$. But $\zeta(s) \cdot \zeta(s)^{-1} = 1$, so that $\sum_{k=1}^{\infty} k^{-s} \sum_{m=1}^{\infty} \mu(m) m^{-s} = 1$. The double sum may be written as $\sum_{k=1}^{\infty} \sum_{m=1}^{\infty} (km)^{-s} \mu(m)$ or, setting $km = n$, as $\sum_{n=1}^{\infty} n^{-s} \sum_{m|n} \mu(m)$, so that

$$\sum_{n=1}^{\infty} n^{-s} \sum_{m|n} \mu(m) = 1. \tag{1}$$

Both sides in equation (1) are *Dirichlet series*, that is, series of the form $\sum_{n=1}^{\infty} a_n n^{-s}$, and (1) may be written as $\sum_{n=1}^{\infty} a_n n^{-s} = \sum_{n=1}^{\infty} b_n n^{-s}$, with $a_n = \sum_{m|n} \mu(m)$; $b_1 = 1$, $b_n = 0$ $(n > 1)$. It is easy to verify that if a Dirichlet series converges for $s = s_0 (\in \mathbf{R})$, then it also converges for $s \geq s_0$.

The following uniqueness theorem holds for Dirichlet series. Assuming it for a moment, the proof of Theorem 9 reduces to the remarks that $a_1 = \sum_{m|1} \mu(m) = b_1 = 1$ and $a_n = \sum_{m|n} \mu(m) = b_n = 0$ for $n > 1$.

Theorem 10. *If the Dirichlet series $\sum_{n=1}^{\infty} a_n n^{-s}$ and $\sum_{n=1}^{\infty} b_n n^{-s}$ converge both for $s \geq s_0$, then they are equal for all $s \geq s_0$ if and only if $a_n = b_n$ for all n.*

PROOF. The "if" part is trivial. To prove the "only if," assume the equality of the two series and let $s \to +\infty$; then all terms with $n > 1$ tend to zero in both series. One easily shows (by first proving, then using the uniformity of the convergence) that $\lim_{s \to \infty} \sum_{n=2}^{\infty} a_n n^{-s} = \lim_{s \to \infty} \sum_{n=2}^{\infty} b_n n^{-s} = 0$ so that $a_1 = b_1$, and the statement is proven for $n = 1$. Having established that $a_n = b_n$ for $n = 1, 2, \ldots, k - 1$, it remains to show that also $a_k = b_k$, and the proof by induction of the Theorem will be complete. From $a_n = b_n$ $(n = 1, 2, \ldots, k - 1)$ and $\sum_{n=1}^{\infty} a_n n^{-s} = \sum_{n=1}^{\infty} b_n n^{-s}$ it follows that $\sum_{n=k}^{\infty} a_n n^{-s} = \sum_{n=k}^{\infty} b_n n^{-s}$, or multiplying by k^s, that $a_k + \sum_{n=k+1}^{\infty} a_n (k/n)^s = b_k + \sum_{n=k+1}^{\infty} b_n (k/n)^s$.

Observing that for $n \geq k + 1$, $k/n < 1$, it follows as before that both infinite sums approach zero when $s \to \infty$; consequently, $a_k = b_k$, and Theorem 10 is proven.

While Theorem 9 may strike the reader as comparatively trivial (and he may well wonder why one should prove it in three different ways), it actually is a very powerful tool. As an illustration we shall use it to prove a corollary that to the best of my knowledge is due to A. F. Möbius (see [16] for a recent presentation), and is surprisingly general and versatile.

Corollary 9.2.   *Consider an arbitrary set of n couples $\{(\alpha_j, d_j)\}$, $1 \leq j \leq n$, where the α_j may be arbitrary real, or even complex, numbers and the d_j are positive integers. For any integer m set*

$$S_m = \sum_{d_j \equiv 0 \,(\mathrm{mod}\ m)} \alpha_j,$$

and let $S' = \sum_{d_j = 1} \alpha_j$; then

$$S' = \sum_{m=1}^{\infty} \mu(m) S_m.$$

COMMENTS. S_m is the sum of those α_j whose companions d_j are multiples of the integer m; clearly $S_m = 0$ unless $m | d_j$ for at least some $j (1 \leq j \leq n)$. S' is the sum of those α_j whose companions satisfy $d_j = 1$.

PROOF.   By the definition of S_m, $\sum_{m=1}^{\infty} \mu(m) S_m = \sum_{m=1}^{\infty} \mu(m) \sum_{d_j \equiv 0(m)} \alpha_j$; rearranging the terms we obtain $\sum_{j=1}^{n} \alpha_j \sum_{m | d_j} \mu(m) = \sum_{d_j=1} \alpha_j = S'$, because if $d_j \neq 1$ the inner sum vanishes on account of Theorem 9. The Corollary is proven.

To illustrate the power of Corollary 9.2, let us prove

Corollary 9.3.   *For every real $x \geq 1$, $\sum_{m=1}^{x} \mu(m)[x/m] = 1$.*

PROOF.   In Corollary 9.2 with $n = [x]$, let $d_j = j$, $\alpha_j = 1 (1 \leq j \leq x)$; then $S' = 1$, $S_m = \sum_{j \leq x, \, j \equiv 0(m)} 1 = [x/m]$, and by Corollary 9.2, $1 = S' = \sum_{m=1}^{\infty} \mu(m) S_m = \sum_{m=1}^{\infty} \mu(m)[x/m] = \sum_{m=1}^{x} \mu(m)[x/m]$ as claimed.

Theorem 11 (Möbius' inversion formula).   *If $f(n)$ is a function defined on $\mathbf{Z}^+$ and $F(n)$ is defined by $F(n) = \sum_{d|n} f(d)$, then $f(n) = \sum_{d|n} \mu(d) F(n/d)$. Conversely, if $F(n)$ is a function on $\mathbf{Z}^+$ and $f(n)$ is defined by $f(n) = \sum_{d|n} \mu(d) F(n/d)$, then $F(n) = \sum_{d|n} f(d)$.*

PROOF.

$$\sum_{d|n} \mu(d) F(n/d) = \sum_{d|n} \mu(d) \sum_{d_1|(n/d)} f(d_1) = \sum_{d_1|n} f(d_1) \sum_{d_1 d|n} \mu(d)$$

$$= \sum_{d_1|n} f(d_1) \sum_{d|(n/d_1)} \mu(d).$$

The last sum vanishes, by Theorem 9, unless $n/d_1 = 1$, that is unless $n = d_1$, when it reduces to one and $\sum_{d|n} \mu(d) F(n/d) = \sum_{d_1 = n} f(d_1) \cdot 1 = f(n)$. For the converse, one has successively $\sum_{d|n} f(d) = \sum_{d|n} \sum_{d_1|d} \mu(d_1) F(d/d_1) =$

$\Sigma_{d|n}\Sigma_{d_1|d}F(d_1)\mu(d/d_1)$; inverting the order of summation, this equals $\Sigma_{d_1|n}F(d_1)\Sigma_{d_1|d|n}\mu(d/d_1)$. Setting $k = d/d_1$, the inner sum equals $\Sigma_{k|(n/d_1)}\mu(k)$ and is zero, except for $d_1 = n$ when $\Sigma_{k|1}\mu(k) = 1$; $d_1 = n$ also implies $F(d_1) = F(n)$, and that finishes the proof of the statement.

Theorem 11 makes no assumption concerning the multiplicative character of the functions involved. However, we have

Theorem 12.

   (i) *If $f(n)$ is multiplicative and $F(n) = \Sigma_{d|n}f(d)$, then $F(n)$ is also multi-plicative, and conversely,*

   (ii) *If $F(n)$ is multiplicative, then so is $f(n)$.*

PROOF.   (i) If $(m, n) = 1$, then it is clear that we obtain each divisor r of mn exactly once if we set $r = dd_1$ and let d and d_1 run independently through the divisors of m and n, respectively. Hence $F(mn) = \Sigma_{r|mn}f(r) = \Sigma_{dd_1|mn}f(dd_1)$ $= \Sigma_{d|m}\Sigma_{d_1|n}f(d)f(d_1)$, by the multiplicativity of $f(n)$. The double sum, how-ever, equals $\Sigma_{d|m}f(d)\Sigma_{d_1|n}f(d_1) = F(m)F(n)$. The proof of (ii) is similar and is left as an exercise for the reader (Hint: Use the Möbius inversion formula).

SECOND PROOF OF THEOREM 7.   In Theorem 11 take $F(n) = n$. Then $f(n) = \Sigma_{d|n}\mu(n/d)d$ and $F(n) = n = \Sigma_{d|n}f(d)$. But $f(n) = \Sigma_{d|n}\mu(d)(n/d) = n\Sigma_{d|n}(\mu(d)/d) = n\Pi_{p|n}(1 - p^{-1}) = \phi(n)$; hence $F(n) = n = \Sigma_{d|n}f(d)$ be-comes $n = \Sigma_{d|n}\phi(d)$.

The "sum function" of the Möbius function $\mu(n)$ has been defined by $M(x) = \Sigma_{n\leq x}\mu(n)$. It is completely trivial to observe that for $x \geq 2$, $|M(x)| \leq \Sigma_{n\leq x}|\mu(x)| < \Sigma_{n\leq x}1 = x$. It seems reasonable, however, to assume that there are about as many squarefree integers with an even as with an odd number of distinct primes; therefore, it would seem plausible that many cancellations occur in the sum $\Sigma_{n\leq x}\mu(n)$ and one would expect $M(x)$ to increase less fast than x. If this guess is correct, then the ratio $M(x)/x$ should decrease to zero as x increases indefinitely. This is in fact so, but the proof is far from simple. The ease with which we just proved

Theorem 13. $|M(x)| < x$

should be contrasted with the long and difficult considerations needed for the apparently simple improvement of Theorem 13 to

Theorem 13'.   *If $\varepsilon > 0$ but arbitrarily small, then for all sufficiently large x, $|M(x)| < \varepsilon x$.*

A proof of Theorem 13' will be given at the end of this chapter, under anticipation of a result to be proven only later.

6   LIOUVILLE'S FUNCTION

As an example of a totally multiplicative function that behaves in many ways like Möbius' μ-function, one may mention Liouville's function $\lambda(n)$.

DEFINITION 6.   Let $n = p_1^{s_1} p_2^{s_2} \cdots p_r^{s_r}$; then $\lambda(n) = (-1)^s$, where $s = \sum_{i=1}^{r} s_i$. The multiplicativity follows directly from the definition.

7   THE FUNCTION $\sigma_k(n)$

DEFINITION 7.   The sum of the kth powers of the divisors of the integer n is denoted by $\sigma_k(n)$. In particular, $\sigma_0(n)$ is the *number of divisors* of n (sometimes, although not in this book, $\sigma_0(n)$ is denoted by $\tau(n)$), and $\sigma_1(n)$ (usually denoted simply by $\sigma(n)$) is the *sum of the divisors* of n. In symbols, $\sigma_k(n) = \sum_{d|n} d^k$.

Theorem 14. $\sigma_k(n)$ *is multiplicative.*

PROOF.   The function $f(n) = n^k$ is trivially (even totally) multiplicative; if we use it in Theorem 12, the result follows from the definition $\sigma_k(n) = \sum_{d|n} d^k$.

Theorem 15.   *For $k \neq 0$,*

$$\sigma_k(n) = \prod_{p|n} \frac{p^{(s_p+1)k} - 1}{p^k - 1},$$

where s_p stands for the highest power of p that divides n.

PROOF.   By the multiplicative property,

$$\sigma_k(n) = \prod_{p|n} \sigma_k(p^{s_p}) = \prod_{p|n} (1 + p^k + p^{2k} + \cdots + p^{s_p k})$$

$$= \prod_{p|n} (p^{(s_p+1)/k} - 1)/(p^k - 1).$$

Corollary 15.1.

$$\sigma(n) = \prod_{p|n} \frac{p^{s_p+1} - 1}{p - 1}.$$

Corollary 15.2.

$$\sigma_0(n) = \prod_{p|n} (s_p + 1).$$

PROOFS. For Corollary 15.1, set $k = 1$ in Theorem 15; for Corollary 15.2, set $k = 0$ in the proof of the theorem.

8 *PERFECT NUMBERS

DEFINITION 8. A number n is called *perfect* if the sum of all its proper divisors (i.e., of divisors $d \neq n$) adds up to n, that is, if $\sigma(n) = 2n$.

Examples of even perfect numbers are $6(= 1 + 2 + 3)$ and $28(= 1 + 2 + 4 + 7 + 14)$; not many are known. No odd perfect numbers are known; their existence is highly unlikely, but has never been disproven. It has been shown, however (see [3], [5 pp. 239–240], [8], [9]) that if any odd perfect numbers exist at all, they must be very large ($> 10^{200}$; see [1]) and must contain at least 8 distinct prime divisors (see [7]), the largest of which cannot be less than 100,129 (see [6]). In fact, the largest cannot be less than 300,000 (see [2]), but this last result (due to an undergraduate student) has not been published.

Except for finding more, and more stringent, necessary conditions for the existence of odd perfect numbers (and thus making their existence so much more unlikely), almost all our present knowledge about perfect numbers goes back to Euclid, who already knew the following

Theorem 16. *If $2^{n+1} - 1 = p$ is a prime, then $m = \frac{1}{2}p(p + 1)$ is a perfect number; and if m is an even perfect number, then $m = 2^n(2^{n+1} - 1)$, with $2^{n+1} - 1 = p$, a prime.*

PROOF. Let $2^{n+1} - 1 = p$ be a prime; then $(p + 1)/2 = 2^n$, so that if we set $m = 2^n p$ we obtain by Corollary 15.1 $\sigma(m) = \sigma(2^n p) = ((2^{n+1} - 1)/(2 - 1))((p^2 - 1)/(p - 1)) = (2^{n+1} - 1)(p + 1) = p(p + 1) = 2m$, and m is perfect.

Conversely, if m is even and perfect, $m = 2^n m_1$, $n > 0$, m_1 odd. By Corollary 15.1

$$\sigma(m) = \sigma(2^n)\sigma(m_1) = (2^{n+1} - 1)\sigma(m_1),$$

and if m is perfect, $\sigma(m) = 2m = 2^{n+1}m_1$. Hence

$$2^{n+1}m_1 = (2^{n+1} - 1)\sigma(m_1).$$

Let $d = (m_1, \sigma(m_1))$; then, observing that $(2^{n+1} - 1, 2^{n+1}) = 1$, it follows that $m_1 = d(2^{n+1} - 1)$ and $\sigma(m_1) = 2^{n+1}d$. We assert that $d = 1$; indeed, if $d \neq 1$, then m_1 would have at least the distinct divisors $m_1, d, (2^{n+1} - 1), 1$, so that

$$\sigma(m_1) \geq m_1 + d + 2^{n+1} - 1 + 1 = d(2^{n+1} - 1) + d + 2^{n+1}$$

$$= 2^{n+1}(d + 1) > 2^{n+1}d = \sigma(m_1),$$

which is a contradiction. Hence $d = 1$, $m_1 = 2^{n+1} - 1$, and $m = 2^n m_1 = 2^n \cdot (2^{n+1} - 1)$. Now $\sigma(m)$ becomes

$$\sigma(m) = \sigma(2^n)\sigma(m_1) = (2^{n+1} - 1)\left(m_1 + 1 + \sum_{\substack{1 < c < m_1 \\ c \mid m_1}} c \right)$$

$$\geq (2^{n+1} - 1)(2^{n+1} - 1 + 1) = 2m,$$

with equality possible only if the sum over the divisors c of $m_1, 1 < c < m_1$ vanishes. But that means precisely that $m_1 = 2^{n+1} - 1$ has to be a prime, and the proof is complete.

Prime numbers of the form $p = 2^{n+1} - 1$ are called *Mersenne primes* and $n + 1 = q$ must then be itself a prime; otherwise, if $q = m \cdot k, m > 1, k > 1, 2^{m \cdot k} - 1 = (2^m)^k - 1$ is divisible at least by $1, 2^m - 1$, and $2^q - 1$, all distinct because $q = mk > m > 1$. It is known that $2^q - 1$ is prime for $q = 2, 3, 5, 7, 13, 17, 19, 31, 67, 127, 257$, and composite for all other primes less than 257. Hence, the condition that q be a prime is a necessary but *not a sufficient* condition for the primality of $2^q - 1$. In some cases it is known that $2^q - 1$ is composite without knowing any factor of it, on the basis of different tests of primality, some of which were devised specially for Mersenne numbers. The oldest seems to be Euler's (1736, see [4]); the most useful is presumably due to Lucas (see [11] p. 310, also [5] pp. 80–81, 223, and 231), who published his paper in Volume 1 of the newly launched *American Journal of Mathematics*, (1878). The largest known prime number is a Mersenne prime, recently discovered by L. Nickel and C. Noll (1979) and corresponds to $p = 21,701$. Its value is $2^p - 1 = 4.48 \cdot 10^{6532}$, so that it has 6,533 digits in decimal notation.[†]

9   *RAMANUJAN SUMS

These sums, important especially in the theory of representation of integers as sums of squares, occur since shortly after 1900 in the work of many mathematicians (for instance, Jensen, Landau, etc.). However, in view of Ramanujan's contribution to their study (1918; see [12], pp. 179–199), Hardy's choice of calling them "Ramanujan's Sums" seems well justified.

[†] A new Mersenne prime has been discovered; it corresponds to $p = 86,243$ so that $M = 2^p - 1 \cong 5.36 \cdot 10^{25961}$. M_p has 25,962 digits in decimal notation; this is a very large integer and is the largest prime number known at present.

We shall list some of their principal properties and give sketches for the proofs. For more details, one may consult [5] and [12]. Concerning notations, there is no universal agreement; here we shall again follow Hardy and denote Ramanujan's sums by $c_n(m)$. Before we define them formally, we recall that a complex number $a + ib$ may be represented by a point in the plane with coordinates $x = a$ and $y = b$. Its distance from the origin is $(a^2 + b^2)^{1/2}$ and is called the absolute value of the complex number, usually denoted by $|a + ib|$. The angle θ formed by the ray from the origin to $a + ib$ with the x axis is called the argument; clearly $\theta = \operatorname{arctg}(b/a)$. The series expansion of the exponential function of real argument, $e^x = \sum_{n=0}^{\infty}(x^n/n!)$, is taken as the definition of the exponential function in the case of a complex argument. In particular, if θ is real, $e^{i\theta} = \sum_{n=0}^{\infty}(i\theta)^n/n!$, and separating the real from the purely imaginary part,

$$e^{i\theta} = \sum_{n=0}^{\infty}(-1)^n\frac{\theta^{2n}}{(2n)!} + i\sum_{n=0}^{\infty}(-1)^n\frac{\theta^{2n+1}}{(2n+1)!}.$$

We now easily identify these well-known series and obtain the formula $e^{i\theta} = \cos\theta + i\sin\theta$. It is easy to show (use the addition formulas for sine and cosine) that with this definition the usual formal rules of computation with exponentials continue to hold, in particular $e^{a+ib} = e^a(\cos b + i\sin b)$, and $e^{ix} \cdot e^{iy} = e^{i(x+y)}$. Also $|e^{i\theta}| = (\cos^2\theta + \sin^2\theta)^{1/2} = 1$ for real θ. Clearly $e^{2\pi ik} = \cos(2\pi k) + i\sin(2\pi k) = 1$ for every integer k.

DEFINITION 9. The sums $c_n(m) = \sum_n e^{2\pi ihm/n}$, where h runs through a complete set of reduced residues modulo n, depend on the two parameters n (the index) and m (the argument), and are called *Ramanujan's sums*.

REMARK 3. Each summand $s = e^{2\pi ihm/n}$ entering the Ramanujan sums satisfies $s^n = e^{2\pi imh} = 1$; hence such summands are called *nth roots of unity*. If in $s = e^{2\pi ihm/n}$ the fraction m/n is in reduced form, then it is clear that $s^k \neq 1$ for $0 < k < n$. Such nth roots of unity that are not roots of unity of any lower order are called *primitive nth roots of unity*.

Theorem 17. *Ramanujan's sums are multiplicative functions of their indices; that is, if $(m, n) = 1$, then $c_m(k)c_n(k) = c_{mn}(k)$ for all integers k.*

PROOF.

$$c_m(k)c_n(k) = \sum_{\substack{h_1 \bmod m \\ (h_1, m)=1}} e^{2\pi ih_1k/m} \sum_{\substack{h_2 \bmod n \\ (h_2, n)=1}} e^{2\pi ih_2k/n} = \sum_{h_1}\sum_{h_2} e^{2\pi ik(h_1n+h_2m)/mn},$$

the summation being over h_1 and h_2 with the same conditions. By Lemma 2 we know that for $(m, n) = 1$, if h_1 runs through a complete reduced set of residues modulo m and h_2 runs independently through a complete set of reduced residues modulo n, then $h = h_1n + h_2m$ runs through a complete set of reduced residues modulo mn; hence the last sum can be written as $\sum\limits_{\substack{h \bmod mn \\ (h, mn)=1}} e^{2\pi ikh/mn}$ and equals, by definition, $c_{mn}(k)$.

Theorem 18. $c_p(m) = -1$ *for every prime p and every integer $m \not\equiv 0 \pmod{p}$.*

PROOF. If h runs through a complete set of reduced residues mod p, so does mh, because $(m, p) = 1$. Hence

$$c_p(m) = \sum_{h=1}^{p-1} e^{2\pi ihm/p} = \sum_{\substack{h' \bmod p \\ (h', p)=1}} e^{2\pi ih'/p} = \sum_{h''=1}^{p-1} e^{2\pi ih''/p} = \sum_{h=1}^{p} e^{2\pi ih/p} - 1.$$

The proof is completed by verifying that the finite sum vanishes. To do that, set $x = e^{2\pi i/p}$, then $\sum_{h=1}^{p} e^{2\pi ih/p} = \sum_{h=1}^{p} x^h = (x^{p+1} - x)/(x - 1) = x(x^p - 1)/(x - 1) = 0$ because, as already observed, $x^p = e^{2\pi i} = \cos 2\pi + i \sin 2\pi = 1$. More generally, we have

Theorem 19. *Let $(m, n) = d$; then $c_n(m) = \sum_{k|d} k\mu(n/k)$.*

REMARK 4. If $n = p$, $(m, p) = 1$, then Theorem 19 yields $c_p(m) = \sum_{k|1} k\mu(p/k) = \mu(p) = -1$, so that Theorem 18 is actually a Corollary of Theorem 19.

PROOF. From Theorem 9 we know that $\sum_{k|u} \mu(k) = 1$ or 0, depending on whether $u = 1$ or $u \neq 1$. This permits us to simplify the summation conditions in the definition of Ramanujan's sums. Instead of letting h range only over a complete set of *reduced* residues mod n, we let h range over *all* residues (thus ignoring the annoying side condition $(h, n) = 1$) but introduce the extra factor $\sum_{d|(h, n)} \mu(d)$, which vanishes when $(h, n) \neq 1$ and reduces conveniently to unity precisely when h is prime to n. Hence

$$c_n(m) = \sum_{h \bmod n} e^{2\pi ihm/n} \sum_{c|(h, n)} \mu(c) = \sum_{c|n} \mu(c) \sum_{\substack{h \equiv 0 \,(\mathrm{mod}\ c) \\ 1 \le h \le n}} e^{2\pi ihm/n}.$$

Setting $h = rc$, $r = 1, 2, \ldots, n/c$, $c_n(m) = \sum_{c|n} \mu(c) \sum_{r=1}^{n/c} e^{2\pi ircm/n} = \sum_{c|n} \mu(c) \sum_{r=1}^{n/c} e^{2\pi irm/(n/c)}$. If $(n/c)|m$, $e^{2\pi irm/(n/c)} = 1$, $\sum_{r=1}^{n/c} e^{2\pi irm/(n/c)} = n/c$; otherwise, as seen in the proof of Theorem 18, the sum vanishes. Hence

$$c_n(m) = \sum_{c|n, (n/c)|m} \mu(c)n/c,$$

or with $k = n/c$, $c = n/k$, $c_n(m) = \sum_{k|n, k|m} k\mu(n/k)$. However $k|n$, $k|m$ if and only if $k|d = (m, n)$, and Theorem 19 is proven.

10 FUNCTIONS RELATED TO PRIME NUMBERS

In the theory of primes, there are several functions that play an important role. Some of these are defined by the following formulas:

DEFINITION 10.

(1) $\pi(x) = \sum\limits_{p \leq x} 1$, the number of primes not exceeding the real number x;

(2) $\theta(x) = \sum\limits_{p \leq x} \log p$;

(3) $\psi(x) = \sum\limits_{p^m \leq x} \log p = \sum\limits_{p \leq x} \left[\dfrac{\log x}{\log p} \right] \log p = \sum\limits_{n \leq x} \Lambda(n)$, with

(4) $\Lambda(n) = \begin{cases} \log p & \text{if } n = p^m, \\ 0 & \text{if } n \neq p^m. \end{cases}$

$\psi(x)$ may be interpreted as the logarithm of the least common multiple of all integers n up to x (including x, if x is an integer).

REMARK 5. The smallest prime is $p = 2$; hence $\pi(x) = \theta(x) = \psi(x) = 0$, for $x < 2$.

REMARK 6. The functions $\pi(x), \theta(x), \psi(x)$ are defined for all real x; the function $\Lambda(n)$ is defined only for integral n. While the function $\pi(x)$ is a very "natural" one, it is not immediately apparent why $\theta(x)$ and $\psi(x)$ had to be introduced (by Tchebycheff, see [6] and [10]). It turns out that $\theta(x)$ and $\psi(x)$ are easier to handle than $\pi(x)$. Indeed, for large x, $\theta(x)$ and $\psi(x)$ behave practically like x itself; in fact, they are *asymptotically equal* to x, which means that $\lim_{x \to \infty}\{\theta(x)/x\} = \lim_{x \to \infty}\{\psi(x)/x\} = 1$. Actually, it is possible to prove the Prime Number Theorem (denoted from here on by PNT) directly in the form $\lim_{x \to \infty}(\pi(x) \cdot (\log x)/x) = 1$, and this is precisely what we are going to do. However, in the classical approach it turned out that a direct proof of the PNT was more difficult than the proof of $\lim_{x \to \infty}(\theta(x)/x) = 1$ or that of $\lim_{x \to \infty}(\psi(x)/x) = 1$, and it is very easy to show that either of these equalities is equivalent to the PNT. Indeed, we have

Theorem 20. *If one of the three functions $\theta(x)/x$, $\psi(x)/x$, $(\pi(x)\log x)/x$ approaches a limit when $x \to \infty$, then all three do, and the three limits are equal.*

Before we prove Theorem 20, it is convenient to introduce some very useful notations, due to E. Landau, which have received widespread acceptance. Let $f(x)$ and $g(x)$ be any two functions, $g(x) > 0$; then if there exist positive constants C and ε such that $|f(x)| < Cg(x)$ holds for all x satisfying $|x - x_0| < \varepsilon$, we write $f(x) = O(g(x))(x \to x_0)$. If $|f(x)| < Cg(x)$ holds for all sufficiently large x, we write $f(x) = O(g(x))(x \to \infty)$. Usually it is unnecessary to mention x_0, which is understood from the context, and then the last bracket is omitted. This notation is particularly useful if $g(x)$ is a much simpler function than $f(x)$. If, for instance, we want to state that $f(x)$ stays bounded, $|f(x)| < C$, we take $g(x) = 1$ and write $f(x) = O(1)$. The following are a few further examples of the use of the O notation: When $x \to +\infty$, then $\sin x = O(1)$; $e^{-x} = O(x^{-n})$ for every n; $[x] = O(x)$; $x/[x] = O(1)$.

If an inequality $|f(x)| < \varepsilon g(x)$ $(x \to x_0)$ holds for every $\varepsilon > 0$, provided only that $|x - x_0|$ is small enough (or, provided that x is sufficiently large, if $x_0 = \infty$), then we write $f(x) = o(g(x))$. In previous examples, for instance, when $x \to \infty$, $\sin x = o(x^{1/2})$, $e^{-x} = o(x^{-n})$, for every n; also $x = o(x^2)$, $x^{1/2} = o(x)$, and so on. If $f(x) \to 0$, we may express this by writing $f(x) = o(1)$.

Theorem 20 will follow easily from a few preliminary theorems.

Theorem 21. $\psi(x) = \theta(x) + \theta(x^{1/2}) + \theta(x^{1/3}) + \theta(x^{1/4}) + \cdots$.

REMARK 7. The sum breaks off after k terms if k is the first integer such that $x^{1/(k+1)} < 2$ (see Remark 5).

PROOF. $\psi(x) = \sum_{p^m \le x} \log p = \sum_{m \ge 1} \sum_{p \le x^{1/m}} \log p = \sum_{m \ge 1} \theta(x^{1/m})$.

Theorem 22. *For $x > 1$, $\theta(x) < x \log x$.*

PROOF.

$$\theta(x) = \sum_{p \le x} \log p \le \sum_{p \le x} \log x = \log x \sum_{p \le x} 1 = \log x \cdot \pi(x) < x \log x.$$

Theorem 23. *For $x > 1$, $\theta(x) \le \psi(x) < \theta(x) + x^{1/2} \log^2 x$, or less precisely, $\psi(x) - \theta(x) = O(x^{1/2} \log^2 x)$.*

PROOF. The first inequality follows from Theorem 21; for the second, let $k = [\log x / \log 2]$ and observe that

$$\psi(x) - \theta(x) = \sum_{m=2}^{k} \theta(x^{1/m}) \le \sum_{m=2}^{k} x^{1/m} \log x^{1/m} = \sum_{m=2}^{k} \frac{1}{m} \log x \cdot x^{1/m}$$

by Theorem 22. Hence, *a fortiori*,

$$\psi(x) - \theta(x) \le \sum_{m=2}^{k} \tfrac{1}{2} x^{1/m} \log x = \tfrac{1}{2} \log x \sum_{m=2}^{k} x^{1/m} \le \tfrac{1}{2} \log x \sum_{m=2}^{k} x^{1/2}$$

$$= \tfrac{1}{2}(k-1) x^{1/2} \log x < \frac{k}{2} x^{1/2} \log x$$

$$\le \frac{1}{2 \log 2} x^{1/2} \log^2 x < x^{1/2} \log^2 x,$$

which proves the theorem.

Theorem 24 (Tchebycheff). *There exist positive constants A, A'; B, B', such that for $x > 2$, $Ax < \theta(x) < A'x$, $Bx < \psi(x) < B'x$.*

PROOF. On account of Theorem 23, it is sufficient to prove $\theta(x) < A'x$ and $\psi(x) > Bx$, because then

$$\psi(x) < \theta(x) + O(x^{1/2}\log^2 x) < A'x + O(x^{1/2}\log^2 x)$$
$$= x(A' + O(x^{-1/2}\log^2 x)) < (A' + 1)x$$

and

$$\theta(x) > \psi(x) - O(x^{1/2}\log^2 x) > Bx(1 - O(x^{-1/2}\log^2 x)) > \tfrac{1}{2}Bx$$

follow immediately for sufficiently large x. We shall actually prove rather more, namely that $\theta(x) < (2\log 2)x$ and $\psi(x) > ((\log 8)/10)x$.

For $n = 1$ and $n = 2$, $\theta(n) < (2\log 2)n$ holds trivially. In order to complete the induction on n, we observe that in the expansion of $(1 + 1)^{2m+1}$, the binomial coefficient[†] $M = (2m + 1)!/m!(m + 1)! = N/D$ occurs twice, so that $M < 2^{2m}$. All primes p satisfying $m + 1 < p \leq 2m + 1$ divide N, but not D; hence $(\prod_{p>m+1}^{2m+1} p)|M$ and, in particular, $(\prod_{p>m+1}^{2m+1} p) \leq M$. On taking logarithms, $\theta(2m + 1) - \theta(m + 1) = \sum_{p>m+1}^{2m+1}\log p \leq \log M < 2m\log 2$. The proof that $\theta(k) < (2\log 2)k$ can now be completed by induction on k. Let us assume that the inequality, already checked for $k = 1$ and $k = 2$, holds up to $k = n - 1$. We show that it then also holds for $k = n$. If n is even this is trivial, because n (being divisible by 2) cannot be a prime; hence

$$\theta(n) = \theta(n - 1) < (2\log 2)(n - 1) < (2\log 2)n.$$

If n is odd, it is of the form $n = 2m + 1$, and we just proved that $\theta(n) = \theta(2m + 1) < \theta(m + 1) + (2\log 2)m$; by the induction hypothesis, this is $< (2\log 2)(m + 1) + (2\log 2)m = (2\log 2)(2m + 1) = (2\log 2)n$, and this finishes the proof of the statement. Indeed, having proven the inequality for integers, it holds for all real numbers, because if $n < x < n + 1$, then

$$\theta(x) = \theta(n) < (2\log 2)n < (2\log 2)x.$$

To prove $\psi(x) > ((\log 8)/10)x$, we proceed in a similar way. One easily checks that

$$\psi(1) = 0,$$
$$\psi(2) = \log 2 > 2(\cdot 346),$$
$$\psi(3) = \log 2 + \log 3 = \log 6 > 3(\cdot 597),$$
$$\psi(4) = 2\log 2 + \log 3 > 4(\cdot 621),$$
$$\psi(5) = \psi(4) + \log 5 > 5(\cdot 818),$$

[†] The binomial coefficient, its numerator, and its denominator are denoted here for convenience by capitals, although they are rational integers.

so that for $x \le 5$, $\psi(x) > ((\log 8)/10)x \cong (\cdot 2079)x$ is amply satisfied, and from here on we may assume that $x > 5$. Now

$$2^n < \frac{2n}{n} \cdot \frac{2n-1}{n-1} \cdot \ldots \cdot \frac{n+1}{1} = \frac{(2n)!}{n!n!}.$$

By Theorem 2,

$$\frac{(2n)!}{n!n!} = \prod_{p \le 2n} p^{k_p}, \quad \text{with} \quad k_p = \sum_{1 \le m \le m_1} \left(\left[\frac{2n}{p^m} \right] - 2 \left[\frac{n}{p^m} \right] \right),$$

where $m_1 = [\log 2n/\log p]$ (clearly, all terms are zero for $m > (\log 2n/\log p)$). By Theorem 1(8) the bracket is either 0 or 1; hence, $k_p \le \sum_{1 \le m \le m_1} 1 = [\log 2n/\log p]$. Consequently,

$$n \log 2 = \log 2^n \le \log \prod_{p \le 2n} p^{k_p}$$

$$= \sum_{p \le 2n} k_p \log p \le \sum_{p \le 2n} \left[\frac{\log 2n}{\log p} \right] \log p = \psi(2n).$$

For given x, set $n = [\tfrac{1}{2}x]$; then

$$\psi(x) \ge \psi(2n) \ge n \log 2 > (\tfrac{1}{2}x - 1)\log 2$$

$$= \frac{\log 2}{2} x \left(1 - \frac{2}{x} \right)$$

$$> \left(1 - \frac{2}{5} \right) \frac{\log 2}{2} x = \frac{3 \log 2}{10} x \qquad \text{for } x > 5.$$

This finishes the proof of Theorem 24.

PROOF OF THEOREM 20.   First, by Theorem 23,

$$\frac{\theta(x)}{x} \le \frac{\psi(x)}{x} \le \frac{\theta(x)}{x} + O(x^{-1/2}\log^2 x),$$

so that

$$\lim_{x \to \infty} \frac{\theta(x)}{x} = \lim_{x \to \infty} \frac{\psi(x)}{x},$$

if either limit exists. Next, as already seen,

$$\theta(x) = \sum_{p \le x} \log p < \sum_{p \le x} \log x = \log x \sum_{p \le x} 1 = \pi(x)\log x,$$

so that $\theta(x)/x \le (\pi(x)\log x)/x$. Finally, for any $\varepsilon > 0$ but arbitrarily small,

$$\theta(x) \ge \sum_{x^{1-\varepsilon} < p \le x} \log p \ge \sum_{x^{1-\varepsilon} < p \le x} \log x^{1-\varepsilon} = (1-\varepsilon)\log x \sum_{x^{1-\varepsilon} < p \le x} 1$$

$$= (1-\varepsilon)(\log x)(\pi(x) - \pi(x^{1-\varepsilon})) > (1-\varepsilon)(\log x)(\pi(x) - x^{1-\varepsilon}).$$

Consequently,

$$\frac{\theta(x)}{x} > (1 - \varepsilon)\frac{\pi(x)\log x}{x}\left(1 - \frac{x^{1-\varepsilon}}{\pi(x)}\right);$$

but we just proved that $\theta(x) < \pi(x)\log x$, so that $\pi(x) > \theta(x)/(\log x)$ and $x^{1-\varepsilon}/\pi(x) < (x^{1-\varepsilon}/\theta(x))\log x$. By Theorem 24, this is less than $(x^{1-\varepsilon}/Ax)\log x = (\log x)/Ax^{\varepsilon}$ with $A > 0$, $\varepsilon > 0$; hence, the fraction becomes arbitrarily small, say $< \varepsilon$, as $x \to \infty$, and

$$\frac{\theta(x)}{x} > (1 - \varepsilon)\frac{\pi(x)\log x}{x}\left(1 - \frac{x^{1-\varepsilon}}{\pi(x)}\right) > (1 - 2\varepsilon)\frac{\pi(x)\log x}{x}.$$

Combining the results it follows that the inequalities

$$\frac{\theta(x)}{x} \leq \frac{\pi(x)\log x}{x} < \frac{1}{1 - 2\varepsilon} \cdot \frac{\theta(x)}{x}$$

hold for arbitrarily small $\varepsilon > 0$. It follows that if $\lim_{x \to \infty}\theta(x)/x$ exists, so does $\lim_{x \to \infty}(\pi(x)\log x)/x$ and conversely, and that if these limits exist, then they are equal. This finishes the proof of Theorem 20.

11   FORMULAE THAT YIELD PRIMES

A question that arises quite naturally at this point is the following: Is there no formula that yields the primes themselves? Or to be more precise, given an integer n, is there a function, say $f(n)$, such that $f(n) = p_n$, the nth prime? In fact, such formulae do exist, but they are disappointing. They are usually of the recursive kind; that means they permit us to compute the primes only one by one, and we need to know the first $n - 1$ primes to use the formula for the computation of the nth prime. Several such formulae have been proposed—one of the simplest is the following, due to J. M. Gandhi, who presented it at the International Congress of Mathematicians, Moscow, 1966.

Let Q_n be the product of the first $n - 1$ primes, and set $\sigma_n = \sum_{d|Q_n}(\mu(d)/(2^d - 1))$. One may show that σ_n decreases with n, and that $1/2 < \sigma_n < 1$, so that $0 < \sigma_n - 1/2 < 1/2$ for all n. If we set $(\sigma_n - 1/2)^{-1} = c_n$, then $c_n > 2$, and between c_n and $2c_n$ there is precisely one integral power of 2, $c_n < 2^{f(n)} < 2c_n$. The statement now is (see [15]) that $f(n) = p_n$, the nth prime, or equivalently, that

$$p_n = \left[\frac{\log 2c_n}{\log 2}\right] = 1 + \left[\frac{\log c_n}{\log 2}\right].$$

So for example if $n = 4$, then $Q_4 = 2 \cdot 3 \cdot 5 = 30$, $\sigma_4 = 1 - 1/3 - 1/7 - 1/31 + 1/63 + 1/1023 + 1/32767 - 1/(2^{30} - 1) = .50844\ldots$, $c_4 =$

$1/.00844\ldots = 118.48\ldots, 2c_4 = 236.98\ldots$, and if $118.48\ldots < 2^{f_4} < 236.98\ldots$, then $2^{f_4} = 128$, so that $f_4 = 7$, the 4th prime. Alternatively, $p_4 = [\log c_4/\log 2] + 1 = [\log 118.48\ldots/\log 2] + 1 = [4.7708\ldots/0.6931\ldots] + 1 = 6 + 1 = 7$, the 4th prime. These formulae are not useful for the effective computation of the primes, nor for the evaluation of the number of primes in given intervals, say between 1 and x. Their main importance is indeed the very fact that such formulae exist at all, something considered for a long time as highly unlikely.

12 **ON THE SUM FUNCTION $M(x)$

At the end of Section 5 we had Theorem 13', which can be restated now in the following neater form.

Theorem 13'. *If $x \to \infty$, then $M(x) = \displaystyle\sum_{n \le x} \mu(n) = o(x)$.*

We made the remark that, while $M(x) = O(x)$ is trivial, the apparently modest sharpening represented by Theorem 13' is far from trivial. In the remaining part of this chapter, we shall prove Theorem 13' by anticipating a result that will be proven only later, namely the PNT; that is,

$$\pi(x) = \frac{x}{\log x} + o\left(\frac{x}{\log x}\right);$$

we know from Theorem 20 that this statement is equivalent to $\psi(x) = x + o(x)$.

The reader may convince himself that this result will be proven without the use of anything developed in the present section, so that no circularity vitiates the reasoning.

Let $H(x) = \sum_{n=1}^{x} \mu(n)\log n$; then $\mu(n) = (H(n) - H(n-1))/(\log n)$, so that $M(x) = 1 + \sum_{n=2}^{x}(H(n) - H(n-1))/(\log n)$. We may regroup the terms by combining those with the same $H(n)$ in consecutive summands, and observing also that $H(1) = 0$, we obtain

$$M(x) = 1 + \sum_{n=2}^{x} H(n)\left(\frac{1}{\log n} - \frac{1}{\log(n+1)}\right) + \frac{H(x)}{\log([x]+1)}.$$

We shall presently show that as $x \to \infty$, $H(x) = o(x \log x)$. Assuming this result for a moment, it follows that given $\varepsilon_0 = 3\varepsilon > 0$ arbitrarily small, there exists an integer $x_1 > 8$, such that $|H(x)| < \varepsilon x \log x$ for $x \ge x_1$. Also let $C = \max_{x \le x_1}|H(x)|$. One observes that for $x \ge x_1$, $|H(x)|/\log([x]+1) \le \varepsilon x$, and that

$$\frac{1}{\log n} - \frac{1}{\log(n+1)} = \frac{\log(n+1) - \log n}{\log n \log(n+1)} < \frac{\log(1+n^{-1})}{\log^2 n} < \frac{1}{n \log^2 n}.$$

It now follows for $x \geq x_1$ that

$$|M(x)| \leq 1 + \sum_{n=2}^{x_1} C \cdot \frac{1}{n \log^2 n} + \sum_{n=x_1+1}^{x} \frac{\varepsilon n \log n}{n \log^2 n} + \varepsilon x$$

$$\leq C' + \varepsilon \sum_{n=x_1+1}^{x} \frac{1}{\log n} + \varepsilon x,$$

where

$$C' \leq 1 + \sum_{n=2}^{\infty} \frac{C}{n \log^2 n}.$$

Also,

$$\sum_{n=x_1+1}^{x} \frac{1}{\log n} < \frac{1}{\log x_1} \sum_{n=x_1+1}^{x} 1 < x/2,$$

because $\log x_1 > \log 8 > 2$, and for sufficiently large x, $C' < \varepsilon x$. Consequently, $|M(x)| \leq 3\varepsilon x = \varepsilon_0 x$, and this is the statement we wanted to prove. Hence in order to finish the proof of Theorem 13' it only remains to show (besides the two anticipated results) that $H(x) = o(x \log x)$, which means that $\lim_{x \to \infty} H(x)/(x \log x) = 0$. In the proof, we shall need the following result.

Lemma 3. *For every integer n_0,*

$$\left| \sum_{n=1}^{n_0} \frac{\mu(n)}{n} \right| \leq 1.$$

We use the result of Corollary 9.3, stating that for every x, $\sum_{n=1}^{x} \mu(n)[x/n] = 1$. We take for x the integer n_0 and denote the fractional part $(n_0/n) - [n_0/n]$ by α_n; clearly, $\alpha_1 = 0$, and $0 \leq \alpha_n < 1$ for $2 \leq n \leq n_0$. Hence

$$1 = \sum_{n=1}^{n_0} \mu(n) \left[\frac{n_0}{n} \right] = \sum_{n=1}^{n_0} \mu(n) \frac{n_0}{n} - \sum_{n=2}^{n_0} \alpha_n \mu(n).$$

The last sum is less in absolute value than $\sum_{n=2}^{n_0} 1 = n_0 - 1$; hence

$$\left| \sum_{n=1}^{n_0} \mu(n) \frac{n_0}{n} \right| = n_0 \cdot \left| \sum_{n=1}^{n_0} \frac{\mu(n)}{n} \right| \leq 1 + (n_0 - 1) = n_0,$$

and the Lemma is proven.

To show that $H(x) = o(x \log x)$ we first find a different expression for $H(x)$. As already seen in the second proof of Theorem 9, $\zeta(s)^{-1} = \sum_{n=1}^{\infty} \mu(n) n^{-s}$, so that

$$\frac{d}{ds} \frac{1}{\zeta(s)} = -\frac{\zeta'(s)}{\zeta^2(s)} = -\sum_{n=1}^{\infty} \mu(n)(\log n) n^{-s}.$$

Also,

$$\log \zeta(s) = -\sum_p \log(1 - p^{-s}) = \sum_p \sum_m \frac{p^{-ms}}{m},$$

by the expansion of the logarithm. Differentiating,

$$\zeta'(s)/\zeta(s) = -\sum_p \sum_m (\log p)/p^{ms},$$

or, using Definition 10(4) of $\Lambda(n)$, $\zeta'(s)/\zeta(s) = -\sum_{n=1}^{\infty}\Lambda(n)/n^s$. The validity of these formal operations can easily be proven, on account of the uniform convergence (for $\sigma \geq 1 + \varepsilon$, $\varepsilon > 0$) of all series involved.

We observe that $-\zeta'(s)/\zeta^2(s) = (1/\zeta(s))(-\zeta'(s)/\zeta(s))$; hence

$$-\sum_{n=1}^{\infty} \frac{\mu(n)\log n}{n^s} = \sum_{k=1}^{\infty} \frac{\mu(k)}{k^s} \sum_{m=1}^{\infty} \frac{\Lambda(m)}{m^s}.$$

Setting $km = n$ in the second member, this can be written as

$$\sum_{n=1}^{\infty} \frac{1}{n^s} \sum_{k|n} \mu(k)\Lambda(n/k) = \sum_{n=1}^{\infty} \frac{a_n}{n^s},$$

so that

$$-\sum_{n=1}^{\infty} \frac{\mu(n)\log n}{n^s} = \sum_{n=1}^{\infty} \frac{a_n}{n^s}, \quad \text{with} \quad a_n = \sum_{k|n} \mu(k)\Lambda(n/k).$$

By Theorem 10, two such Dirichlet series can coincide for all values of s with $\sigma \geq 1 + \varepsilon$ only if they are equal termwise; that is, if they have the same coefficients. Hence $a_n = \sum_{k|n}\mu(k)\Lambda(n/k) = -\mu(n)\log n$, and $H(x)$ becomes

$$H(x) = \sum_{n=1}^{x} \mu(n)\log n = -\sum_{n=1}^{x} \sum_{k|n} \mu(k)\Lambda(n/k)$$

$$= -\sum_{k=1}^{x} \mu(k) \sum_{\substack{n=mk \\ 1 \leq m \leq x/k}} \Lambda(n/k) = -\sum_{k=1}^{x} \mu(k) \sum_{m=1}^{[x/k]} \Lambda(m)$$

$$= -\sum_{k=1}^{x} \mu(k)\psi(x/k)$$

by the definition of $\psi(x)$. Until now we have made no use of the PNT; now we shall use it by observing that, according to it, the following is true: Given $\varepsilon > 0$ but arbitrarily small, we can find a constant y (depending only on ε, but not on x) such that for $x \geq y$, $|\psi(x) - x| < \varepsilon x$. If we set $n_0 = [x/y]$, then

$$-H(x) = \sum_{n=1}^{x} \mu(n)\psi(x/n) = \sum_{n=1}^{n_0} \mu(n)\psi(x/n) + \sum_{n=n_0+1}^{x} \mu(n)\psi(x/n).$$

In the first sum, $x/n \geq x/n_0 \geq y$; hence

$$\left| \sum_{n=1}^{n_0} \mu(n)\psi\left(\frac{x}{n}\right) \right| = \left| \sum_{n=1}^{n_0} \mu(n)\left(\frac{x}{n} + \left(\psi\left(\frac{x}{n}\right) - \left(\frac{x}{n}\right)\right)\right) \right|$$

$$\leq \left| \sum_{n=1}^{n_0} \mu(n)\frac{x}{n} \right| + \varepsilon \cdot \sum_{n=1}^{n_0} \left| \frac{\mu(n)x}{n} \right|$$

$$\leq x \sum_{n=1}^{n_0} \frac{\mu(n)}{n} + \varepsilon \cdot x \cdot \sum_{n=1}^{n_0} \frac{1}{n}.$$

Using Lemma 3 and the fact that $\sum_{n=1}^{N}(1/n) < \log N + 1$, it follows that the last sums add up to less than $x + \varepsilon x(\log n_0 + 1) < x + \varepsilon x \log x$ (for $y > e$; the other alternative is trivial). In the second sum occurring in $-H(x)$, $\psi(x/n) \leq \psi(x/(n_0 + 1)) < \psi(y)$, a constant, independent of x. Hence the second sum is less in absolute value than $\sum_{n=n_0}^{x} |\mu(n)\psi(y)| < \psi(y)\sum_{n=n_0}^{x} 1 < x\psi(y)$, so that finally,

$$|H(x)| < x + \varepsilon x \log x + x\psi(y).$$

Consequently, $|H(x)| < x(1 + \psi(y)) + \varepsilon x \log x$, and $|H(x)|/(x \log x) < \varepsilon + (1 + \psi(y))/(\log x) < 2\varepsilon$ if x is sufficiently large. Hence $H(x) = o(x \log x)$, as asserted, and (except for the two anticipated results) our proofs are complete.

PROBLEMS

1. Prove Theorem 1 in detail.
2. Let k_p be the highest power of the prime p that divides $(2n)!/(n!)^2$. Show that

$$k_p = \sum_{m \geq 1}\left(\left[\frac{2n}{p^m}\right] - 2\left[\frac{n}{p^m}\right]\right).$$

3. Find the highest power of 5 that divides 3,000!
4. Find $\sigma_0(24)$, $\sigma_1(24)$, $\sigma_2(24)$, (a) directly, and (b) using Theorem 15.
5. Find $\sigma(1728)$.
6. Prove: $\sum_{d|n}\lambda(d) = \begin{cases} 1 & \text{if } n = m^2 \text{ for some integer } m, \\ 0 & \text{otherwise.} \end{cases}$
7. Prove the second part of Theorem 12: If $F(n) = \sum_{d|n}f(d)$ and $F(n)$ is multiplicative, then so is $f(n)$.
8. Let $\phi(n)$ be the Euler function. Prove: $\phi(n) = \sum_{dc=n} d\mu(c)$. (Hint: $n = \sum_{d|n}\phi(d)$; use the Möbius inversion formula.)

9. Compute $\sum_{d|6000}\phi(d)$.

10. Let $F(x) = \sum_{n \le x}\phi(n)$. By anticipating the result of Section 8.7 (See also Chapter 7, Problem 19) that $\zeta(2) = \pi^2/6$, prove that $F(x) = (3x^2/\pi^2)(1 + o(1))$. (Hint: Use Problem 8, change order of summation; use Corollary 9.3 and replace $[x/d]^2$ by $((x/d) - \alpha)^2$. In estimating the result, use $\zeta(2) = \pi^2/6$.)

11. (Landau). Compute $(\phi(5186), \phi(5187), \phi(5188))$, and $[\phi(5186), \phi(5187), \phi(5188)]$.

12. Show that n cannot be a perfect number if it is a prime or the product of two odd primes; also, find all exceptions to the statement that n cannot be a perfect number if it is the product of two primes.

13. Prove in detail the equalities occurring in the definition of $\psi(x)$.

14. Prove Theorem 3 in detail.

15. Define the Dedekind sum (see [9], where the notation is slightly different) $s(h, k) = \sum_{m \bmod k}((m/k))((mh/k))$. Prove:

    (a) $h_1 \equiv h_2 \pmod{k} \Rightarrow s(h_1, k) = s(h_2, k)$,
    (b) $h\bar{h} \equiv 1 \pmod{k} \Rightarrow s(h, k) = s(\bar{h}, k)$,
    (c) $s(-h, k) = -s(h, k)$,
    (d) $hh^* \equiv -1 \pmod{k} \Rightarrow s(h^*, k) = -s(h, k)$,
    (e) $h^2 \equiv -1 \pmod{k} \Rightarrow s(h, k) = 0$.

16. Find the perfect numbers corresponding to the Mersenne primes $2^q - 1$ for $q = 5$ and $q = 7$; check the result.

17. Compute the Ramanujan sums $c_3(5)$ and $c_6(10)$ directly and by Theorems 18 and 19.

18. Prove $c_n(1) = \mu(n)$.

19. Prove $\sum_{d|n}\mu(d)\sigma_0(n/d) = 1$.

20. Prove $\sum_{n \le x}\phi(n)/n = (6/\pi^2)x + O(\log x)$. (Hint: Use Problem 8).

BIBLIOGRAPHY

1. M. Buxton and S. R. Elmore, Abstract 731-10-40 *Notices AMS*, **23** (1976) A-55.

2. J. Condict, Unpublished senior thesis, Middlebury College, Middlebury, Vermont, June 1979.

3. L. E. Dickson, *History of the Theory of Numbers*, Vol. 1. Washington, D.C.: Carnegie Inst. of Washington, 1919–1923.

4. L. Euler, *Comm. Acad. Petrop.* **6** (1732–3) 103–107; *Opera Omnia* (1) **2**, 1–5.

5. G. H. Hardy and E. M. Wright, *An Introduction to the Theory of Numbers*, 3rd ed. Oxford: The Clarendon Press, 1954.

6. P. Hagis and W. L. McDaniel, *Mathem. of Computation* **29** (1975) 922–924.

7. P. Hagis, *Mathematics of Computation*, **35** (1980) pp. 1027–1032.

8. H. J. Kanold, *Journal für die reine und angewandte Mathem.* **186** (1944) 25–29; **197** (1957) 82–96.

9. U. Kühnel, *Mathem. Zeitschrift*, **52** (1949) 202–211.

10. E. Landau, *Handbuch der Lehre von der Verteilung der Primzahlen*, 2nd ed. New York: Chelsea Publishing Co., 1953.

11. E. A. Lucas, *Amer. Journal of Mathem.* **1** (1878) 185–240, 289–321.

12. S. Ramanujan, *Collected Papers*, edited by G. H. Hardy, P. V. Sashu Aiyar, and B. M. Wilson. Cambridge: Cambridge University Press, 1927.

13. B. Riemann and R. Dedekind, Erläuterung zu den Fragmenten XXVIII, *Collected Works of B. Riemann*, edited by H. Weber, 2nd ed. (1892–1902). New York: Dover Publications, 1953.

14. P. Tchebycheff, Several memoirs, see in partic. *Journal des Math. pures et appliquées* (1) **17** (1852) 366–390.

15. C. van den Eynden, *Amer. Math. Monthly* **79** (1972), 625.

16. I. M. Vinogradov, *Elements of Number Theory*. New York: Dover Publications, 1954.

Chapter 7

The Theory of Partitions

1   INTRODUCTION

Since the 18th century the theory of partitions has interested some of the best minds. While it seems to have little or no practical application, it has, in a certain sense, just the right degree of difficulty. The problems are far from trivial, but at the same time they are not so hard as to discourage any attempt at a solution. Besides, through the introduction by Euler (1707–1783) of generating functions, the highly developed apparatus of the theory of functions became available for the study of partitions. A further circumstance of great help in this study is the fact that the generating functions which occur in the theory of partitions and functions closely related to them belong to two important classes of functions, namely the theta functions and the modular functions, both of which have received much attention and have been most thoroughly investigated since the time of Jacobi (1804–1851).

In the following pages it will be possible to consider only a few, mostly classical, aspects of the theory of partitions. It is my hope that the reader's curiosity will not have been satisfied with (or dulled by) this limited selection of topics. In case his interest has been sufficiently stimulated that he may want to learn something about the many facets of the theory that, unfortunately, had to be ignored here, then I would suggest as a starting point some of the survey papers, like [13] or [6], where more extensive bibliographies may also be found.

2   DEFINITIONS AND NOTATIONS

The following definitions will clarify the meaning of some of the technical terms used or to be used; some others will be explained later. Throughout this chapter we shall often have to refer to the set $\mathbf{Z}_0$ of non-negative integers $0, 1, 2, \ldots$; clearly $\mathbf{Z}_0 = \{0\} \cup \mathbf{Z}^+$ and $\mathbf{Z}_0 \subset \mathbf{Z}$.

DEFINITION 1. Let $f(n)$ be a function defined for $n \in \mathbf{Z}_0$ and let $F(x) = \sum_{n=0}^{\infty} f(n) x^n$; then $F(x)$ is said to be the *generating function* of $f(n)$.

REMARK 1. The definition does not imply that the series converges anywhere if $x \neq 0$; but as a matter of fact the generating functions we shall meet with will all converge in some finite circle, usually the unit circle.

Although the reader already knows from our introductory remarks what is meant by the word "partition," this concept will be further clarified and made more precise by the following definition.

DEFINITION 2. Let $\mathbf{A} = \{a_1, a_2, \ldots, a_r, \ldots\}$ be a finite or infinite set of positive integers. If $a_{i_1} + a_{i_2} + \cdots + a_{i_r} = n$, with $a_{i_j} \in \mathbf{A}(j = 1, 2, \ldots, r)$, then we say that the sum $a_{i_1} + a_{i_2} + \cdots + a_{i_r}$ is a *partition* of n into summands (or parts) belonging to the set $\mathbf{A}$.

REMARK 2. If $n \in \mathbf{A}$, then n itself has to be counted as a partition of itself (see the examples with $n = 5$ at the end of this section).

REMARK 3. According to the definition of partitions, the summands do not have to be distinct; an explicit restriction to have all summands distinct may, however, be added.

The order of the summands is irrelevant; two partitions of n that differ only by the order of the summands are not considered distinct.[†] Hence, rearranging the summands if necessary and grouping together summands that are equal, every partition can be written uniquely in the form $n = k_1 a_1 + k_2 a_2 + \cdots + k_i a_i + \cdots$, where $a_1, a_2, \ldots$ are the distinct elements of $\mathbf{A}$ in increasing order, and where the k_i are non-negative integers, only finitely many of which are different from zero. Each coefficient k_i indicates how often a given summand $a_i \in \mathbf{A}$ occurs in the partition under consideration, and it is called the *frequency* of a_i in that partition of n. It should be clear that any given partition is completely determined by the set of its frequencies.

DEFINITION 3. Let $\mathbf{A}$ be a finite or infinite set of positive integers. The number of distinct partitions of a natural integer n into summands (or parts) belonging to $\mathbf{A}$ is denoted by $p_{\mathbf{A}}(n)$ and is called the *partition function* relative to the set $\mathbf{A}$. If no further conditions are imposed, we also call $p_{\mathbf{A}}(n)$ the number of *unrestricted partitions* of n into parts that belong to $\mathbf{A}$. If $\mathbf{A}$ is the set $\mathbf{Z}^+$ of all natural integers, the partition function is denoted simply by $p(n)$. If

---

[†] If we prefer to distinguish between partitions with identical summands taken in different order, then we call them *representations* rather than partitions.

other conditions are imposed (such as to have all summands distinct, or no more than a fixed number m of summands), then the partitions are called partitions with restrictions, or *restricted partitions*.

NOTATIONS. We shall use the following notations:

$p(n)$ stands for the number of partitions of n into natural integers, without any restrictions.

$p_A(n)$ stands for the number of partitions of n into parts belonging to **A**, without other restrictions.

$p_{A,m}(n)$ stands for the number of partitions of n into parts not exceeding m and belonging to **A**.

$p_A^{(m)}(n)$ stands for the number of partitions of n into at most m summands, all belonging to **A**.

$p_A^{(d)}(n)$ stands for the number of partitions of n into distinct summands, all belonging to **A**.

$p_{A,k}^{(d,m)}(n)$ stands for the number of partitions of n into at most m distinct parts, not exceeding k, of the set **A**.

$p_A^{(o)}(n)$ stands for the number of partitions of n into an odd number of parts belonging to **A**.

$p_A^{(e)}$ stands for the number of partitions of n into an even number of parts, all belonging to **A**.

$p_A^{(o,d)}$ and $p_A^{(e,d)}$ stand for the number of partitions of n into an odd (respectively an even) number of distinct parts belonging to **A**.

We denote the set of positive odd integers by **O** and the set of non-negative even integers by **E**; clearly $\mathbf{O} \cup \mathbf{E} = \mathbf{Z}_0$. Observe that $p_E(n)$, the number of partitions of n into *even summands*, and $p^{(e)}(n)$, the number of partitions of n into an *even number of summands*, are not the same thing (see examples below). Occasionally, a restriction that need not be specified will be denoted by an*. Generating functions of partitions will be denoted by $F(x)$, with the same subscripts or superscripts as the partitions they generate. So for instance,

$$F_A^{(m)}(x) = \sum_{n=0}^{\infty} p_A^{(m)}(n)x^n,$$

and so on.

REMARK 4. As already observed, each partition is completely determined by the set of its frequencies $k_1, k_2, \ldots, k_r, \ldots$; hence the number $p_A(n)$ of partitions of n is precisely the number of distinct solutions of the Diophantine equation $k_1 a_1 + k_2 a_2 + \cdots + k_i a_i + \cdots = n$, in positive integers k_i and elements $a_i \in \mathbf{A}$; in symbols,

$$p_\mathbf{A}(n) = \sum_{\substack{\Sigma k_i d_i = n \\ a_i \in \mathbf{A},\, k_i \in \mathbf{Z}^+}} 1.$$

EXAMPLES

For $n = 5$, we observe that $5 = 4 + 1 = 3 + 2 = 3 + 1 + 1 = 2 + 2 + 1$ $= 2 + 1 + 1 + 1 = 1 + 1 + 1 + 1 + 1$. Hence $p_{\mathbf{Z}^+}(5) = p(5) = 7$; $p_\mathbf{E}(5) = 0$; $p_\mathbf{O}(5) = 3$; $p_3(5) = 5$, $p^{(3)}(5) = 5$; $p_\mathbf{O}^{(d)}(5) = 1$; $p_3^{(d)} = 1$; $p_5(5) = 7$; $p_8(5) = 7$; $p^{(e)}(5) = 3 (\neq p_\mathbf{E}(5) = 0)$ $p^{(d,2)}(5) = p^{(d,3)}(5) = 3$.

In what precedes, we required the elements of $\mathbf{A}$ to be natural integers. Applying this condition strictly, it follows that zero has no partitions; that is, $p(0) = 0$. However, usually (not always! see, e.g., the proof of Theorem 10) it is more convenient to define $p(0) = 1$ (and occasionally also $p_{\mathbf{A},\cdot}^{(\cdot)}(0) = 1$); whenever this is the case, we shall not hesitate to do so.

3   SURVEY OF METHODS

Several methods have been used successfully in the theory of partitions. We already mentioned the method of *generating functions*, used brilliantly by Euler and his successors. Next are the purely *combinatorial methods*, whose usefulness is greatly enhanced by the use of *graphs*. Besides the use of generating functions as a formal device, they may also be used in connection with algebraic methods such as decomposition into *partial fractions* (see [2], [9], and [14]), or in connection with analytic methods such as Cauchy's *theorem on residues*, *contour integration*, or *Tauberian theorems*, to quote only a few. In what follows we shall have occasion to exemplify the use of several of these methods, but shall remain throughout on an elementary level; that is, we shall make no appeal to any deeper theorems of either algebra or analysis.

4   GENERATING FUNCTIONS

Let $F_\mathbf{A}(x) = \sum_{n=0}^{\infty} p_\mathbf{A}(n) x^n$ and $F_\mathbf{A}^{(d)}(x) = \sum_{n=0}^{\infty} p_\mathbf{A}^{(d)}(n) x^n$ be the generating functions for $p_\mathbf{A}(n)$ and $p_\mathbf{A}^{(d)}(n)$, respectively; then the following theorem holds.

Theorem 1. $F_\mathbf{A}(x) = \prod_{a \in \mathbf{A}} (1 - x^a)^{-1}$ *and* $F_\mathbf{A}^{(d)}(x) = \prod_{a \in \mathbf{A}} (1 + x^a)$.

PROOF. Taking the product of $(1 - x^a)^{-1} = 1 + x^a + x^{2a} + \cdots + x^{ka} + \cdots$ over all $a \in A$, we obtain

$$\prod_{a \in A} \left(1 + x^a + x^{2a} + \cdots + x^{ka} + \cdots \right) = \sum_{n=0}^{\infty} b_n x^n,$$

where b_n is the number of times we obtain n as a sum of terms of the form ka, all as being distinct; that is, b_n is the number of solutions of the Diophantine equation $k_1 a_1 + k_2 a_2 + \cdots = n$, with $a_j \in A$, $a_i \neq a_j$ if $i \neq j$, and $k_j \in Z_0$. Hence by Remark 4, $b_n = p_A(n)$, proving the first statement. Similarly, $\prod_{a \in A}(1 + x^a) = \sum_{n=0}^{\infty} c_n x^n$, where c_n is the number of times we obtain n as a sum of distinct exponents $a \in A$, that is, $c_n = p_A^{(d)}(n)$.

REMARK 5. In both statements of Theorem 1, the coefficient of x^0 (that is, the constant term) is 1; hence we are led to make the announced convention and set $p_A(0) = p_A^{(d)}(0) = 1$.

From Theorem 1 and Definitions 1, 2, and 3 we obtain almost immediately the following statements, which we list as

Corollary 1.1.

(1) $F(x) = \displaystyle\prod_{k=1}^{\infty} (1 - x^k)^{-1}$ is the generating function of $p(n)$.

(2) $F_E(x) = \displaystyle\prod_{k=1}^{\infty} (1 - x^{2k})^{-1}$ is the generating function of $p_E(n)$.

(3) $p_E(2n + 1) = 0$.

(4) $F_O(x) = \displaystyle\prod_{k=1}^{\infty} (1 - x^{2k-1})^{-1}$ is the generating function of $p_O(n)$.

(5) $F_m(x) = \displaystyle\prod_{k=1}^{m} (1 - x^k)^{-1}$ is the generating function of $p_m(n)$.

(6) $F^{(d)}(x) = \displaystyle\prod_{k=1}^{\infty} (1 + x^k)$ is the generating function of $p^{(d)}(n)$.

(7) $p_{A,m}^{(*)}(n) \leq p_A^{(*)}(n)$ for all $m \geq 1$.

(8) $p_{A,m}^{(*)}(n) = p_A^{(*)}(n)$ for $m \geq n$.

(9) $p^{(d)}(n) = p_O(n)$.

PROOFS. Statements (1), (2), (4), (5), and (6) are particular cases of Theorem 1. Statement (3) simply restates that $F_E(x)$ has only even powers of x. Statement (7) follows from the remark that adding restrictions (in this case, concerning the size of admissible summands) cannot increase the number of partitions. However, no parts in excess of n occur in any of its partitions; hence if $m \geq n$, to restrict the parts not to exceed m is not to restrict them at all and statement (8) follows. It remains to prove statement (9). We do it by showing that $p^{(d)}(n)$ and $p_O(n)$ have the same generating function, that is, that $\prod_{k=1}^{\infty}(1 + x^k) = \prod_{k=1}^{\infty}(1 - x^{2k-1})^{-1}$. We observe that $\prod_{k=1}^{\infty}(1 + x^k) = \prod_{k=1}^{\infty}((1 - x^{2k})/(1 - x^k))$. All factors $1 - x^k$ occur in the denominator, both with k even and with k odd. In the numerator occur only (but all) factors $1 - x^{2k}$ with even exponent; these cancel the corresponding factors in the denominator, leaving 1 in the numerator and only (but all) factors $1 - x^{2k-1}$ in the denominator. The product has become precisely $\prod_{k=1}^{\infty}(1 - x^{2k-1})^{-1}$, as we wanted to prove.

The reader will have observed that so far we have not raised the question of convergence of the generating function, represented either as a series or as a product. This is not an oversight. Indeed, all that we are interested in at present are formal manipulations that insure the identity of the coefficients of the formal power series expansions of the functions appearing on the right and on the left of the $=$ sign. So for instance the formal series $\sum_{n=0}^{\infty} n! x^n$ is the generating function for the arithmetical function $f(n) = n!$; but except for $x = 0$ the series does not converge. For more on formal power series, the reader may want to consult [11]. As a matter of fact, the series and products met in this chapter do converge, at least for $|x| < 1$; hence they actually represent functions analytic at least inside the unit circle. Although this consideration, of great importance in some of the analytic methods, will hardly ever be used here (see however the proof of Theorem 8), we shall prove

Theorem 2.   *The generating functions of the partition functions converge inside the unit circle.*

PROOF.   The number of partitions of n, with or without restrictions, is a non-negative integer; also, as already observed, the number of partitions can only decrease if restrictions are added. Hence the coefficients of the generating function of the partition function are non-negative integers, nondecreasing when we remove restrictions so that if z is a complex variable and $r = |z|$, then

$$|F_A^{(\cdot,\cdot,\cdot)}(z)| \le F_A^{(\cdot,\cdot,\cdot)}(r) \le F(r)$$

$$= \prod_{k=1}^{\infty}(1 - r^k)^{-1} = \prod_{k=1}^{\infty}(1 + f_k(r)),$$

with

$$f_k(r) = r^k + r^{2k} + \cdots = \frac{r^k}{1 - r^k}(> 0) \quad \text{for} \quad 0 < r < 1.$$

The infinite product of factors $1 + f_k(r)$ (with $f_k(r) > 0$) converges (see [1], p. 382), provided that the series $\sum_{k=1}^{\infty} f_k(r)$ converges and this is actually the case. Indeed on account of $0 < r < 1$, $r^k \to 0$ as $k \to \infty$, and the series $\sum_{k=1}^{\infty} f_k(r)$ converges by comparison with the series $\sum_{k=1}^{\infty} r^k$.

REMARK 6. The convergence is uniform for $|z| \leq 1 - \varepsilon(\varepsilon > 0$, arbitrarily small).

REMARK 7. The series and products, representing the generating functions mentioned, do not converge outside the unit circle; this can be seen easily if we let $x \to 1^-$ in the product representation and observe that the functions cannot stay bounded.

Corollary 2.1. $\displaystyle\lim_{n \to \infty} \frac{p(n + 1)}{p(n)} = 1.$

PROOF. We can make a distinct partition of $p(n + 1)$ correspond to each partition of n counted by $p(n)$, simply by increasing the largest (or one of the largest, if there are several equal summands larger than the others) summand occurring in a given partition of n by one unity. In addition, $n + 1$ has the partition $1 + 1 + 1 + \cdots + 1$ ($n + 1$ parts each equal to unity), which has no "mate" among the partitions of n, under the correspondence we set up; hence $p(n + 1) \geq p(n) + 1$. On the other hand, by Theorem 2 $F(x) = \sum_{n=0}^{\infty} p(n)x^n$ converges for any $0 \leq x < 1$; hence

$$\lim_{n \to \infty} \left| \frac{p(n + 1)x^{n+1}}{p(n)x^n} \right| = \lim_{n \to \infty} \frac{p(n + 1)}{p(n)} |x| \leq 1.$$

Therefore,

$$\lim_{n \to \infty} \frac{p(n + 1)}{p(n)} \leq \frac{1}{|x|} \quad \text{for any} \quad |x| < 1.$$

It follows that $\lim_{n \to \infty} p(n + 1)/p(n) \leq 1$. However $p(n + 1) > p(n)$ implies $\lim_{n \to \infty} p(n + 1)/p(n) \geq 1$; hence $\lim_{n \to \infty} p(n + 1)/p(n) = 1$, as asserted.

It is often easier to work with partitions where the size of the largest summand is limited than with unrestricted partitions. It is for this reason that statements (7) and (8) of Corollary 1.1 are often useful. One reason why $p_m(n)$ is often easy to handle is the following theorem, which permits us to reduce $p_m(n)$ to $p_{m-1}(n)$.

Theorem 3.

$$p_m(n) = p_{m-1}(n) + p_m(n - m) \quad \text{for} \quad n \geq m, m \geq 1; \tag{1}$$

more generally,

$$p_m(n) = p_{m-1}(n) + p_{m-1}(n - m) + p_{m-1}(n - 2m) + \cdots$$
$$+ p_{m-1}(n - km) + p_m(n - (k + 1)m)$$
$$= p_{m-1}(n) + p_{m-1}(n - m) + \cdots + p_{m-1}(n - rm), \quad r = \left[\frac{n}{m}\right].$$

PROOF.   Either a given partition of n does not contain the summand m, and then it is counted (exactly once) by $p_{m-1}(n)$; or else it does contain m, and then it is of the form $n = a_1 + a_2 + \cdots + a_r + m(1 \le a_j \le m, j = 1, 2, \ldots, r)$. Here $a_1 + \cdots + a_r$ is a partition of $n - m$ and to each of the $p_m(n - m)$ such partitions of $n - m$ into summands not in excess of m corresponds exactly one partition of n, effectively containing m, which is precisely $a_1 + a_2 + \cdots + a_r + m$. This proves the first equality and the other two follow by induction on k and the remark that for $v > 0$, $p_m(-v) = 0$.

Among the easiest to handle and most important generating functions is of course

$$F(x) = \sum_{n=0}^{\infty} p(n)x^n = \prod_{k=1}^{\infty} (1 - x^k)^{-1}.$$

Still simpler is the reciprocal $F(x)^{-1}$, which we shall denote by $\Phi(x)$. This function $\Phi(x) = \prod_{n=1}^{\infty}(1 - x^n)$, introduced by Euler, seems to be the first appearance in the literature of a *theta function*. The functions of this class have two important properties: (i) They may be represented by a product (which we know, in this instance); and (ii) they may be represented by a power series such that the exponents of the independent variable are polynomials of second degree in the summation index. In the particular case of the function $\Phi(x)$, these properties form the content of the following famous theorem of Euler, generally known as the "pentagonal numbers" theorem.

Theorem 4.

$$\Phi(x) = \prod_{k=1}^{\infty} (1 - x^k) = \sum_{n=-\infty}^{\infty} (-1)^n x^{n(3n+1)/2}$$

$$= 1 + \sum_{n=1}^{\infty} (-1)^n x^{n(3n+1)/2} + \sum_{n=1}^{\infty} (-1)^n x^{n(3n-1)/2}.$$

Before we prove this theorem, let us make a few applications, which will illustrate the power of the method of generating functions introduced by Euler.

The obvious identity $F(x)\Phi(x) = 1$ leads to the following interesting *recurrence formula*, which permits the computation of $p(n)$ if $p(k)$ is already known for $k < n$.

Theorem 5. $p(n) = p(n - 1) + p(n - 2) - p(n - 5) - p(n - 7) + \cdots + (-1)^{j+1}p(n - n_j) + \cdots$ *where* $n_j = \frac{1}{2}j(3j \pm 1)$ *are the so-called pentagonal numbers.*

REMARK 8.   If we border a regular pentagon, marked by 5 dots, so as to obtain successively pentagons with $3, 4, \ldots, j, \ldots$ dots on each side, then the total number of dots is $\frac{1}{2}j(3j - 1)$ (see Fig. 1).

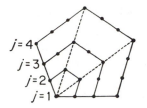

Figure 1.   Pentagonal Numbers

PROOF OF THEOREM 5.

$$1 = F(x)\Phi(x) = \sum_{k=0}^{\infty} p(k)x^k \sum_{m=-\infty}^{\infty} (-1)^m x^{m(3m+1)/2}$$

$$= \sum_{n=0}^{\infty} x^n \sum_{k+\frac{m(3m+1)}{2}=n} (-1)^m p(k)$$

$$= \sum_{n=0}^{\infty} x^n \sum_{\frac{m(3m+1)}{2} \le n} (-1)^m p\left(n - \frac{m(3m+1)}{2}\right).$$

Here the first and last terms of this sequence of equalities are formal power series; indeed $1 = 1 + 0 \cdot x + 0 \cdot x^2 + \cdots$. Two power series (just as two polynomials) can be equal for all values of the variable x only if they have the same coefficients. Hence except for $n = 0$ (which implies $m = 0$ in the inner sum), all coefficients of x^m have to vanish. Successively making

$$m = 0, +1, -1, +2, -2,\ldots,$$

we obtain

$$0 = \sum_{m}(-1)^m p\left(n - \frac{m(3m+1)}{2}\right) = p(n) - p(n-1) - p(n-2)$$

$$+p(n-5) + p(n-7) - \cdots .$$

The series breaks off when $m(3m+1)/2 > n$, and Theorem 5 is proven.

Only slightly more complicated is the proof of

Theorem 6.

$$np(n) = p(0)\sigma(n) + p(1)\sigma(n-1) + \cdots + p(r)\sigma(n-r) + \cdots$$

$$+p(n-1)\sigma(1).$$

PROOF.   From $F(x) = 1/\Phi(x)$ we obtain by differentiation

$$F'(x) = -\frac{\Phi'(x)}{\Phi^2(x)} = -\frac{\Phi'(x)}{\Phi(x)}\frac{1}{\Phi(x)} = -\frac{\Phi'(x)}{\Phi(x)}F(x).$$

We now proceed to replace each function by its power series expansion. From $\Phi(x) = \prod_{k=1}^{\infty}(1 - x^k)$ follows

$$\log \Phi(x) = \sum_{k=1}^{\infty} \log(1 - x^k) \quad \text{and} \quad \frac{\Phi'(x)}{\Phi(x)} = -\sum_{k=1}^{\infty} \frac{kx^{k-1}}{1 - x^k}.$$

Hence

$$F'(x) = \sum_{n=1}^{\infty} np(n)x^{n-1} = -F(x)\frac{\Phi'(x)}{\Phi(x)} = -\left(\sum_{m=0}^{\infty} p(m)x^m\right)\left(-\sum_{k=1}^{\infty} \frac{kx^{k-1}}{1 - x^k}\right),$$

or successively,

$$\sum_{n=1}^{\infty} np(n)x^n = \sum_{m=0}^{\infty} p(m)x^m \sum_{k=1}^{\infty} k\left(x^k + x^{2k} + \cdots + x^{rk} + \cdots\right)$$

$$= \sum_{n=1}^{\infty} x^n \sum_{m+rk=n} kp(m)$$

$$= \sum_{n=1}^{\infty} x^n \sum_{m=0}^{n} p(m) \sum_{k \mid n-m} k = \sum_{n=1}^{\infty} x^n \sum_{m=0}^{n} p(m)\sigma(n - m),$$

and we obtain the theorem if we identify the coefficients of the first and last member.

ALTERNATIVE PROOF (see [3]).   Let us write out explicitly all partitions of n, say $n = a_1 + a_2 + \cdots + a_r$. Adding them all up, on the left we obtain $np(n)$. On the right, we regroup the summands according to their size. We have to determine how often a specific summand, say m, occurs. As we already had an opportunity to observe, if m appears in a given partition of n, the other summands form one of the partitions of $n - m$; hence so far, m occurred $p(n - m)$ times. However, m may also occur in some of these partitions of $n - m$ (because it could occur more than once in a given partition of n). Reasoning as before, we find that if m actually occurs in a partition of $n - m$, then the other summands form a partition of $n - 2m$, of which there are exactly $p(n - 2m)$, and so forth. Altogether, if we add up all summands m that occur, we obtain $\sum_{k=1}^{[n/m]} mp(n - km)$. And if we add up *all* summands, we obtain $\sum_{m=1}^{n}\sum_{k=1}^{n/m} mp(n - km)$. We formulate the result obtained as

Theorem 7.

$$np(n) = \sum_{m=1}^{n} \sum_{k=1}^{n/m} mp(n - km).$$

4 Generating Functions 119

From Theorem 7 we can immediately obtain again Theorem 6, namely if we set $km = r$,

$$np(n) = \sum_{m=1}^{n} \sum_{k=1}^{n/m} mp(n - km) = \sum_{r=1}^{n} p(n - r) \sum_{m|r} m = \sum_{r=1}^{n} p(n - r)\sigma(r).$$

We still owe the reader a proof of Theorem 4. There are several proofs of this important theorem. The simplest, conceptually, is Euler's own. This consists in a systematic multiplication of the product $(1 - x)(1 - x^2)(1 - x^3)\dots$, with a skillful arrangement of the cancellations and an easy induction.

Another, very elegant, proof is due to Professor N. Fine [4]; it consists in formal work with power series, involving additional parameters. It has, however, several rather subtle points. Actually, Theorem 4 follows as a simple Corollary from an important identity due to Jacobi ([8], Section 64), which we shall discuss next. For the reader who does not want to get involved with Jacobi's identity, a direct combinatorial proof of Theorem 4, using graphs, will also be presented.

Theorem 8 (Jacobi). *For $z \neq 0$ and $|x| < 1$,*

$$\prod_{n=1}^{\infty} \{(1 - x^{2n})(1 + x^{2n-1}z)(1 + x^{2n-1}z^{-1})\} = \sum_{n=-\infty}^{\infty} x^{n^2}z^n. \qquad (2)$$

Assuming for a moment the validity of Theorem 8, let us prove Theorem 4. In (2) we replace x by $u^{3/2}$ and z by $-u^{1/2}$; then the first member becomes

$$\prod_{n=1}^{\infty} \{(1 - u^{3n})(1 - u^{3n-3/2+1/2})(1 - u^{3n-3/2-1/2})\}$$

$$= \prod_{n=1}^{\infty} \{(1 - u^{3n-2})(1 - u^{3n-1})(1 - u^{3n})\} = \prod_{m=1}^{\infty} (1 - u^m) = \Phi(u).$$

The second member becomes

$$\sum_{n=-\infty}^{\infty} u^{3n^2/2}(-u^{1/2})^n = \sum_{n=-\infty}^{\infty} (-1)^n u^{n(3n+1)/2},$$

and Theorem 4 is proven.

Besides Theorem 4, some other important corollaries of Theorem 8 are the following:

Corollary 8.1. $\prod_{n=0}^{\infty}(\{1 - x^{2n+2})(1 + x^n)\} = \sum_{n=-\infty}^{\infty} x^{n(n+1)/2}.$

REMARK 9. $t_n = n(n + 1)/2$ is called the nth triangular number.

PROOF.   In (2), set $x = z = u^{1/2}$, obtaining in the first member

$$\prod_{n=1}^{\infty} \left\{ (1 - u^n)(1 + u^{n-\frac{1}{2}+\frac{1}{2}})(1 + u^{n-\frac{1}{2}-\frac{1}{2}}) \right\}$$

$$= \prod_{n=1}^{\infty} (1 - u^{2n})(1 + u^{n-1}) = \prod_{n=0}^{\infty} (1 - u^{2n+2})(1 + u^n),$$

and in the second member

$$\sum_{n=-\infty}^{\infty} u^{n^2/2 + n/2} = \sum_{n=-\infty}^{\infty} u^{n(n+1)/2},$$

proving the corollary.

Furthermore, by (9) of Corollary 1.1 (see its proof),

$$\prod_{n=1}^{\infty} (1 + x^n) = \prod_{n=1}^{\infty} \frac{1}{1 - x^{2n-1}};$$

hence

$$\prod_{n=0}^{\infty} (1 + x^n) = 2 \prod_{n=1}^{\infty} (1 + x^n) = 2 \prod_{n=1}^{\infty} \frac{1}{1 - x^{2n-1}}$$

$$= 2 \frac{1}{(1 - x)(1 - x^3)(1 - x^5) \cdots},$$

and the first member in Corollary 8.1 may be written as

$$2 \frac{(1 - x^2)(1 - x^4)(1 - x^6)}{(1 - x)(1 - x^3)(1 - x^5)} \cdots .$$

We record this result as

Corollary 8.2. $\displaystyle \prod_{n=1}^{\infty} \frac{1 - x^{2n}}{1 - x^{2n-1}} = \sum_{n=0}^{\infty} x^{n(n+1)/2}.$

Corollary 8.3. $\displaystyle \Phi^3(x) = \prod_{n=1}^{\infty} (1 - x^n)^3 = \sum_{n=1}^{\infty} (-1)^n (2n + 1) x^{n(n+1)/2}.$

PROOF.   We set $z = -xy$ in (2) and obtain the first member $\prod_{n=1}^{\infty}(1 - x^{2n})$. $(1 - yx^{2n})(1 - y^{-1}x^{2n-2})$. If we write the last factor as $1 - y^{-1}x^{2n}$, then as n runs from 1 to ∞ the product is the same, except that we lose the factor $1 - y^{-1}$ corresponding to $n = 1$ in the original form of the product. Hence the first member becomes

$$(1 - y^{-1}) \prod_{n=1}^{\infty} (1 - x^{2n})(1 - yx^{2n})(1 - y^{-1}x^{2n}).$$

The second member of (2) becomes

$$\sum_{n=-\infty}^{\infty} (-1)^n x^{n^2+n} y^n,$$

and we obtain by (2), after division by $(1 - y^{-1})$, that

$$\prod_{n=1}^{\infty} (1 - x^{2n})(1 - yx^{2n})(1 - y^{-1}x^{2n}) = \frac{1}{1 - y^{-1}} \sum_{n=-\infty}^{\infty} (-1)^n x^{n^2+n} y^n.$$

The second member may also be written

$$\sum_{n=0}^{\infty} (-1)^n \frac{x^{n(n+1)}(y^n - y^{-n-1})}{1 - y^{-1}}$$

$$= \sum_{n=0}^{\infty} (-1)^n x^{n(n+1)} (y^n + y^{n-1} + y^{n-2} + \cdots + 1 + y^{-1} + \cdots + y^{-n}).$$

If we write the equality of the first and second member for $y = 1$, we obtain

$$\prod_{n=1}^{\infty} (1 - x^{2n})^3 = \sum_{n=0}^{\infty} (-1)^n (2n + 1) x^{n(n+1)},$$

as asserted.

Theorem 8 and its corollaries, including Theorem 4, present examples of theta functions and state the identity between their representations as infinite products and as power series.

*PROOF OF THEOREM 8.   Let[†]

$$\Phi_M(x, z) = \prod_{m=1}^{M} (1 + zx^{2m-1})(1 + z^{-1}x^{2m-1}).$$

Then

$$\Phi_M(x, x^2 z) = \prod_{m=1}^{M} (1 + zx^{2m+1})(1 + z^{-1}x^{2m-3})$$

$$= \Phi_M(x, z) \frac{(1 + zx^{2M+1})(1 + z^{-1}x^{-1})}{(1 + zx)(1 + z^{-1}x^{2M-1})}$$

$$= \Phi_M(x, z) \frac{1 + zx^{2M+1}}{xz + x^{2M}},$$

or

$$(xz + x^{2M})\Phi_M(x, x^2 z) = (1 + zx^{2M+1})\Phi_M(x, z). \tag{3}$$

† Exceptionally a rational integer is denoted here by capital M.

If we expand the (finite) product defining $\Phi_M(x, z)$, the coefficient of z^m will be a certain polynomial $P_m(x)$. The product Φ_M does not change if we replace z by z^{-1}; hence the coefficient of z^{-m} is the same polynomial, or $P_m(x) = P_{-m}(x)$. The highest power of z that occurs is clearly z^M, and its coefficient is

$$P_M(x) = x^{1+3+\cdots+(2M-1)} = x^{M^2}.$$

Hence $\Phi_M(x, z) = \sum_{m=-M}^{M} P_m(x)z^m$ with

$$P_M(x) = P_{-M}(x) = x^{M^2} \quad \text{and} \quad P_m(x) = P_{-m}(x).$$

Substituting this expansion in the recurrence relation (3), we obtain

$$(xz + x^{2M}) \sum_{m=-M}^{M} P_m(x)x^{2m}z^m = (1 + zx^{2M+1}) \sum_{m=-M}^{M} P_m(x)z^m. \quad (4)$$

Equating the coefficient of z^m on both sides,

$$x^{2m-1}P_{m-1} + x^{2(M+m)}P_m = P_m + x^{2M+1}P_{m-1} \quad (5)$$

for $-M \leq m - 1 < m \leq M$. We observe that in (4) z^{-M} and z^{M+1} also occur, to which the recurrence does not apply; but we also easily check that the corresponding equalities of coefficients still hold, and are as a matter of fact the trivial $P_M(x) = P_M(x)$ and $P_{-M}(x) = P_{-M}(x)$; actually, we already know their common value x^{M^2}. From (5) we obtain

$$P_m(x) = x^{2m-1}\frac{1 - x^{2(M-m+1)}}{1 - x^{2(M+m)}}P_{m-1}(x) \quad (6)$$

or

$$P_{m-1}(x) = P_m(x) \cdot x^{-2m+1}\frac{1 - x^{2(M+m)}}{1 - x^{2(M-m+1)}}.$$

In particular, for $m = M$:

$$P_{M-1}(x) = P_M(x) \cdot x^{-2M+1}\frac{1 - x^{4M}}{1 - x^2} = x^{(M-1)^2}\frac{1 - x^{4M}}{1 - x^2};$$

similarly by (6),

$$P_{M-2}(x) = P_{M-1}(x) \cdot x^{-2M+3}\frac{1 - x^{4M-2}}{1 - x^4}$$

$$= x^{(M-1)^2 - 2(M-1)+1} \cdot \frac{1 - x^{4M}}{1 - x^2}\frac{1 - x^{4M-2}}{1 - x^4}$$

$$= x^{(M-2)^2} \cdot \frac{(1 - x^{4M})(1 - x^{4M-2})}{(1 - x^2)(1 - x^4)},$$

and by induction,

$$P_{M-k}(x) = x^{(M-k)^2} \frac{(1 - x^{4M})(1 - x^{4M-2}) \cdots (1 - x^{4M-2k+2})}{(1 - x^2)(1 - x^4) \cdots (1 - x^{2k})}.$$

In particular,

$$P_0(x) = \frac{(1 - x^{4M})(1 - x^{4M-2}) \cdots (1 - x^{2M+2})}{(1 - x^2)(1 - x^4) \cdots (1 - x^{2M})}.$$

The careful reader will have observed the strong analogy between these rational expressions and the binomial coefficients. As a matter of fact, if the factors $1 - x^k$ are replaced by the exponents k, we obtain ordinary binomial coefficients. Just as the ordinary binomial coefficients are actually integers although they are written as fractions, so also our "Gaussian binomial coefficients" $P_m(x)$ are really polynomials, in spite of their appearance. Having obtained the polynomials $P_m(x)$, we can now write, using (6),

$$\Phi_M(x, z) = P_0(x) \left\{ 1 + x \frac{1 - x^{2M}}{1 - x^{2M+2}} (z + z^{-1}) \right.$$

$$+ x^4 \frac{(1 - x^{2M})(1 - x^{2M-2})}{(1 - x^{2M+2})(1 - x^{2M+4})} (z^2 + z^{-2}) + \cdots$$

$$\left. + x^{M^2} \frac{(1 - x^{2M}) \cdots (1 - x^2)}{(1 - x^{2M+2}) \cdots (1 - x^{4M})} (z^M + z^{-M}) \right\}.$$

If we multiply both sides by $\prod_{m=1}^{M}(1 - x^{2m})$, this cancels the denominator of $P_0(x)$ and we obtain

$$\prod_{m=1}^{M} (1 - x^{2m})(1 + zx^{2m-1})(1 + z^{-1}x^{2m-1})$$

$$= (1 - x^{2M+2})(1 - x^{2M+4}) \cdots (1 - x^{4M})$$

$$\times \left\{ 1 + x \frac{1 - x^{2M}}{1 - x^{2M+2}} (z + z^{-1}) + \cdots \right.$$

$$\left. + x^{M^2} \frac{(1 - x^{2M}) \cdots (1 - x^2)}{(1 - x^{2M+2}) \cdots (1 - x^{4M})} (z^M + z^{-M}) \right\}.$$

We observe that in the denominator occur only factors $(1 - x^k)$ with $k \geq 2M + 2$; hence in the expansion of

$$\frac{1}{1 - x^k} = 1 + x^k + x^{2k} + \cdots,$$

also, only powers of x with exponent larger than $2M$ will occur. Furthermore, if we look only at the coefficients of $(z^n + z^{-n})$ with $n < M/2$, then the numerators have factors $(1 - x^{2M})(1 - x^{2M-2}) \cdots (1 - x^{2M-2n})$ and the lowest power of x that occurs has an exponent $2M - 2n > M$. Therefore

$$\prod_{m=1}^{M} (1 - x^{2m})(1 + zx^{2m-1})(1 + z^{-1}x^{2m-1})$$

$$= 1 + x(z + z^{-1})(1 + x^{k_1} + \cdots) + x^{2^2}(z^2 + z^{-2})(1 + x^{k_2} + \cdots) + \cdots$$

$$+ x^{n^2}(z^n + z^{-n})(1 - x^{k_n} + \cdots) + \cdots$$

with $k_j > M$. As $M \to \infty$, so do the exponents k_j and $x^{k_j} \to 0$, because $|x| < 1$. Therefore, the limit for $M \to \infty$ of the nth summand (n fixed) is $x^{n^2}(z^n + z^{-n})$, and we have proved that

$$\prod_{m=1}^{\infty} (1 - x^{2m})(1 + zx^{2m-1})(1 + z^{-1}x^{2m-1})$$

$$= \lim_{M \to \infty} \prod_{m=1}^{M} (1 - x^{2m})(1 + zx^{2m-1})(1 + z^{-1}x^{2m-1})$$

$$= 1 + \sum_{n=1}^{\infty} x^{n^2}(z^n + z^{-n}) = \sum_{n=-\infty}^{\infty} z^n x^{n^2},$$

that is, Theorem 8.

As we saw, Theorem 8 (Jacobi's identity) yields a large number of results. However, if our aim is only to prove Theorem 4, we can obtain it with far less complicated machinery. For that reason, and also to introduce a new approach, we shall present another extremely simple proof of Theorem 4. Our main tool will be the consideration of graphs.

5   GRAPHS

Graphs are pictorial representations of partitions. Each summand is represented as a number of dots arranged, say horizontally. The partition $3 + 2$ of five has the graph:

$$\circ \quad \circ \quad \circ$$
$$\circ \quad \circ \quad .$$

The same graph may be interpreted, however, so that dots of the same vertical line correspond to a summand. Looked upon that way, the graph stands for the partition $2 + 2 + 1$. Two partitions related like $3 + 2$ and $2 + 2 + 1$, i.e., that are representable by the same graph, are called *conjugate partitions*. This concept of "conjugacy" establishes a pairing off, or correspondence between partitions; under this correspondence some partitions, like

$$\begin{matrix} \circ & \circ & \circ \\ \circ & & \\ \circ & & \end{matrix} \quad ,$$

correspond to themselves; these are called *self-conjugate* partitions. If m is the largest summand in the graph of a partition, then the conjugate partitions contains exactly m summands, and conversely. Hence we infer that the number of partitions of n having the largest summand m is equal to the number of partitions of n into exactly m summands. It now follows, considering all partitions with largest summand not in excess of m, that this equals the number of partitions into no more than m summands and we have proven

Theorem 9. *The number of partitions into parts not in excess of m equals the number of partitions into at most m summands, or in symbols,*

$$p_m(n) = p^{(m)}(n).$$

The method of graphs will permit us to prove Theorem 4 by very elementary considerations. First, however, we shall transform the problem slightly. If we multiply out the product $(1 - x)(1 - x^2)(1 - x^3) \cdots$, it turns out that most powers of x have coefficient zero. This means that the products

$$(-x^{a_1})(-x^{a_2}) \cdots (-x^{a_r}) \quad \text{with} \quad a_1 + a_2 + \cdots + a_r = n$$

and an even number of factors each with a different exponent, which lead to $+x^n$, occur in general exactly the same number of times as the products $(-x^{b_1}) \cdots (-x^{b_s})$ with $b_1 + b_2 + \cdots + b_s = n$, distinct bs and an odd number of terms, leading to $-x^n$. An exception occurs, according to Theorem 4, if and only if $n = j(3j \pm 1)/2$. In this case the number of partitions of n into an odd number of distinct parts exceeds the number of partitions into an even number of distinct parts by exactly one if j is odd; the number of partitions of n into an even number of distinct parts exceeds the number of partitions into an odd number of distinct parts (again by exactly one) if j is even. It is in this equivalent formulation,

$$p^{(e,\,d)}(n) - p^{(o,\,d)}(n) = \begin{cases} (-1)^j & \text{if } n = j(3j \pm 1)/2 \\ 0 & \text{otherwise,} \end{cases}$$

that we are going to prove Theorem 4.

This remarkable proof is due to Franklin, one of the first American mathematicians [5]. In order to prove the theorem, we shall try to construct a one-to-one correspondence between the partitions of n into an even and an odd number of distinct summands; we shall find that the correspondence will break down, and that one or the other kind of partitions will exceed the other kind by exactly one, precisely when n is of the form $j(3j \pm 1)/2$. In order to be certain that each partition is considered exactly once, we shall insist that all summands should be arranged in decreasing order. There will be no ambiguity, because all summands involved are distinct. Consider then such a partition of n, with summands all unequal and arranged in decreasing order, and for simplicity let us place the dots at equal distance from each other, so that they lie on the vertices of little squares:

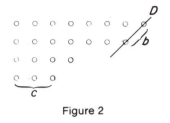

Figure 2

Denote the size of the smallest summand by c; consider also a line D from the upper right corner, going down at 45°. We denote by b the number of dots of the graph on this line. If $b < c$ (as in Figure 2) we may remove the b dots of the line D and add them at the bottom of the graph. In this way the number of summands has been increased by one; hence, its parity has been changed. If $c < b$, or even if $c = b$, the previous operation does not work, because the summand added at the bottom would be either larger than the preceding one, thus violating the rule of decreasing summands, or equal to it, while we insist on having distinct summands (see Fig. 3). But in this case we can, in general, change the parity by the operation of distributing the last summand, consisting of c dots, among the c largest summands; this requires adding a line D', consisting of $c \leq b$ dots, parallel to D. Again, we obtain a legitimate partition with one summand less; hence the parity of the number of summands has been changed. Clearly, either $b < c$ or $c \leq b$; hence only one of the operations is possible. It is clear that whenever one or the other of these operations can be performed, we can make correspond to each partition with an even number of summands one with an odd number of summands; hence each kind of partition occurs the same number of times and $p^{(o,d)}(n) = p^{(e,d)}(n)$. However, there are exactly two cases when *both* above operations become impossible.

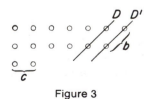

Figure 3

(i)   Let $b = c$ with D passing through the last dot of the smallest summand (see Fig. 4). If we try to remove the last summand, there are only $b - 1 = c - 1$ summands left, and we cannot distribute the c dots among them, one to each. If we try to remove the $b = c$ dots of D and form a new summand, we leave only $c - 1$ dots in the last summand and are not allowed to add underneath a new summand of $b = c > c - 1$ dots.

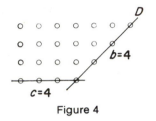

Figure 4

(ii)   A similar situation exists if $c = b + 1$ and D contains again the last dot of the smallest summand. The reasoning being similar to the preceding one, we leave it up to the reader to convince himself using a graph like Fig. 5.

In case (i),

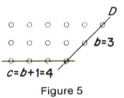

Figure 5

$$n = c + (c + 1) + (c + 2) + \cdots + (c + (b - 1)) = b(2c + b - 1)/2,$$

or, having $b = c$, $n = b(3b - 1)/2$. The excess partition (that without a match) has $b = c$ summands. Hence if b, the number of summands, is even, there is an excess of one partition with an even number of summands over the number of partitions with an odd number of summands; if b is odd, then the lone extra partition has an odd number of summands. Similarly, in case (ii), with $c = b + 1$ as in Fig. 5, $n = b(2c + b - 1)/2 = b(3b + 1)/2$, with the same result. In either case, $p^{(e)}(n) - p^{(o)}(n) = (-1)^b$, and that finishes the proof of the theorem.

6   THE SIZE OF $p(n)$

Euler, Jacobi, and their immediate successors were mainly interested in pretty relations of the kind seen. They do not seem to have tried to estimate the size of $p_A(n)$. Clearly, Theorem 5 permits us to compute recursively every

$p(n)$ exactly, by simple additions. This may have seemed sufficient in the 18th and early 19th century, but the numerical values of $p(n)$ increase so fantastically fast that one cannot go really far in this way and must look for other methods of computing or at least estimating $p(n)$. Now methods for at least a rough appraisal of $p(n)$ were definitely accessible to Euler; therefore, the fact that no theorems in this direction are known earlier than the middle of the 19th century suggests that before that time mathematicians were simply not interested in knowing how large $p(n)$ gets. Instead, they studied the theta function aspect of $F(x)^{-1} = \Phi(x)$ and obtained from it a whole host of identities involving the partition function, some of which we saw. Then around 1880 Sylvester [14] took up the study of $p(n)$ from a somewhat different point of view. His procedures are rather complicated, but are based on a simple idea, which goes back to Laguerre [9]. Let us consider $p_m(x)$, which by (5) of Corollary 1.1 has the generating function

$$F_m(x) = \frac{1}{(1 - x)(1 - x^2)(1 - x^3) \cdots (1 - x^m)}.$$

Then $F_m(x)$ may be decomposed into "partial fractions" of the form $C/(x - \alpha)^r$, with α a root of unity (that is, $\alpha^h = 1$ for some integer h). The coefficients in the power series expansion of such simple "partial fractions" are easily written down, and equating their sum for a given n with $p_m(n)$, we obtain relations of a new type. The roots of unity can be expressed by trigonometric functions, and these formulas then show that, in addition to a principal term which increases monotonically with n, $p_m(n)$ also has a periodic or oscillating part that somehow reflects the arithmetic character of the problem. From $p_m(n)$ we are able to draw conclusions concerning $p(n)$ itself, with the help of relations of which (7) and (8) in Corollary 1.1 are examples. From consideration of the principal increasing term, we are able to estimate rather closely how large $p(n)$ can be for a given n. Actually, we can now formulate the problem of determining such a "principal term" precisely: Having learned that $p(n)$, as a function of n, is rather complicated, we want to split it into two parts, so that

$$p(n) = P(n) + r(n), \tag{7}$$

where $P(n)$ should be a function as simple as possible, and at the same time should represent $p(n)$ well. By this we usually mean that $r(n)$ stays small with respect to $P(n)$ or $p(n)$, i.e., that $\lim_{n \to \infty} r(n)/P(n) = 0$. It is clear that in this case, $\lim_{n \to \infty} p(n)/P(n) = 1$. As pointed out in Chapter 6,[†] whenever we succeed in finding a relation like (7) we say that $p(n)$ is "asymptotically equal" to $P(n)$; and unless there is some special reason to focus our attention on $r(n)$,

---

[†] See pp. 97–8; see also p. 8.

we neglect it completely and write (7) as $p(n) \sim P(n)$. This as well as (7) may clearly also be written as $p(n) = P(n)(1 + o(1))$. It was mainly during the early part of the 20th century that real progress was made in the determination of good asymptotic formulas for $p(n)$, and most of the credit is due to Hardy and Ramanujan [7]. First, by elementary reasonings they showed that

$$\log p(n) = \pi\sqrt{\tfrac{2}{3}}\,\sqrt{n} + o(\sqrt{n}); \tag{8}$$

next, by the use of a Tauberian argument they could show that

$$p(n) = \frac{1}{4\sqrt{3}\,n}e^{\pi\sqrt{2/3}\,\sqrt{n}}(1 + o(1)). \tag{9}$$

Finally, they showed that $F(x)$, the generating function of $p(n)$, is essentially a *modular form*. By that is meant the following: If we make the change of variable $x = e^{2\pi i \tau}$, then the denominator of $F(x)$ differs only by a simple factor from

$$\eta(\tau) = e^{\pi i \tau/12} \prod_{m=1}^{\infty} \left(1 - e^{2\pi i m \tau}\right).$$

If $|x| < 1$, then the imaginary part of τ is positive. It is clear that if here we replace τ by $\tau + 1$, we obtain $\eta(\tau + 1) = e^{\pi i/12}e^{\pi i \tau/12}\prod_{m=1}^{\infty}(1 - e^{2\pi i m \tau} \cdot e^{2\pi i m}) = \omega \cdot \eta(\tau)$, with $|\omega| = |e^{\pi i/12}| = 1$. In fact it turns out that, if we set $\tau' = (a\tau + b)/(c\tau + d)$, where a, b, c, d are integers with $ad - bc = 1$, then $\eta(\tau') = \omega\{-i(c\tau + d)^{1/2}\}\eta(\tau)$, with ω a rather complicated function of a, b, c, d, but with $|\omega| = 1$. The case $\tau' = \tau + 1$ is clearly the particular case $a = b = d = 1$, $c = 0$. In general, if a function $f(\tau)$ has the property that $f((a\tau + b)/(c\tau + d)) = \omega(c\tau + d)^r f(\tau)$, with $|\omega| = 1$, we say that $f(\tau)$ is a *modular form of dimension* $-r$. Exploiting the modular character of $F(x)$, Hardy and Ramanujan were able to apply to $F(x)$ the general theory of Cauchy (residues! see [15]), concerning the determination of the coefficients in the power series expansion of a known function; in this way they found a representation of $p(n)$ by a series. At that time (1918) it was not known whether the series does or does not converge (it turned out that the series does *not* converge), but Hardy and Ramanujan could show that if one stopped at some small term, the sum obtained differed from the true value of $p(n)$ by less than $\frac{1}{2}$; hence $p(n)$, being an integer, was the nearest integer to the partial sum of said series. Later (1937), Rademacher (see [12]) modified the Hardy-Ramanujan method and obtained an explicit representation of $p(n)$ by a convergent series. This fascinating work is unfortunately beyond the scope of these notes. Somewhat later, Erdös (1942, see [3]) proved by entirely elementary considerations that a formula of the type $p(n) = An^{-1}e^{\pi\sqrt{2/3}\,\sqrt{n}}(1 + o(1))$ holds and soon afterwards (1951), D. J. Newman [10] showed, also by elementary reasonings, that Erdös' constant A was in fact $1/4\sqrt{3}$. Actually, Erdös' "elementary" paper is not really easy and we shall not attempt here to give a proof of (9). We shall, however, still following Erdös, give a proof of the following theorem, which while weaker than (9) is stronger than and implies (8):

Theorem 10.   *For every $\varepsilon > 0$,*

$$e^{(\pi\sqrt{2/3} - \varepsilon)\sqrt{n}} < p(n) < e^{\pi\sqrt{2/3}\sqrt{n}} \tag{10}$$

holds, provided that n is sufficiently large.

7   SOME LEMMAS

In the proof of Theorem 10, we shall make use of some simple Lemmas, which we state and prove here.

Lemma 1.

(1) *For $\alpha \leq n$,*

$$\sqrt{n - \alpha} < \sqrt{n}\left(1 - \frac{\alpha}{2n}\right).$$

(2) *Given $0 < \varepsilon_1 < 1$, for $n > \alpha/\varepsilon_1$, one has*

$$\sqrt{n - \alpha} > \sqrt{n}\left(1 - \frac{\alpha}{2n}(1 + \varepsilon_1)\right).$$

PROOFS.

(1) $\sqrt{n - \alpha} = \sqrt{n}\left(1 - \frac{\alpha}{n}\right)^{1/2} < \sqrt{n}\left(1 - \frac{\alpha}{2n}\right),$

because

$$1 - \frac{\alpha}{n} < \left(1 - \frac{\alpha}{2n}\right)^2 = 1 - \frac{\alpha}{n} + \frac{\alpha^2}{4n^2}.$$

(2) $\sqrt{n - \alpha} = \sqrt{n}\left(1 - \frac{\alpha}{n}\right)^{1/2} \geq \sqrt{n}\left(1 - \frac{\alpha}{2n}(1 + \varepsilon_1)\right),$

provided that

$$1 - \frac{\alpha}{n} \geq \left(1 - \frac{\alpha}{2n}(1 + \varepsilon_1)\right)^2,$$

which, solved for n, gives

$$n \geq \frac{\alpha}{\varepsilon_1}\left(\frac{1 + \varepsilon_1}{2}\right)^2.$$

The result follows, observing that $(1 + \varepsilon_1)/2 \leq 1$.

Lemma 2. *For $x > 1$,*

$$\sum_{v=1}^{\infty} v x^{-v} = \frac{x^{-1}}{(1 - x^{-1})^2} = \frac{x}{(x - 1)^2}.$$

PROOF.

$$(x - 1)^{-2} = x^{-2}(1 - x^{-1})^{-2}$$

$$= x^{-2}\left(1 + \frac{2}{1}x^{-1} + \frac{2 \cdot 3}{1 \cdot 2}x^{-2} + \frac{2 \cdot 3 \cdot 4}{1 \cdot 2 \cdot 3}x^{-3} + \cdots\right)$$

$$= x^{-2}\sum_{v=1}^{\infty} v x^{-v+1} = \sum_{v=1}^{\infty} v x^{-v-1},$$

and multiplying by x, we obtain the lemma.

Lemma 3.

(1) *For $x > 0$,* $\quad \dfrac{e^{-x}}{(1 - e^{-x})^2} < \dfrac{1}{x^2};$

(2) *for $0 < \varepsilon_2 < 1$ and $0 < x < \sqrt{10 \cdot \varepsilon_2} \leq 6$,*

$$\frac{e^{-x}}{(1 - e^{-x})^2} > \frac{1}{x^2}(1 - \varepsilon_2).$$

PROOFS.

(1) We show that $(x^2 e^{-x})/(1 - e^{-x})^2 < 1$. Indeed,

$$\frac{x^2 e^{-x}}{(1 - e^{-x})^2} = \frac{x^2}{(e^{x/2} - e^{-x/2})^2} = \frac{(x/2)^2}{\left(\dfrac{e^{x/2} - e^{-x/2}}{2}\right)^2}$$

$$= \frac{(x/2)^2}{\left(\dfrac{x}{2} + \dfrac{1}{3!}\left(\dfrac{x}{2}\right)^3 + \cdots\right)^2}$$

$$= \frac{1}{(1 + A)^2} < 1, \quad \text{where} \quad A = \frac{1}{3!}\left(\frac{x}{2}\right)^2 + \frac{1}{5!}\left(\frac{x}{4}\right)^4 + \cdots.$$

(2) From $((x^2 e^{-x})/(1 - e^{-x})^2)) = ((1/(1 + A)^2))$ it follows that (2) of the Lemma holds, provided that $(1 + A)^{-2} > 1 - \varepsilon_2$. This is indeed the case for $A < \varepsilon_2/2$, as then $(1 + A)^2 < 1 + \varepsilon_2$, so that $(1 + A)^{-2} > (1 + \varepsilon_2)^{-1}$, and this exceeds $1 - \varepsilon_2$ because $(1 - \varepsilon_2)(1 + \varepsilon_2) = 1 - \varepsilon_2^2 < 1$.

On the other hand,

$$A = \frac{x^2}{24}\left(1 + \frac{1}{4\cdot 5}\left(\frac{x}{2}\right)^2 + \frac{1}{4\cdot 5\cdot 6\cdot 7}\left(\frac{x}{2}\right)^4 + \cdot\right)$$

$$< \frac{x^2}{24}\left(1 + \frac{x^2}{80} + \left(\frac{x^2}{80}\right)^2 + \cdots\right)$$

$$= \frac{x^2}{24}\frac{1}{1 - x^2/80},$$

and it is sufficient to require that $((x^2/24)/(1 - x^2/80)) < \varepsilon_2/2$. Here the left side increases monotonically with x, and for $x^2 = 10\varepsilon_2$ its value is $((10\varepsilon_2/24)/(1 - \varepsilon_2/8)) = ((10\varepsilon_2)/(24 - 3\varepsilon_2)) < \varepsilon_2/2$, the last inequality being justified by $\varepsilon_2 < 1$. Hence, (2) of the Lemma also holds for all smaller x, i.e., for $x \le \sqrt{10\varepsilon_2}$, which was assumed. This finishes the proof of the lemma.

8   PROOF OF THE THEOREM.

Let us denote the constant $\sqrt{2/3}\,\pi$ simply by α. We prove separately the two inequalities of (10) and start with

$$p(n) < e^{\alpha\sqrt{n}}. \tag{10'}$$

Exceptionally we find it convenient to define here $p(0) = p(-m) = 0$. We prove (10') by induction on n. For $n = 1$, (10') holds because it reduces to $p(1) = 1 < e^{\alpha}$, which is true. Assuming now that (10') holds for $p(k)$ with $k = 1, 2, \ldots, n - 1$, we shall prove that (10') also holds for $p(n)$. For that we make use of Theorem 7, which by a change in the order of summation may be written more conveniently in the form $np(n) = \sum_{k=1}^{n}\sum_{m=1}^{n/k}mp(n - km)$. By the induction assumption, all partition functions on the right satisfy (10'), so that $np(n) < \sum_{k=1}^{n}\sum_{m=1}^{n/k}me^{\alpha\sqrt{n-km}}$, or by Lemma 1(1),

$$np(n) < \sum_{k=1}^{n}\sum_{m=1}^{n/k}me^{\alpha\sqrt{n}(1-km/2n)} = \sum_{k=1}^{n}\sum_{m=1}^{n/k}me^{\alpha\sqrt{n}}e^{-\alpha(km/2\sqrt{n})}$$

$$< e^{\alpha\sqrt{n}}\sum_{k=1}^{\infty}\sum_{m=1}^{\infty}me^{-(\alpha k/2\sqrt{n})\cdot m}.$$

By Lemma 2, the inner sum equals $(e^{-(\alpha k/2\sqrt{n})})/((1 - e^{-(\alpha k/2\sqrt{n})})^2)$; this in turn, by Lemma 3, is less than $(2\sqrt{n}/\alpha k)^2$. Hence

$$np(n) < e^{\alpha\sqrt{n}}\sum_{k=1}^{\infty}\frac{4n}{\alpha^2 k^2} = \frac{4n}{\alpha^2}e^{\alpha\sqrt{n}}\sum_{k=1}^{\infty}\frac{1}{k^2}.$$

It is known (here we anticipate a result of Section 8.7; see also Problem 19) that the last sum is equal to $\pi^2/6$; hence by also replacing α^2 by its value $2\pi^2/3$, we obtain

$$np(n) < \left(4n/(2\pi^2/3)\right)\left(\pi^2/6\right)e^{\alpha\sqrt{n}} = ne^{\alpha\sqrt{n}},$$

or

$$p(n) < e^{\alpha\sqrt{n}},$$

what we wanted to prove.

The opposite inequality,

$$p(n) > e^{(\alpha-\varepsilon)e\sqrt{n}}, \tag{10''}$$

for sufficiently large n, is somewhat more difficult to prove. Let us assume that we can prove first that $p(n) > Ae^{(\alpha-\varepsilon/2)\sqrt{n}}$ holds for given $\varepsilon > 0$ and *all* n, provided that we have selected A small enough (A may actually depend on ε). Then we observe that if we keep ε and A fixed and let n increase, for sufficiently large n we have that $p(n) > Ae^{(\alpha-\varepsilon/2)\sqrt{n}} = Ae^{\varepsilon\sqrt{n}/2}e^{(\alpha-\varepsilon)\sqrt{n}}$, and for sufficiently large n this is larger than $e^{(\alpha-\varepsilon)\sqrt{n}}$. Hence the theorem will be completely proven if we can show that for a given $\varepsilon > 0$ we can find a constant A such that

$$p(n) > Ae^{(\alpha-\varepsilon)\sqrt{n}} \tag{10'''}$$

holds for all n. The proof is, as before, by induction. Besides $\varepsilon > 0$, which occurs in (10'''), let also $\varepsilon_1, \varepsilon_2, \varepsilon_3$ be given arbitrarily small positive quantities, chosen so as to satisfy

$$\varepsilon_1 + \varepsilon_2 + \frac{6}{\pi^2}\varepsilon_3 < \frac{\varepsilon}{\alpha}. \tag{11}$$

Furthermore, in order to avoid trivial complications, we shall assume from the start that $0 < \varepsilon, \varepsilon_1, \varepsilon_2, \varepsilon_3 < (1/50)$. For reasons that will become clear later, we start by determining $n_i (i = 1, 2, 3, 4)$ as the smallest integers satisfying the inequalities

$$n_1^\varepsilon \geq \frac{1}{\varepsilon_1}, \quad n_2 \geq \left(\frac{\alpha^2}{10 \cdot \varepsilon_2}\right)^3, \quad n_3 \geq \left(\frac{2}{\varepsilon_3}\right)^3, \quad n_4 \geq \left(\frac{2500\alpha}{\varepsilon}\right)^{3/2} \tag{12}$$

and then define[†]

$$N = \max\left(\left(\max_{1\leq i\leq 4} n_i\right), 2^7\right). \tag{13}$$

[†] For convenience, a rational integer is once more denoted by a capital letter, N.

Next, for all $n \le N$, consider the ratios $p(n)/e^{(\alpha - \varepsilon)\sqrt{n}} = A_n$ and select $A = \min_{1 \le n \le N} A_n$. Then by the selection of A, (10''') holds for all $n \le N$. We now claim that (10''') holds also for $n > N$. The proof is by induction and uses the fact that by the induction assumption (10''') holds for all integers up to and including $n - 1$. This will finish the proof by induction of (10''') and hence that of the theorem.

REMARK 10.   One observes that here the induction starts not, as usually, with $n = 1$, but with $n = N + 1$.

By Theorem 7, $np(n) = \sum_{k=1}^{n}\sum_{m=1}^{n/k} mp(n - mk)$; by the choice of A and the induction assumption, $np(n) > A\sum_{k=1}^{n}\sum_{m=1}^{n/k} me^{(\alpha - \varepsilon)\sqrt{n - km}}$. We further reduce the second member by summing only for $k \le n^{1/3}$ and $m \le n^{2/3 - \varepsilon}$. Then setting $c = km$, we have $c \le n^{1-\varepsilon}$; hence $(c/n) \le n^{-\varepsilon} < \varepsilon_1$, because $n > N \ge n_1$. By Lemma 1(2), $\sqrt{n - c} > \sqrt{n}\,(1 - c(1 + \varepsilon_1)/2n)$; therefore

$$(\alpha - \varepsilon)\sqrt{n - mk} > (\alpha - \varepsilon)\sqrt{n}\left(1 - \frac{km}{2n}(1 + \varepsilon_1)\right),$$

so that

$$np(n) > A \sum_{k=1}^{n^{1/3}} \sum_{m=1}^{n^{2/3 - \varepsilon}} me^{(\alpha - \varepsilon)\sqrt{n}} \cdot \exp\left\{-\frac{(\alpha - \varepsilon)k(1 + \varepsilon_1)}{2\sqrt{n}} m\right\}$$

$$= Ae^{(\alpha - \varepsilon)\sqrt{n}} \sum_{k=1}^{n^{1/3}} \sum_{m=1}^{n^{2/3 - \varepsilon}} m \exp\left\{-\frac{(\alpha - \varepsilon)(1 + \varepsilon_1)k}{2\sqrt{n}} \cdot m\right\}.$$

We set $(\alpha - \varepsilon)(1 + \varepsilon_1) = \beta$ and observe, using $\alpha = \pi\sqrt{2/3}$ and $0 < \varepsilon, \varepsilon_1 < 1/50$, that $2.4 < \beta < 2.5$. The inner sum may be written as

$$\sum_{m=1}^{\infty} me^{-\beta(k/2\sqrt{n})m} - \sum_{m > n^{2/3 - \varepsilon}} me^{-\beta(k/2\sqrt{n})m}.$$

As before, the first sum equals $e^{-\beta(k/2\sqrt{n})}/(1 - e^{-\beta(k/2\sqrt{n})})^2$ by Lemma 2. The second sum is actually bounded and may be estimated ("integral test" for series) as follows: Clearly, $me^{-(\beta k/2\sqrt{n})m} \le me^{-(\beta/2\sqrt{n})m}$, and the function $xe^{-(\beta/2\sqrt{n})x}$ is monotonically decreasing for $x > 2\sqrt{n}/\beta$; hence if

$$x > n^{(2/3) - \varepsilon}\left(= n^{1/2} \cdot n^{(1/6) - \varepsilon} > n^{1/2}\frac{2}{2 \cdot 4}n^{(1/6) - \varepsilon} > n^{1/2}\frac{2}{\beta}\right),$$

then

$$me^{-(\beta/2\sqrt{n})m} < \int_{m-1}^{m} xe^{-(\beta/2\sqrt{n})x}\, dx.$$

Setting $m_0 = [n^{(2/3)-\varepsilon}] + 1$, the desired sum satisfies

$$\sum_{m \geq m_0} m e^{-\beta(k/2\sqrt{n})m} \leq \sum_{m \geq m_0} m e^{-(\beta/2\sqrt{n})m} \leq m_0 e^{-(\beta/2\sqrt{n})m_0}$$

$$+ \sum_{m=m_0+1}^{\infty} \int_{m-1}^{m} x e^{-(\beta/2\sqrt{n})x}\, dx$$

$$= m_0 e^{-(\beta/2\sqrt{n})m_0} + \int_{m_0}^{\infty} x e^{-(\beta/2\sqrt{n})x}\, dx.$$

The integral can be computed explicitly (integration by parts), the result is further increased if we replace m_0 by $n^{(2/3)-\varepsilon}$, and also using $\varepsilon < 1/50$, we obtain

$$n^{(2/3)-\varepsilon} e^{-(\beta/2)n^{(1/6)-\varepsilon}} + \left(\frac{2}{\beta} n^{(7/6)-\varepsilon} + \frac{4}{\beta^2} n \right) e^{-(\beta/2)n^{(1/6)-\varepsilon}}$$

$$= \frac{2}{\beta} n^{(7/6)-\varepsilon} e^{-(\beta/2)n^{(1/6)-\varepsilon}} \left(1 + \frac{2}{\beta} n^{-(1/6)+\varepsilon} + \frac{\beta}{2} n^{-1/2} \right)$$

$$< 1 \cdot 3\, n^{(7/6)-\varepsilon} e^{-(\beta/2)n^{(1/6)-\varepsilon}}$$

for $n > 2^7$.

We increase this value still further by replacing the exponent $(1/6) - \varepsilon$ by $1/7$ and by dropping ε in the exponent of n. Hence, the second sum is less than $1 \cdot 3\, n^{7/6} e^{-(\beta/2)n^{1/7}}$. The function $x^{7/6} e^{-(\beta/2)x^{1/7}}$ increases from zero to a maximum of $(49/3\beta e)^{49/6}$ (attained for $x = (49/3\beta)^7$) and then decreases monotonically to zero as $x \to \infty$. Hence

$$1 \cdot 3\, n^{7/6} e^{-(\beta/2)n^{1/7}} \leq 1 \cdot 3 \left(\frac{49}{3\beta e} \right)^{49/6} < 1 \cdot 3 \left(\frac{49}{3(2 \cdot 4)e} \right)^{49/6} < 2500,$$

so that

$$np(n) > A e^{(\alpha-\varepsilon)\sqrt{n}} \sum_{k=1}^{n^{1/3}} \left\{ \frac{e^{-(\beta k/2\sqrt{n})}}{(1 - e^{-(\beta k/2\sqrt{n})})^2} - 2500 \right\}. \qquad (14)$$

Now for $k < n^{1/3}$, $k/\sqrt{n} < n^{-1/6}$ so that the exponent

$$\frac{\beta k}{2\sqrt{n}} < \frac{\alpha}{2} \left(1 - \frac{\varepsilon}{\alpha} \right)(1 + \varepsilon_1) n^{-1/6} < \alpha n^{-1/6}.$$

However, by (12) and (13)

$$\alpha n^{-1/6} \leq \alpha N^{-1/6} \leq \alpha n_2^{-1/6} \leq \sqrt{6 \cdot \varepsilon_2},$$

so that by Lemma 3(2),

$$\frac{e^{-(\beta/2\sqrt{n})k}}{(1 - e^{-(\beta/2\sqrt{n})k})^2} > \frac{1 - \varepsilon_2}{(\beta k/2\sqrt{n})^2}.$$

Substituting this in (14), we obtain

$$np(n) > Ae^{(\alpha-\varepsilon)\sqrt{n}}(1 - \varepsilon_2)\frac{4n}{\beta^2}\sum_{k=1}^{n^{1/3}}\frac{1}{k^2} - 2500Ae^{(\alpha-\varepsilon)\sqrt{n}}n^{1/3}.$$

The sum

$$\sum_{k=1}^{n^{1/3}}\frac{1}{k^2} = \sum_{k=1}^{\infty}\frac{1}{k^2} - \sum_{k>n^{1/3}}\frac{1}{k^2} = \frac{\pi^2}{6} - \delta.$$

Here

$$\delta = \sum_{k>n^{1/3}}\frac{1}{k^2} \le \sum_{k>N^{1/3}}\frac{1}{k^2} < \frac{1}{N^{2/3}} + \int_{N^{1/3}}^{\infty}\frac{dx}{x^2}$$

$$= \frac{1}{N^{2/3}} + \frac{1}{N^{1/3}} < \frac{2}{N^{1/3}} \le \varepsilon_3$$

because $N \ge n_3 \ge \left(\dfrac{2}{\varepsilon_3}\right)^3$. Hence

$$np(n) > Ae^{(\alpha-\varepsilon)\sqrt{n}}(1 - \varepsilon_2)\frac{4n}{\frac{2}{3}\pi^2\left(1 - \dfrac{\varepsilon}{\alpha}\right)^2(1 + \varepsilon_1)^2}\left(\frac{\pi^2}{6} - \varepsilon_3\right)$$

$$- 2500Ae^{(\alpha-\varepsilon)\sqrt{n}}n^{1/3},$$

and

$$p(n) > Ae^{(\alpha-\varepsilon)\sqrt{n}}\frac{\left(1 - \dfrac{6}{\pi^2}\cdot\varepsilon_3\right)(1 - \varepsilon_2)}{\left(1 - \dfrac{\varepsilon}{\alpha}\right)^2(1 + \varepsilon_1)^2} - \frac{2500}{n^{2/3}}Ae^{(\alpha-\varepsilon)\sqrt{n}}$$

$$= Ae^{(\alpha-\varepsilon)\sqrt{n}}\left(\frac{\left(1 - \dfrac{6}{\pi^2}\varepsilon_3\right)(1 - \varepsilon_2)}{\left(1 - \dfrac{\varepsilon}{\alpha}\right)^2(1 + \varepsilon_1)} - \frac{2500}{n^{2/3}}\right)$$

$$> Ae^{(\alpha-\varepsilon)\sqrt{n}}\left(1 - \frac{6}{\pi^2}\varepsilon_3 - \varepsilon_2 + \frac{2\varepsilon}{\alpha} - \varepsilon_1 - \frac{2500}{n^{2/3}}\right).$$

By (11), $\varepsilon_1 + \varepsilon_2 + (6/\pi^2)\varepsilon_3 < \varepsilon/\alpha$; by (12) and (13),

$$\frac{1}{n^{2/3}} \leq \frac{1}{N^{2/3}} \leq \frac{1}{n_4^{2/3}} \leq \frac{\varepsilon}{2500\alpha}.$$

Hence the last bracket exceeds 1 and $p(n) > Ae^{(\alpha-\varepsilon)\sqrt{n}}$ holds for $n > N$, assuming that $(10''')$ held for all integers up to $n - 1$. This finishes the proof that $(10''')$ holds for all integers n, and hence that $(10'')$ holds for all sufficiently large integers n. This together with $(10')$ finishes the proof of Theorem 10.

The Theory of Partitions does not end here, of course. The study of partition functions $p_{A,\cdots}^{(\cdots)}(n)$ with all kinds of restrictions and relative to many types of sets has continued and led to most interesting results. Among the sets $\mathbf{A}$ considered we may mention "congruence sets," that is, sets of the form

$$\mathbf{A} = \{n|n \equiv a_i(\bmod k), i = 1, 2, \ldots, r\},$$

where $a_1, \ldots, a_r$ are r distinct residues modulo k. Also $\mathbf{A} = \{n|n = p^m\}$, that is, $\mathbf{A}$ is the set of all mth powers of primes (m fixed or arbitrary) has been considered. Some of the most fascinating results have to do with the ratios $(p_{\mathbf{A}}(n))/(p_{\mathbf{B}}(n))$, where $\mathbf{A}$ and $\mathbf{B}$ are "related sets," for instance the sets of quadratic residues and of nonresidues modulo some fixed prime p, respectively. But these involve concepts like class-number and fundamental unit of algebraic number fields, and properly belong to the theory of algebraic numbers. Therefore I stop here, in the hope that the reader has become sufficiently interested *not* to stop here, but to go on.

PROBLEMS

1. (a) Find $p(n)$ for $n = 1, 2, 3, 4,$ and 5, by explicitly listing all partitions.
   (b) Verify the results of Part (a) using Theorem 5.
   (c) Use the results of Parts (a) and (b) with Theorem 5 to compute $p(n)$ for $6 \leq n \leq 10$.

2. Verify the results of Problem 1 with the help of Theorem 6.

3. Let $\mathbf{P}$ be the set of prime numbers; find $p_{\mathbf{P}}(5)$, $p_{\mathbf{P},3}(5)$, $p_{\mathbf{P}}^{(3)}(5)$, $p_{\mathbf{P}}^{(o)}(5)$, and $p_{\mathbf{P}}^{(e)}(5)$.

4. Let $\mathbf{R}(5)$ and $\mathbf{N}(5)$ stand for the set of quadratic residues modulo 5 and the set of quadratic nonresidues modulo 5, respectively; find the product representations of the generating functions for $p_{\mathbf{R}(5)}(n)$ and $p_{\mathbf{N}(5)}(n)$.

5. Complete all details in the proof of Corollary 8.2.

6. Complete all details in the proof of Corollary 8.3.

7. (a) Show that for every $n \in \mathbf{Z}^+$, $p_1(n) = 1$.

   (b) Show by a direct argument (that is, by enumeration of the partitions) that $p_2(n) = \frac{1}{2}n + 1$ if n is even and $p_2(n) = \frac{1}{2}(n + 1)$ if n is odd.

   (c) Obtain the result of Part (b) using Theorem 3.

8. Consider

$$F_2(x) = \frac{1}{(1 - x)(1 - x^2)} = \sum_{n=0}^{\infty} p_2(n)x^n.$$

Set up the decomposition into partial fractions

$$F_2(x) = \frac{1}{4}\left\{ \frac{2}{(1 - x)^2} + \frac{1}{1 - x} + \frac{1}{1 + x} \right\},$$

replace each simple fraction by its Taylor series, then combine similar powers of x to obtain

$$F_2(x) = 1 + x + 2x^2 + \cdots + \left(\frac{1}{2}(n + 1) + \frac{1}{4}(1 + (-1)^n)\right)x^n + \cdots.$$

Compare the value of $p_2(n)$ you find by this method with the result of Problem 7.

*9. (a) Treat $F_3(x) = \{(1 - x)(1 - x^2)(1 - x^3)\}^{-1}$ the way $F_2(x)$ was treated in Problem 8; obtain a formula for $p_3(n)$.

   (b) Verify the result of Part (a) using Theorem 3. (Answer: for $n \equiv 1$ or $n \equiv 5 \pmod 6$, $p_3(n) = \frac{1}{12}(n + 1)(n + 5)$; otherwise it is the smallest integer larger than $\frac{1}{12}(n + 1)(n + 5)$).

**10. Use the method of Problems 8 and 9 to show that in general, for fixed m and $n \to \infty$,

$$p_m(n) = \frac{n^{m-1}}{m!(m - 1)!}\left(1 + O\left(\frac{1}{n}\right)\right).$$

(Hints: Observe that

$$F_m(x) = \prod_{k=1}^{m} (1 - x^k)^{-1} = \sum_{k=1}^{m} A_k(1 - x)^{-k} + \cdots$$

with $A_m = 1/m!$. Also,

$$(1 - x)^k = \sum_{n=0}^{\infty} \frac{(n + 1) \cdots (n + k - 1)}{(k - 1)!} x^n$$

$$= \sum_{n=0}^{\infty} c_{k,n} x^k \quad \text{with} \quad c_{k,n} = \frac{n^{k-1}}{(k - 1)!}\left(1 + O\left(\frac{1}{n}\right)\right).$$

Summing over k, one finds that the summands $\sum_{k=1}^{m} A_k (1-x)^{-k}$ of $F_m(x)$ contribute

$$\sum_{k=1}^{m} A_k c_{k,n} = \frac{n^{m-1}}{m!(m-1)!}\left(1 + O\left(\frac{1}{n}\right)\right)$$

to the coefficient of x^n, and it only remains to show that the contribution of the neglected fractions of $F_m(x)$ is at most $O(n^{m-2})$.)

11. Consider the rational fraction

$$\begin{bmatrix} n \\ m \end{bmatrix} = \frac{(1-x^n)(1-x^{n-1})\cdots(1-x^{n-m+1})}{(1-x)(1-x^2)\cdots(1-x^m)};$$

prove that $\begin{bmatrix} n \\ m \end{bmatrix}$ is actually a polynomial.

12. Prove:

$$\prod_{n=0}^{\infty}\left\{(1-x^{2kn+k-m})(1-x^{2kn+k+m})(1-x^{2kn+2k})\right\}$$

$$= \sum_{n=-\infty}^{\infty}(-1)^n x^{kn^2+nm}$$

(Hint: Use Theorem 8.)

13. Prove:

$$\prod_{\substack{m\in\mathbf{Z}^+ \\ m\notin\mathbf{R}(5)}}(1-x^m) = \sum_{n=-\infty}^{\infty}(-1)^n x^{n(5n+1)/2}$$

(Hint: Choose appropriate numerical values for k and m in Problem 12.)

14. Find a series expansion for $\displaystyle\prod_{\substack{m\in\mathbf{Z}^+ \\ m\notin\mathbf{N}(5)}}(1-x^m)$ analogous to that occurring in Problem 13.

15. As an illustration of Sylvester's point of view, identify for $m=2$ (if possible also for $m=3$), the terms of $p_m(n)$ that increase monotonically with n and the periodic part of $p_m(n)$ that reflects the arithmetic nature of n; in particular show that the monotonic terms are a polynomial of degree $m-1$ in n.

16. Give a proof of Lemma 2 by differentiating the identity $\sum_{v=0}^{\infty} z^v = (1-z)^{-1}$ and replacing z by x^{-1}.

17. Write out in full detail the proof of $(10''')$.

18. Modify the proof of Theorem 10 so as to obtain the stronger result: Given $C > 0$ arbitrarily large and $\varepsilon > 0$ arbitrarily small, there exists an integer N such that for $n \geq N$ one has $Ce^{(\sqrt{2/3}\pi - \varepsilon)\sqrt{n}} < p(n) < e^{\sqrt{2/3}\pi\sqrt{n}}$.

*19. Let $f(x) = x^2$ for $|x| \leq \pi$, and for $|x| > \pi$ define $f(x)$ by periodicity, i.e., $f(x) = f(x + 2\pi)$. Consequently, if $x = x_0 + 2k\pi$, $|x_0| \leq \pi$, then $f(x) = f(x_0) = x_0^2$. Clearly $f(x) = f(-x)$, and it follows (see, e.g.,

[15]) that

$$f(x) = \tfrac{1}{2}a_0 + \sum_{n=1}^{\infty} a_n \cos nx. \tag{15}$$

(i) Show that $a_0 = 2\pi^2/3$ and $a_n = (-1)^n \cdot 4/n^2$ for $n > 0$.

(ii) By replacing the coefficients in (15) by their values, prove the identity

$$x^2 = \frac{\pi^2}{3} - 4\left\{\cos x - \frac{\cos 2x}{2^2} + \frac{\cos 4x}{4^2} - \cdots\right\}$$

valid for $|x| \le \pi$.

(iii) Set $x = \pi$ in the above identity and show that

$$\sum_{k=1}^{\infty} \frac{1}{k^2} = \frac{\pi^2}{6}.$$

BIBLIOGRAPHY

1. T. M. Apostol, *Mathematical Analysis*, Reading, Mass.: Addison-Wesley Publishing Co., 1957.

2. P. T. Bateman and P. Erdös, *Mathematika*, **3**, 1956, pp. 1–14.

3. P. Erdös, *Annals of Mathematics* (2), **43**, 1942, pp. 437–450.

4. N. Fine, *Hypergeometric Series*, (to be published by U. of Pa. Press).

5. F. Franklin, *Comptes Rendus de l'Acad. des Sciences* (Paris), **92**, 1881, pp. 448–450.

6. E. Grosswald, *Revista de la Union Mat. Argentina*, **20**, 1962, pp. 48–57.

7. G. H. Hardy and S. Ramanujan, *Proceedings, London Math. Soc.* (2), **17**, 1918, pp. 75–115.

8. G. J. Jacobi, *Fundamenta Nova Theoriae Functionum Ellipticarum*, *Collected Works*, Vol. 1. Berlin: G. Reimer, 1881, pp. 49–239.

9. E. Laguerre, *Bull. Soc. Math. France*, **5**, 1877, pp. 76–78.

10. D. J. Newman, *Amer. Journal of Math.*, **73**, 1951, pp. 599–601.

11. I. Niven, *Amer. Mathem. Monthly*, **76**, 1969, pp. 871–889.

12. H. Rademacher, *Proceedings, London Math. Soc.* (2), **43**, 1937, pp. 241–254.

13. H. Rademacher, Survey article in the *Bull. Amer. Math. Soc.*, **46**, 1940, pp. 59–73.

14. J. J. Sylvester, *Amer. Journal of Math.*, **5**, 1882, pp. 119–136. *Collected Papers*, Vol. 3. Camb. Univ. Press, 1904 (edited by H. F. Baker), pp. 605–622 and pp. 658–660.

15. E. C. Titchmarsh, *The Theory of Functions*, 2nd ed., Oxford: The Clarendon Press, 1939.

Part Three

Topics from Analytic and Algebraic Number Theory

Chapter 8

The Distribution of Primes and the Riemann Zeta Function

1 THE DISTRIBUTION OF PRIMES AND THE SIEVE METHOD

We recall that the problem of the distribution of primes had been raised at least as far back as the Greek antiquity. The proof of our Theorem 3.9, that there are infinitely many primes, appears in Euclid (Book 9, Section 20), and Eratosthenes[†] devised a systematic method for obtaining all primes up to any given number x. This is called the *sieve method* and is easily described. First, it is clear that if n is not a prime then there exists a prime $p \leq \sqrt{n}$ that divides n. Indeed n not being a prime, it can be factored, $n = a \cdot b$; if $a > \sqrt{n}$ and $b > \sqrt{n}$, then $n = a \cdot b > (\sqrt{n})^2 = n$, a contradiction. Hence if $a \leq b$, then $a \leq \sqrt{n}$ and every prime $p|a$ satisfies $p \leq a \leq \sqrt{n}$. This remark suggests the following procedure for the construction of a list of all primes not in excess of x, if the primes up to $\sqrt{x}$ are already known.

Consider all integers from 2 to x, listed in their natural order; let us keep 2, but cancel (by a slash /) all its multiples, starting with 2^2, i.e., $2 \cdot 2, 3 \cdot 2$, $4 \cdot 2, \dots, n \cdot 2, \dots$ for all $n \leq \frac{1}{2}x$. Next we keep the first number after 2 that has not yet been canceled, that is 3, but cancel (by an inverted slash \) all multiples of 3, starting with 3^2, i.e., $3 \cdot 3, 4 \cdot 3, 5 \cdot 3, \dots, n \cdot 3, \dots$ up to $n \leq x/3$. The next integer remaining in our list after 3 is 5; we keep it, but cancel (by a double slash \\) all its multiples, starting with 5^2, i.e., $5 \cdot 5, 6 \cdot 5, \dots$, and so on. From the previous remark, we know that any integer remaining in our set must be either a prime or larger than $5^2 = 25$. In general, having removed the multiples of some prime p_r, the first remaining integer after p_r is also a prime, say p_{r+1}; otherwise it would have to be divisible by some prime less than itself (actually, less than its square root) and would have been removed. We also do

---

[†] School of Alexandria, born 276 B.C., died, apparently of voluntary starvation, at 80 years of age, ca. 196 B.C.

not have to look at integers less than p_{r+1}^2, because we observed that the composite ones among them had been removed already, all having a prime factor less than p_{r+1}; hence it is sufficient to start removing the multiples of p_{r+1}, starting with p_{r+1}^2. An example of the way in which we proceed follows:

$$2, 3, \not{4}, 5, \not{6}, 7, \not{8}, \not{9}, \not{10}, 11, \not{12}, 13, \not{14}, \not{15}, \not{16}, 17, \not{18}, 19, \not{20}, \not{21}, \not{22}, 23, \not{24},$$
$$\not{25}, \not{26}, \not{27}, \not{28}, 29, \not{30}, 31, \ldots.$$

We realize of course that some of the integers have been crossed out more than once; for instance, 12 had to be canceled as a multiple both of 2 and of 3; 30 as a multiple of 2, 3, and 5, and so on. The sieve method is quite effective for obtaining a list of primes up to a reasonably small limit; but if we try to keep track of the number of integers canceled, in order to find a formula for the (exact or approximate) number of primes $p \le x$, we run into a serious difficulty. The number of integers canceled is expressed by square brackets; for instance, when we cancel the multiples of 2 up to x, we cancel $\left[\frac{1}{2}x\right] - 1$ integers. When we now try to replace $\left[\frac{1}{2}x\right]$ by $\frac{1}{2}x$ (or, more generally, $[x/n]$ by x/n), we introduce an error that may be as big as unity. It soon appears that these errors accumulate so fast that they completely swamp the principal term we try to compute.

The simple approach of Eratosthenes has been much improved by Viggo Brun (see [7]), later by H. Rademacher [21] and especially A. Selberg [26], and also earlier, in a different direction, by Meissel [19].

Some old problems can now be solved with the help of these improved "sieve methods"; for instance, one can show that the sum $\Sigma(1/p)$ extended over all twin primes converges (see [8]), or that every even number is the sum of two odd numbers each containing only a small number of primes (see [5], [22], [30], and [9]). But the sieve method has some natural limitations that seem by now fairly well established (see Selberg [27], last paragraph). In particular, there seems to be no way to use it for a counting of the primes sufficiently accurate to yield the PNT (prime number theorem).

2   FROM TCHEBYCHEFF TO LANDAU

Around the middle of the 19th century, Tchebycheff approached the problem differently. As already seen, he introduced the functions[†] $\theta(x) = \Sigma_{p \le x} \log p$ and $\psi(x) = \Sigma_{p^m \le x} \log p$, and proved that the PNT is equivalent to either of the relations $\lim_{x \to \infty} \theta(x)/x = 1$, $\lim_{x \to \infty} \psi(x)/x = 1$; furthermore, he

---

[†] In order to avoid breaking with an old tradition, variable quantities will be denoted by lower case italic letters taken from the end of the alphabet, such as x, y, z, and t (in addition to σ). Up to this point, lower case italic letters have represented rational integers only, but the context should make any confusion or misinterpretation highly unlikely.

also showed that there exist numerical constants A and A', with $\log(2^{1/2}3^{1/3}5^{1/5}/30^{1/30})$ $(\approx 0 \cdot 921) \leq A$, $1 \leq A' \leq 6A/5$ $(\approx 1 \cdot 105)$, and such that for sufficiently large x all three ratios, $\theta(x)/x$, $\psi(x)/x$, and $(\pi(x)\log x)/x$ stay between A and A'. Tchebycheff's method can be refined to give values of A and A' still closer to unity, but no direct way seems to lead to the conclusion (needed for the PNT) that for increasing x a sequence of couples (A_n, A'_n) can be found, such that for $x > n$, $A_n x < \theta(x) < A'_n x$, with $\lim_{n \to \infty} A_n = \lim_{n \to \infty} A'_n = 1$. A few years after Tchebycheff's work, Riemann published his famous memoir "Ueber die Anzahl der Primzahlen unter einer gegebenen Grösse" (see [23] pp. 145–155), in which he approaches the problem from a completely different angle. He considers the function $\zeta(x) = \sum_{n=1}^{\infty} n^{-s}$, whose connection with the theory of primes had already been recognized by Euler. Riemann does not restrict himself as Euler did to real values of s, but studies the behavior of $\zeta(s)$ in the whole plane of the complex variable $s = \sigma + it$. Riemann's paper is relatively short (less than 10 pages) and is beautifully written. Still, we are not going to follow it here. Indeed as it stands, it is incomplete. While at first reading it might strike us as quite clear and convincing, some reflection reveals several gaps. These could not be filled until some 37 years later (and more than 30 years after Riemann's early death in 1866), through the work of Hadamard on the theory of entire functions and that of Hadamard and de la Vallée-Poussin on the zeta function. At present, it is possible to give a completely satisfactory proof along the lines of Riemann's approach. However, this requires several rather deep theorems of analysis, and we prefer instead to present here one along somewhat different lines.

Up to this point, in addition to the reader's eagerness to know and a certain amount of mathematical maturity, little more than high school mathematics was needed or assumed. For here on, however, we shall have to start using some analysis. An effort has been made to keep this requirement as low as possible. For almost everything that follows, a knowledge of advanced calculus and of the theory of functions of one complex variable, on the level of, say Ahlfors [1], is more than adequate. In one or two places, elements of Fourier analysis will be invoked. The reader willing to accept those simple results should have no difficulty in following the subject matter of interest here; the reader who wants to gain a deeper understanding of the analytic tools used should refer to the sources indicated. Indeed, whenever use is made of any deeper analytic results, at least one easily accessible text is quoted.

3   THE RIEMANN ZETA FUNCTION

Theorem 1.   *Let $s = \sigma + it$ be a complex variable. Then for $\sigma > 1$, the series $\sum_{n=1}^{\infty} n^{-s}$ converges. The convergence is uniform for $\sigma \geq 1 + \varepsilon$, and any $\varepsilon > 0$*

arbitrarily small. Hence in the half plane $\sigma > 1$, $\sum_{n=1}^{\infty} n^{-s}$ *represents an analytic function of s* (*see, e.g.,* [3] *p.* 394 *or* [28] *p.* 95).

PROOF. For any two positive integers $m < k$ one has

$$\left| \sum_{n=m+1}^{k} n^{-s} \right| \leq \sum_{n=m+1}^{k} |n^{-s}| = \sum_{n=m+1}^{k} n^{-\sigma}$$

$$\leq \int_{m}^{k} x^{-\sigma} \, dx = x^{1-\sigma}/(1-\sigma)|_{m}^{k} \leq m^{1-\sigma}/(\sigma - 1) < m^{-\varepsilon}/\varepsilon,$$

because $\sigma \geq 1 + \varepsilon$. For fixed $\varepsilon > 0$, $m^{-\varepsilon}/\varepsilon$ can be made arbitrarily small by taking m sufficiently large, and this proves the uniform convergence of the series. Differentiating termwise, we observe that the differentiated series $-\sum_{n=2}^{\infty} n^{-s} \log n$ also converges for $\sigma > 1$, and uniformly so for $\sigma \geq 1 + \varepsilon$ ($\varepsilon > 0$), the proof being essentially the same as for $\sum_{n=1}^{\infty} n^{-s}$; hence for $\sigma > 1$, by the termwise differentiability of uniformly convergent series (see [3] p. 451 or [31] p. 305), $-\sum_{n=2}^{\infty} n^{-s} \log n$ is the derivative of the function $\sum_{n=1}^{\infty} n^{-s}$, which is consequently analytic for $\sigma > 1$.

DEFINITION 1 (Riemann). The function defined for $\sigma > 1$ by $\sum_{n=1}^{\infty} n^{-s}$ is denoted, following Riemann, by $\zeta(s)$.

REMARK 1. If $s > 1$, then $\sum_{n=1}^{k} n^{-s} > \int_{1}^{k} x^{-s} \, dx = (k^{1-s} - 1)/(1 - s) = (1 - k^{1-s})/(s - 1)$; hence it follows that $\zeta(s) = \lim_{k \to \infty} \sum_{n=1}^{k} n^{-s} \geq 1/(s - 1)$. If we now let s approach 1 from the right along the real axis (in symbols $s \to 1^{+}$), then $1/(s - 1) \to +\infty$. Hence $\zeta(s)$ does *not* stay bounded as $s \to 1^{+}$; the series $\sum_{n=1}^{\infty} n^{-s}$ diverges for $s = 1$, and Theorem 1 cannot be improved even to state the convergence of the series $\sum_{n=1}^{\infty} n^{-s}$ for $\sigma \geq 1$.

Euler, while considering only real values of s, proved the following theorem, equally valid for complex s.

Theorem 2. *For* $\sigma > 1$, $\zeta(s) = \prod_{p} (1 - p^{-s})^{-1}$.

Here and in what follows, the letter p under the symbol for sum or product means that the corresponding sum or product is extended over all primes.

PROOF. We consider the finite product $P_k(s) = \prod_{p \leq k} (1 - p^{-s})^{-1} = \prod_{p \leq k} (1 + 1/p^s + \cdots + 1/p^{ns} + \cdots)$. Multiplying out, the right side is of the form $\sum^{1} n^{-s}$. All and only those integers n that contain no prime divisor larger than k occur in this sum. Also, on account of the uniqueness of factorization, each such integer occurs exactly once. Hence $\sum^{1} n^{-s} = \sum_{n=1}^{\infty} n^{-s} - \sum^{2} n^{-s}$, where the sum $\sum^{2} n^{-s}$ is extended over all integers n containing at least one prime factor $p > k$. For $\sigma > 1$, however, $|\sum^{2} n^{-s}| \leq \sum^{2} n^{-\sigma} < \sum_{n=k+1}^{\infty} n^{-\sigma}$, and as $k \to \infty$ this sum approaches zero, because the series

$\sum_{n=1}^{\infty} n^{-\sigma}$ converges. Hence $\prod_p (1 - p^{-s})^{-1} = \lim_{k \to \infty} P_k(s) = \lim_{k \to \infty} \prod_{p \le k} (1 - p^{-s})^{-1} = \zeta(s) - \lim_{k \to \infty} \sum^2 n^{-s} = \zeta(s)$, proving the theorem.

Corollary 2.1 (Euler). *There exist infinitely many primes.*

PROOF. We saw that for $s \to 1^+$, $\prod_p (1 - p^{-s})^{-1} = \zeta(s)$ does not stay bounded, while if the number of primes were finite, $\lim_{s \to 1^+} \prod_p (1 - p^{-s})^{-1} = \prod_p (1 - p^{-1})^{-1}$ would be a finite, actually even a rational number.

Corollary 2.2. *The sum $\sum \dfrac{1}{p}$ extended over all primes diverges.*

PROOF. Let $x \ge 3$ and observe that, just as in the proof of Theorem 2, $\sum_{n \le x} 1/n < \prod_{p \le x} (1 - (p^{-1}))^{-1}$; hence

$$\log \sum_{n \le x} \frac{1}{n} < - \log \prod_{p \le x} \left(1 - \frac{1}{p}\right)$$

$$= - \sum_{p \le x} \log\left(1 - \frac{1}{p}\right) = \sum_{p \le x} \left\{ \frac{1}{p} + \sum_{m=2}^{\infty} \frac{1}{mp^m} \right\} = \sum_{p \le x} \frac{1}{p} + R(x),$$

with

$$0 < R(x) = \sum_{p \le x} \sum_{m=2}^{\infty} \frac{1}{mp^m} < \sum_{p \le x} \sum_{m=2}^{\infty} \frac{1}{p^m} = \sum_{p \le x} \frac{1}{p^2 - p}$$

$$< 2 \sum_{p \le x} \frac{1}{p^2} < 2 \sum_{n \le x} \frac{1}{n^2} < 2 \sum_{n=1}^{\infty} \frac{1}{n^2} = 2\zeta(2)$$

$(= \pi^2/3$ (see Section 7 or Problem 7.19), but the exact value of this constant is of course irrelevant). It follows that $\sum_{p \le x} p^{-1} > \log \sum_{n \le x} n^{-1} - 2\zeta(2)$, and the result now follows from the divergence of the harmonic series $\sum_{n=1}^{\infty} n^{-1}$.

Corollary 2.3. $\zeta(s) \ne 0$ *for $\sigma > 1$.*

PROOF. For real $s > 0$, $\log(1 - p^{-s})$ is real and negative. For $\sigma > 0$, $|p^{-s}| = p^{-\sigma} < 1$; hence $1 - p^{-s} \ne 0$, and therefore we can define a single-valued branch of the function $\log(1 - p^{-s})$ in $\sigma > 0$, *a fortiori* in $\sigma > 1$. Expanding, $\log(1 - p^{-s})^{-1} = -\log(1 - p^{-s}) = p^{-s} + p^{-2s}/2 + \cdots + (p^{-ms}/m) + \cdots$ and

$$\left|\log(1 - p^{-s})^{-1}\right| = \left| \sum_{m=1}^{\infty} p^{-sm} m^{-1} \right| \le \sum_{m=1}^{\infty} p^{-\sigma m}$$

$$= p^{-\sigma}(1 - p^{-\sigma})^{-1} = (p^{\sigma} - 1)^{-1} < 2p^{-\sigma}$$

because $p \geq 2$ and $\sigma > 1$. Hence

$$\left| \log \prod_{p \leq x} (1 - p^{-s})^{-1} \right| = \left| \sum_{p \leq x} \log(1 - p^{-s})^{-1} \right| < 2 \sum_{p \leq x} p^{-\sigma} < 2 \sum_{n=1}^{\infty} n^{-\sigma}$$

$$= 2\zeta(\sigma) = A,$$

say (A being some constant), because the series converges by Theorem 1. Taking the limit for $x \to \infty$, the infinite product on the left converges to $\zeta(s)$, on account of Theorem 2 (because $\sigma > 1$). We conclude that $|\log \zeta(s)| < A$; hence $\log|\zeta(s)| = $ Real part of $\log \zeta(s) > -A$, so that $|\zeta(s)| > e^{-A} > 0$, as asserted.

*ALTERNATIVE PROOF.　For $\sigma > 1$, $\zeta(s) \neq 0$ as absolutely convergent infinite product of nonvanishing factors.

4　THE ZETA FUNCTION AND $\pi(x)$

Before we continue with the study of the zeta function, it might be of interest to stop for a moment to show the close connection that $\zeta(s)$ has with the function $\pi(x)$; this is illustrated by the following theorem.

Theorem 3.　*For $\sigma > 1$, $\log \zeta(s) = s \int_{2}^{\infty} \dfrac{\pi(x)}{x(x^s - 1)} dx$.*

PROOF.　For $\sigma > 1$, $\zeta(s) \neq 0$ by Corollary 2.3, and for real $s > 1$, $\zeta(s) > 1$. Hence we can define a branch of $\log \zeta(s)$, real and positive for $s > 1$ and single valued for $\sigma > 1$.
Then by Theorem 2,

$$\log \zeta(s) = - \sum_{p} \log(1 - p^{-s}) = - \lim_{k \to \infty} \sum_{n=2}^{k} (\pi(n) - \pi(n - 1))\log(1 - n^{-s}).$$

By *partial summation* (that is, by a regrouping of the finitely many terms of the sum; see Section 5 for a detailed example),

$$\log \zeta(s) = - \lim_{k \to \infty} \left\{ \sum_{n=2}^{k} \pi(n)\left[\log(1 - n^{-s}) - \log(1 - (n + 1)^{-s})\right] \right.$$

$$\left. - \pi(1)\log(1 - 2^{-s}) + \pi(k)\log(1 - (k + 1)^{-s}) \right\}$$

$$= \lim_{k \to \infty} \left\{ \sum_{n=2}^{k} \pi(n)\left[\log(1 - (n + 1)^{-s}) - \log(1 - n^{-s})\right] \right\}$$

$$- \lim_{k \to \infty} \pi(k)\log(1 - (k + 1)^{-s}).$$

For $k \to \infty$, the last term is less than

$$\left| k \left\{ \frac{1}{(k+1)^s} + \frac{1}{2(k+1)^{2s}} + \cdots \right\} \right| < \frac{2k}{|(k+1)^s|} \to 0$$

because $\sigma > 1$. Hence

$$\log \zeta(s) = \sum_{n=2}^{\infty} \pi(n) \int_n^{n+1} \frac{d}{dx} \log(1 - x^{-s}) \, dx$$

$$= \sum_{n=2}^{\infty} \pi(n) \int_n^{n+1} \frac{s}{x(x^s - 1)} \, dx = s \sum_{n=2}^{\infty} \int_n^{n+1} \frac{\pi(x)}{x(x^s - 1)} \, dx$$

$$= s \int_2^{\infty} \frac{\pi(x)}{x(x^s - 1)} \, dx,$$

and the theorem is proven.

The PNT may now be obtained by "inverting" the relation given by Theorem 3, that is, by "solving" the equation of Theorem 3 for $\pi(x)$. Then $\pi(x)$ appears as an integral over $(\log \zeta(s))/s$; this integral is somewhat difficult to evaluate, and before we attempt this calculation we have to return to the study of the zeta function.

5   FURTHER THEORY OF THE ZETA FUNCTION

Several interesting theorems concerning the zeta function may be obtained as particular cases of the following simple lemma, which is concerned only with finite sums:

Lemma 1.   *Let m and k be positive integers, while $s = \sigma + it$ is a complex number. Let $S = S(s; m, k) = \sum_{n=m+1}^{k} n^{-s}$. Then*

$$S = -s \int_m^k \frac{x - [x]}{x^{s+1}} \, dx + \frac{1}{s-1} (m^{1-s} - k^{1-s}).$$

PROOF.   By partial summation we have

$$S = \frac{(m+1) - m}{(m+1)^s} + \frac{(m+2) - (m+1)}{(m+2)^s} + \cdots + \frac{(m+a) - (m+a-1)}{(m+a)^s}$$

$$+ \frac{(m+a+1) - (m+a)}{(m+a+1)^s} + \cdots + \frac{k - (k-1)}{k^s}$$

$$= \frac{-m}{(m+1)^s} + (m+1)\left(\frac{1}{(m+1)^s} - \frac{1}{(m+2)^s}\right) + \cdots$$

$$+ (m+a)\left(\frac{1}{(m+a)^s} - \frac{1}{(m+a+1)^s}\right) + \cdots$$

$$+ (k-1)\left(\frac{1}{(k-1)^s} - \frac{1}{k^s}\right) + \frac{k}{k^s}$$

$$= \sum_{n=m+1}^{k-1} n\left(\frac{1}{n^s} - \frac{1}{(n+1)^s}\right) - \frac{m}{(m+1)^s} + \frac{k}{k^s}$$

$$= \sum_{n=m}^{k-1} n\left(\frac{1}{n^s} - \frac{1}{(n+1)^s}\right) - m^{1-s} + k^{1-s}.$$

As in the proof of Theorem 3, we observe that the general term

$$n\left(\frac{1}{n^s} - \frac{1}{(n+1)^s}\right) = sn\int_n^{n+1} \frac{dx}{x^{s+1}} = s\int_n^{n+1} \frac{[x]}{x^{s+1}}dx$$

and the sum on n becomes a sum of abutting integrals; hence it can be written as a single integral, $s\int_m^k ([x]/x^{s+1})\,dx$, and

$$S = s\int_m^k \frac{[x]}{x^{s+1}}dx - m^{1-s} + k^{1-s}. \tag{1}$$

In some cases, in order to improve the convergence it is desirable to have a smaller numerator under the integral sign; we therefore subtract

$$s\int_m^k \frac{x}{x^{s+1}}dx = s\int_m^k x^{-s}\,dx = \frac{s}{1-s}(k^{1-s} - m^{1-s})$$

and then add back the same quantity, obtaining

$$S = s\int_m^k \frac{[x] - x}{x^{s+1}}dx + \frac{1}{1-s}(k^{1-s} - m^{1-s}),$$

which is precisely the lemma.

Theorem 4. *For* $\sigma > 1$, $\zeta(s)$ *is regular analytic and has the representation*

$$\zeta(s) = s\int_1^\infty \frac{[x]}{x^{s+1}}dx. \tag{2}$$

PROOF. Take $m = 1$ and let $k \to \infty$ in (1); then for $\sigma > 1$,

$$\lim_{k\to\infty} S = \zeta(s) - 1, \quad m^{1-s} = 1, \quad \lim_{k\to\infty} |k^{1-s}| = \lim_{k\to\infty} k^{1-\sigma} = 0,$$

and the improper integral in (1) converges for $\sigma > 1$.

Theorem 5. *For $\sigma > 0$, $\zeta(s)$ is regular analytic, except for a pole of first order at $s = 1$, with residue $a_{-1} = 1$, and has the representation*

$$\zeta(s) = \frac{1}{s-1} + 1 - s \int_1^\infty \frac{x - [x]}{x^{s+1}} \, dx. \qquad (3)$$

PROOF. Take $m = 1$ and let $k \to \infty$ in Lemma 1. Then the term k^{1-s} vanishes for $\sigma > 1$ and $k \to \infty$ and we obtain (3) formally. This representation holds, therefore, at least for $\sigma > 1$. However, the integral already converges for $\sigma > 0$ (because $x - [x]$ stays bounded), and uniformly so for $\sigma \geq \varepsilon$ ($\varepsilon > 0$ arbitrarily small). If we formally take the derivative of the definite integral in (3), by differentiating under the integral sign we obtain $\int_1^\infty (x - [x]) x^{-s-1} \log x \, dx$, again uniformly convergent for $\sigma \geq \varepsilon > 0$. One easily infers from this that the last term in (3) is an analytic function of s, at least in the half plane $\sigma > 0$. This function coincides there with $f(s) = \zeta(s) - (1/(s-1)) - 1$; it therefore represents the analytic continuation of $f(s)$ for $\sigma > 0$. The assertion concerning the pole of $\zeta(s)$ at $s = 1$ can now be read off from (3).

Actually, we can do still better. If in (3) we also introduce the term $-\frac{1}{2}$ in the numerator of the integrand, the integral becomes $\int_1^\infty ((x)) x^{-s-1} dx$. To compensate for the change, we have to subtract $\frac{1}{2} s \int_1^\infty x^{-s-1} dx = \frac{1}{2}$, and then we obtain from (3)

Theorem 6. *For $\sigma > -1$, $\zeta(s)$ admits the representation*

$$\zeta(s) = \frac{1}{s-1} + \frac{1}{2} - s \int_1^\infty \frac{((x))}{x^{s+1}} \, dx. \qquad (4)$$

PROOF. We already know that (4) holds for $\sigma > 0$. In order to prove Theorem 6, all we still have to show is that the integral converges for $\sigma > -1$. Indeed, let $y = \int_1^x ((t)) \, dt$; then $y(1) = 0$ and (see Theorem 6.3(6)) $|y(x)| \leq 1/8$ for all x. Hence, integrating by parts,

$$\int_1^\infty \frac{((x))}{x^{s+1}} \, dx = y(x) x^{-s-1} \Big|_1^\infty + (s+1) \int_1^\infty \frac{y(x)}{x^{s+2}} \, dx.$$

For $\sigma > -1$, the first term vanishes and the last integral converges because $y(x)$ is bounded.

Corollary 6.1. $\lim\limits_{s \to 1} (s - 1)\zeta(s) = 1.$

PROOF is obvious from (4).

Corollary 6.2. *For $\sigma > 0$, $\zeta(s) = (1/(s-1)) + \gamma + O(s-1)$, where $\gamma = \lim_{k \to \infty}(\sum_{n=1}^k n^{-1} - \log k)$ is the Euler-Mascheroni constant.*

PROOF.

$$\int_1^k \frac{x - [x]}{x^2}\, dx = \int_1^k \frac{dx}{x} - \sum_{n=1}^{k-1} n \int_n^{n+1} \frac{dx}{x^2}$$

$$= \log k - \sum_{n=1}^{k-1} n\left(\frac{1}{n} - \frac{1}{n+1}\right)$$

$$= \log k - \sum_{n=1}^{k-1} \frac{1}{n+1} = 1 - \left(\sum_{n=1}^{k} \frac{1}{n} - \log k\right).$$

For $k \to \infty$ the bracket has a finite limit (see Problem 5) denoted by γ; hence (3) may be written as

$$\zeta(s) = \frac{1}{s-1} + 1 - \int_1^{\infty} \frac{x - [x]}{x^2}\, dx - \left\{s\int_1^{\infty} \frac{x - [x]}{x^{s+1}}\, dx - \int_1^{\infty} \frac{x - [x]}{x^2}\, dx\right\}$$

$$= \frac{1}{s-1} + \gamma - I(s);$$

here

$$I(s) = s\int_1^{\infty} \frac{x - [x]}{x^{s+1}}\, dx - \int_1^{\infty} \frac{x - [x]}{x^2}\, dx$$

$$= (s - 1)\int_1^{\infty} \frac{x - [x]}{x^{s+1}}\, dx + \int_1^{\infty} (x^{-s-1} - x^{-2})(x - [x])\, dx$$

$$= (s - 1)\left\{\int_1^{\infty} \frac{x - [x]}{x^{s+1}}\, dx + \int_1^{\infty} \frac{x - [x]}{x^{s+1}} \frac{1 - x^{s-1}}{s - 1}\, dx\right\}$$

$$= (s - 1)G(s),$$

say.

The proof of the corollary will be complete if we show that $G(s)$ stays bounded as $s \to 1$. For $\sigma \geq \varepsilon$, the first integral in the brackets is majorized by $\int_1^{\infty} x^{-1-\varepsilon}\, dx = \varepsilon^{-1}$. For $\sigma \geq \varepsilon$ and s bounded away from 1, say for $|s - 1| \geq \delta > 0$, the second integral is majorized by $\int_1^{\infty}((1 + x^{\sigma-1})/(x^{\sigma+1}\delta))\, dx = \delta^{-1}(\sigma^{-1} + 1) < (1 + \varepsilon)/\varepsilon\delta$. If, on the other hand, $s \to 1$, then $\lim_{s\to 1}((x^{s-1} - x^0)/((s - 1) - 0)) = \{(d/dx)(x^{s-1})\}_{s=1} = \log x$; hence the second integral is majorized by $\int_1^{\infty} x^{-2+\varepsilon'}(\log x + \eta)\, dx < \int_1^{\infty} x^{-3/2}\, dx = 2$. To justify the last inequality, one has to keep ε' and η appropriately small; the conditions $((1 - 4\varepsilon')/(1 - 2\varepsilon')) > \log 2$ and $\eta < 2(((1 - 4\varepsilon')/(1 - 2\varepsilon')) - \log 2)$ are sufficient and easily realized (see Problem 6). It follows that in all cases $|G(s)|$ has a bound that is independent of s, and this finishes the proof of the corollary.

6   *THE FUNCTIONAL EQUATION

Let us look back for a moment to see what we have accomplished: Starting with the representation $\zeta(s) = \sum_{n=1}^{\infty} n^{-s}$, valid only for $\sigma > 1$, we first obtained the representation $\zeta(s) = (1/(s-1)) + 1 - s\int_1^{\infty}((x - [x])/(x^{s+1}))\,dx$, which extends the domain of definition of $\zeta(s)$ from $\sigma > 1$ to $\sigma > 0$; next, going one step further, we proved $\zeta(s) = (1/(s-1)) + \frac{1}{2} - s\int_1^{\infty}(((x))/x^{s+1})\,dx$, which extends the domain of definition of $\zeta(s)$ still further, to $\sigma > -1$. It is possible to continue step by step and it can be proven that we may continue doing this indefinitely, with the result that $\zeta(s)$ can be defined in the whole plane of the complex variable $s = \sigma + it$ and is analytic everywhere except at $s = 1$, where it has a pole of first order. However, instead of doing this piecemeal, it is both easier and more elegant to do it at one stroke, by proving the justly celebrated functional equation of Riemann (which, however, will not be needed in the proof of the PNT).

Theorem 7 (Riemann).   *The zeta function satisfies the functional equation*

$$\zeta(s) = 2^s \pi^{s-1} \zeta(1-s)\Gamma(1-s)\sin\frac{\pi s}{2}. \tag{5}$$

SKETCH OF A PROOF.   For $\sigma < 0$, $s\int_0^1 ((x))x^{-s-1}\,dx$ converges and has the value

$$s\int_0^1 \frac{x - \frac{1}{2}}{x^{s+1}}\,dx = s\left\{\frac{x^{1-s}}{1-s} + \frac{1}{2}\frac{x^{-s}}{s}\right\}_0^1 = s\left(\frac{1}{1-s} + \frac{1}{2s}\right)$$

$$= \frac{s}{1-s} + \frac{1}{2} = \frac{1}{1-s} - 1 + \frac{1}{2} = -\frac{1}{s-1} - \frac{1}{2}.$$

Hence by (4), for $-1 < \sigma < 0$,

$$\zeta(s) = -s\int_0^1 \frac{((x))}{x^{s+1}}\,dx - s\int_1^{\infty} \frac{((x))}{x^{s+1}}\,dx = -s\int_0^{\infty} \frac{((x))}{x^{s+1}}\,dx.$$

Here we replace $((x))$ by its Fourier series $-\pi^{-1}\sum_{n=1}^{\infty} n^{-1}\sin(2\pi nx)$ (see [15] p. 209), integrate termwise (a nontrivial justification is called for, because the series is *not* uniformly convergent; it is mainly because we here suppress this justification, which may be found, e.g., in [28], that the present section is entitled "sketch of a proof," rather than "proof"), and obtain formally:

$$\zeta(s) = -s\int_0^{\infty} x^{-s-1}\left\{-\frac{1}{\pi}\sum_{n=1}^{\infty} \frac{\sin(2\pi nx)}{n}\right\}\,dx$$

$$= \frac{s}{\pi}\sum_{n=1}^{\infty} \frac{1}{n}\int_0^{\infty} x^{-s-1}\sin(2\pi nx)\,dx.$$

We make the change of variables $y = 2\pi nx$ and obtain

$$\zeta(s) = \frac{s}{\pi} \sum_{n=1}^{\infty} \frac{(2\pi n)^s}{n} \int_0^{\infty} y^{-s-1} \sin y \, dy$$

$$= \frac{s}{\pi} (2\pi)^s \sum_{n=1}^{\infty} n^{-(1-s)} \int_0^{\infty} y^{-s-1} \sin y \, dy.$$

On account of $\sigma < 0$, the convergence of the infinite series is guaranteed, and its value is, by Definition 1, $\zeta(1 - s)$. The integral converges at least for $-1 < \sigma < 0$; remembering (see, e.g., [28] pp. 107, 162) that for $0 < \operatorname{Re} z < 1$,

$$\int_0^{\infty} y^{z-1} \sin y \, dy = \sin(\pi z/2)\Gamma(z),$$

we have with $s = -z$ for $-1 < \sigma = \operatorname{Re} s < 0$ that

$$\int_0^{\infty} y^{-s-1} \sin y \, dy = -\sin(\pi s/2)\Gamma(-s).$$

Hence $\zeta(s) = (s/\pi)(2\pi)^s \zeta(1 - s)(-\Gamma(-s))\sin(\pi s/2)$. Observing that (see [28] p. 55) $s(-\Gamma(-s)) = (-s)\Gamma(-s) = \Gamma(1 - s)$, we finally obtain (5). While the formula has been established only for $-1 < \sigma < 0$, it now appears that the right side is well defined and analytic for all s with $\sigma < 0$; hence by analytic continuation (see [28], pp. 138–141), $\zeta(s)$ is now also defined (by (5)) for all s with $\sigma < 0$. Knowing some of the properties of the Γ-function and of $\sin(\pi s/2)$, we can read off formula (5) many interesting properties of $\zeta(s)$. So, for instance, it is clear that the factor $2^s \pi^{s-1} \Gamma(1 - s)\zeta(1 - s)$ is analytic and different from zero for $\sigma < 0$. Hence any zero or singularity of $\zeta(s)$ for $\sigma < 0$ must come from $\sin(\pi s/2)$. But $\sin(\pi s/2)$ has no singularities, and all its zeros s with $\sigma < 0$ are the even negative integers $-2, -4, \ldots$. It follows that $\zeta(s)$ is analytic for $\sigma < 0$, and that $\zeta(s) = 0$ has the roots $s = -2n(n \in \mathbf{Z}^+)$. These roots are usually called the "trivial roots" of the zeta function, to distinguish them from some other harder to come by roots, located in the strip $0 < \sigma < 1$. Next we use (5) to show how one can compute, for instance, $\zeta(0)$. We have $2^0 \pi^{0-1}\Gamma(1 - 0) = 1/\pi$; hence

$$\zeta(0) = \lim_{s \to 0} 2^s \pi^{s-1} \Gamma(1 - s)\zeta(1 - s)\sin\frac{\pi s}{2} = \frac{1}{\pi} \lim_{s \to 0} (-s)\zeta(1 - s)\frac{\sin(\pi s/2)}{-s}$$

$$= -\frac{1}{2} \cdot \lim_{s \to 0} \frac{\sin(\pi s/2)}{\pi s/2}$$

by Corollary 6.1. Therefore, finally, $\zeta(0) = -\frac{1}{2}$.

7   THE VALUES OF $\zeta(s)$ AT INTEGRAL ARGUMENTS

To make another interesting application of the functional equation (5), we shall need some additional preparation. The reader of this section will no doubt increase his awe for the achievements of Euler, who had already obtained the values of $\zeta(2n)$ $(n = 1,2,3,\ldots)$ 225 years ago [11].

Lemma 2.

$$\pi \cotg \pi z = \frac{1}{z} + \sum_{n=1}^{\infty} (-1)^n \frac{(2\pi)^{2n}}{(2n)!} B_{2n} z^{2n-1},$$

where the B_{2n} are rational numbers, and where the series converges at least for $|z| < 1$.

PROOF.

$$\cotg x = \frac{\cos x}{\sin x} = \left(1 - \frac{x^2}{2!} + \frac{x^4}{4!} - \cdots\right)\left(x - \frac{x^3}{3!} + \frac{x^5}{5!} - \cdots\right)^{-1}$$

$$= x^{-1} \sum_{n=1}^{\infty} (-1)^n \frac{x^{2n}}{(2n)!} \sum_{m=0}^{\infty} c_m x^m.$$

Here the coefficients c_m of $x/\sin x = \sum_{m=0}^{\infty} c_m x^m$ may be obtained, e.g., by writing that $\sum_{n=0}^{\infty}(-1)^n x^{2n}/(2n + 1)! \sum_{m=0}^{\infty} c_m x^m = 1$ holds, or equivalently, $\sum_{k=0}^{\infty} x^k \sum_{m+2n=k}((-1)^n c_m)/((2n + 1)!) = 1$. The constant term c_0 equals 1. The coefficients of all other powers x^k have to vanish, whence $c_1 = 0$, $c_2 - c_0/3! = 0$, or $c_2 = 1/3!$, etc., with all coefficients c_m rational and all $c_{2q+1} = 0$ (because $x/\sin x$ is an even function). It follows that $\cotg x = x^{-1} + \sum_{n=1}^{\infty} b_{2n} x^{2n-1}$. If one here sets $x = \pi z$ and multiplies by π, one obtains $\pi \cotg \pi z = z^{-1} + \sum_{n=1}^{\infty} \pi^{2n} b_{2n} z^{2n-1}$, with rational b_{2n}s. Finally, to conform to tradition set $B_{2n} = (-1)^n (2n)! b_{2n}$. The series in the second member is the Taylor series of the function $F(z) = \pi \cotg \pi z - z^{-1}$. As $\lim_{z \to 0} \pi z \cotg \pi z = 1$, the origin is a regular point for $F(z)$ and its Taylor series converges for $|z| < r$ where r is the smallest modulus of a singular point (see [28]). Here this is $z = \pm 1$, the closest zero $z \neq 0$ of $\sin \pi z$, so that the series converges for $|z| < 1$, and Lemma 2 is proved.

REMARK 2.   The coefficients B_k are called Bernoulli numbers; they are usually defined as the coefficients of $z^n/n!$ in the expansion

$$\frac{z}{e^z - 1} = \sum_{n=0}^{\infty} B_n \frac{z^n}{n!}.$$

The left side equals

$$\frac{z}{z + \dfrac{z^2}{2!} + \dfrac{z^3}{3!} + \cdots} = \frac{1}{1 + \dfrac{z}{2} + \dfrac{z^2}{6} + \cdots} = 1 - \frac{z}{2} + \cdots,$$

and we immediately find that $B_0 = 1$ and $B_1 = -1/2$. Next, $z/(e^z - 1) + z/2 = (z/2)((e^z + 1)/(e^z - 1)) = f(z)$, say, is an even function, i.e., $f(-z) = f(z)$. Indeed, $2f(-z) = -z((e^{-z} + 1)/(e^{-z} - 1)) = -z((1 + e^z)/(1 - e^z)) = z((e^z + 1)/(e^z - 1)) = 2f(z)$. It follows as before that all odd powers of $f(z) = 1 + \sum_{n=2}^{\infty} B_n(z^n/n!)$ vanish, so that $B_{2m+1} = 0$ $(0 < m \in \mathbf{Z})$.

Lemma 3. $\pi \operatorname{cotg} \pi z = 1/z + \sum_{n=1}^{\infty} 2z/(z^2 - n^2)$.

SKETCH OF A PROOF.   Lemma 3 is essentially the partial fraction (or Mittag-Leffler, see [28]) decomposition of $g(z) = \pi \operatorname{cotg} \pi z$. It is clear that $g(z) = z(\cos \pi z / \sin \pi z)$ is analytic, except at zeros of the denominator, i.e., for $z \in \mathbf{Z}$. When $z \to n$, then by l'Hôpital's Rule (see [3]),

$$\lim_{z \to n} (z - n)g(z) = \lim_{z \to n} \pi \cos \pi z \frac{z - n}{\sin \pi z} = (-1)^n \lim_{z \to n} (\cos \pi z)^{-1} = 1.$$

The series $\sum_{i=-\infty}^{\infty} 1/(z - n)$ is not convergent, but for $z \notin \mathbf{Z}$, $S(z) = \sum_{n=-\infty, n \neq 0}^{\infty}(1/(z - n) + 1/n) + 1/z = 1/z + \sum_{n=-\infty, n \neq 0}^{\infty} z/n(z - n)$ converges like $\sum_{n \neq 0} 1/n^2$, and also $\lim_{z \to n}(z - n)S(z) = 1$. Hence, $\lim_{z \to n}(g(z) - S(z)) = 0$ and $G(z) = g(z) - S(z)$ is analytic everywhere in the finite complex plane, i.e., it is an entire function. Furthermore, we verify that $g(z + 1) = g(z)$, $S(z + 1) = S(z)$, so that $G(z)$ is periodic. If $G(z)$ is bounded in a strip of width 1, say for $|\operatorname{Re} z| \leq 1/2$, then $G(z)$ is bounded in the whole plane. By Liouville's Theorem it will then follow that $G(z)$ is a constant. We can evaluate that constant, e.g., for $z = 0$, and find that $G(0) = 0$. We conclude that

$$g(z) = S(z) = \lim_{N \to \infty} \left\{ \frac{1}{z} + \sum_{\substack{n=-N \\ n \neq 0}}^{N} \left(\frac{1}{z - n} + \frac{1}{n} \right) \right\}$$

$$= \frac{1}{z} + \lim_{N \to \infty} \sum_{n=1}^{N} \frac{2z}{z^2 - n^2} = \frac{1}{z} + \sum_{n=1}^{\infty} \frac{2z}{z^2 - n^2},$$

as claimed, because the last series again converges like $\sum_{n=1}^{\infty} 1/n^2$.
    To show the boundedness of $G(z)$, we observe that for $z = \sigma + it$,

$$\pi \operatorname{cotg} \pi z = \pi \frac{\cos \pi(\sigma + it)}{\sin \pi(\sigma + it)} = \pi i \frac{e^{-\pi t + i\pi\sigma} + e^{\pi t - i\pi\sigma}}{e^{-\pi t + i\pi\sigma} - e^{\pi t - i\pi\sigma}}$$

$$= -i\pi \frac{1 + e^{-2\pi t + 2i\pi\sigma}}{1 - e^{-2\pi t + 2i\pi\sigma}} \to -i\pi,$$

as $t \to +\infty$, and the limit is $+i\pi$ if $t \to -\infty$. This actually holds for arbitrary σ, not only for $|\sigma| \leq 1/2$. As for the sum,

$$\lim_{t \to +\infty} S(\sigma + it) = \lim_{t \to +\infty} \left\{ \frac{1}{\sigma + it} - 2i \sum_{n=1}^{\infty} \frac{t - i\sigma}{n^2 + t^2 - 2it\sigma - \sigma^2} \right\}$$

$$= -2i \lim_{t \to +\infty} \sum_{n=1}^{\infty} \frac{t(1 - i\sigma/t)}{n^2 + t^2 \left(1 - \frac{2i\sigma}{t} - \frac{\sigma^2}{t^2}\right)}$$

$$= -2i \lim_{t \to +\infty} \sum_{n=1}^{\infty} \frac{t}{n^2 + t^2} .$$

By splitting the sum into two parts, $\sum_{n=1}^{\infty} = \sum_{n \leq t} + \sum_{n > t}$, it is easy to verify that it is bounded (uniformly in t), and this finishes the proof that $G(z)$ is a constant. However,

$$\lim_{z \to 0} G(z) = \lim_{z \to 0} \left\{ \pi z \cot g\, \pi z - 1 + 2z^2 \sum_{n=1}^{\infty} \frac{1}{n^2} \right\} \quad \text{and} \quad \lim_{z \to 0} \pi z \cot g\, \pi z = 1,$$

so that $\lim_{z \to 0} G(z) = G(0) = 0$, and this finishes the proof of Lemma 3. One can do even better (see Problem 10) and show that $\lim_{t \to \infty} \sum_{n=1}^{\infty} (t/(n^2 + t^2)) = \pi/2$. This shows directly that $\lim_{t \to \infty} G(\sigma + it) = 0$, and we don't even have to verify that $G(0) = 0$.

The fraction $2z/(z^2 - n^2)$ that occurs in the sum of Lemma 3 may be written as

$$-\frac{2z}{n^2} \frac{1}{1 - (z/n)^2} = -\frac{2}{z} \left(\frac{z}{n}\right)^2 \left(1 + \left(\frac{z}{n}\right)^2 + \left(\frac{z}{n}\right)^4 + \cdots\right) = -\frac{2}{z} \sum_{m=1}^{\infty} \left(\frac{z}{n}\right)^{2m},$$

and Lemma 3 is equivalent to $\pi \cot g\, \pi z = z^{-1} - 2z^{-1}\sum_{n=1}^{\infty}\sum_{m=1}^{\infty}(z/n)^{2m}$. For $|z| \leq 1 - \varepsilon$, $n \neq 1$, the inner sum is less in absolute value than $1/(n^2 - 1)$; hence the double sum converges absolutely and uniformly, at least for $|z| \leq 1 - \varepsilon$, and we may invert the order of summation. We obtain the result

$$F(z) = \pi \cot g\, \pi z - \frac{1}{z} = -\frac{2}{z} \sum_{m=1}^{\infty} z^{2m} \sum_{n=1}^{\infty} \frac{1}{n^{2m}} = -2 \sum_{m=1}^{\infty} z^{2m-1}\zeta(2m).$$

This is the Taylor expansion of the function $F(z)$ and converges at least in $|z| < 1$. However we already know that Taylor expansion from Lemma 2. If we equate the coefficients of identical powers of the variable in the two series, we obtain

$$\zeta(2n) = (-1)^{n-1} \frac{(2\pi)^{2n}}{2(2n)!} B_{2n}. \tag{6}$$

We may now use the functional equation (5) to obtain the values of $\zeta(s)$ at odd negative integers, $s = 1 - 2n$, $0 < n \in \mathbf{Z}$. We obtain

$$\zeta(1 - 2n) = 2^{1-2n}\pi^{-2n}\zeta(2n)\Gamma(2n)\sin\left(\frac{\pi}{2} - n\pi\right)$$

$$= 2^{1-2n}\pi^{-2n}(-1)^{n-1}\frac{2^{2n}\pi^{2n}}{2(2n)!}B_{2n}(2n-1)!(-1)^n = -\frac{B_{2n}}{2n},$$

a remarkably simple formula.

We formalize the results obtained so far concerning the values of $\zeta(s)$ for s an integer, in a theorem.

Theorem 8. *Let n be a positive integer and denote the Bernoulli numbers by B_k; then $\zeta(-2n) = 0$, $\zeta(1 - 2n) = -B_{2n}/2n$, $\zeta(0) = -1/2$, $\zeta(2n) = (-1)^{n-1}(2\pi)^{2n}B_{2n}/2(2n)!$*

The reader may well wonder about the asymmetry of this statement. In particular, he will be puzzled by the fact that the values of $\zeta(2n + 1)$, $n > 0$, seem to have been left out. This is indeed the case and is no accident; in fact, we hardly know anything about these values. While Euler determined the values of $\zeta(2n)$ over 200 years ago, so far the values of $\zeta(2n + 1)$ are still a complete mystery. We do know that $\zeta(2n)/\pi^{2n}$ is rational, that $\zeta(1 - 2n)$ is rational, and also that $\zeta(0) = -1/2$ is rational. Also $\zeta(-2n)$ (or should we perhaps say $\zeta(-2n)/\pi^{-2n}$?) is rational in a trivial way (namely zero). Is $\zeta(2n + 1)/\pi^{2n+1}$ also rational? What evidence we have shows that this is very unlikely. If $\zeta(2n + 1)/\pi^{2n+1}$ is irrational, what kind of irrational number is it? Does it satisfy some algebraic equation (in which case we would call it an algebraic number) or not (in which case we would call it transcendental)?

There is an enormous literature that discusses this problem; nevertheless, the only known result concerning $\zeta(2n + 1)$, $0 < n \in \mathbf{Z}$, is that $\zeta(3)$ is irrational (see [2]). Weak as this result appears (it says nothing about $\zeta(3)/\pi^3$), it appeared sensational when it was first announced. The method does not even permit asserting the irrationality of $\zeta(5)$.

8 THE ZETA FUNCTION AND ITS DERIVATIVES FOR σ CLOSE TO 1

Theorem 9. *For $\sigma > 1$,*

$$\zeta'(s) = -\sum_{n=1}^{\infty}\frac{\log n}{n^s} \tag{7}$$

and

$$\frac{\zeta'(s)}{\zeta(s)} = -\sum_{n=1}^{\infty} \frac{\Lambda(n)}{n^s}, \tag{8}$$

and if ε > 0, both series converge uniformly for σ ≥ 1 + ε.

PROOF.   The uniform convergence of both series follows by observing that $\Lambda(n) \le \log n < n^{\varepsilon/2}$ for sufficiently large n; hence the terms of both series are less in absolute value than $n^{-(\sigma-\varepsilon/2)} \le n^{-(1+\varepsilon/2)}$ and the series converge by Theorem 1. On account of the uniform convergence of (7), this formal derivative of the series defining the zeta function is actually its derivative for $\sigma > 1$. Next observe that $\sum_{d|n}\Lambda(d) = \sum_{p^m|n}\log p = \log n$. Hence

$$\zeta(s) \sum_{n=1}^{\infty} \frac{\Lambda(n)}{n^s} = \sum_{m=1}^{\infty} \frac{1}{m^s} \sum_{d=1}^{\infty} \frac{\Lambda(d)}{d^s} = \sum_{n=1}^{\infty} \frac{1}{n^s} \sum_{d|n} \Lambda(d) = \sum_{n=1}^{\infty} \frac{\log n}{n^s} = -\zeta'(s),$$

and because $\zeta(s) \ne 0$ when $\sigma > 1$, we may divide by $\zeta(s)$ and obtain (8).

ALTERNATIVE PROOF.   For $\sigma > 1$, $\zeta(s) \ne 0$; hence a single-valued branch of $\log \zeta(s)$ can be defined, real and positive for $s > 1$ (where $\zeta(s) > 1$). For this branch, $\log \zeta(s) = -\sum_p \log(1 - p^{-s}) = \sum_p \sum_{m \ge 1} p^{-ms}/m$; differentiating, $\zeta'(s)/\zeta(s) = -\sum_p \sum_{m \ge 1} p^{-ms} \log p = -\sum_{n=1}^{\infty} \Lambda(n)/n^{-s}$.

Theorem 10 (Hadamard–de la Vallée-Poussin).

$$\zeta(1 + it) \ne 0.$$

PROOF.   For $\sigma > 1$,

$$\log \zeta(s) = -\sum_p \log(1 - p^{-s}) = \sum_p \sum_{m \ge 1} \frac{p^{-ms}}{m}$$

$$= \sum_p \sum_{m \ge 1} \frac{1}{m} e^{-ms \log p} = \sum_{p,m} \frac{1}{m} e^{-m(\sigma + it)\log p}$$

$$= \sum_{p,m} \frac{1}{m} e^{-m\sigma \log p} e^{-mit \log p}$$

$$= \sum_{p,m} \frac{1}{m} e^{-m\sigma \log p} \{\cos(mt \log p) - i \sin(mt \log p)\}.$$

Hence

$$\log|\zeta(s)| = \operatorname{Re} \log \zeta(s) = \sum_{p,m} \frac{1}{m} e^{-m\sigma \log p} \cos(mt \log p) = \sum_{p,m} \frac{\cos(mt \log p)}{mp^{m\sigma}}.$$

Consider now the positive expression $S = |\zeta^3(\sigma)\zeta^4(\sigma + it)\zeta(\sigma + 2it)|$. Using the above result for $\log|\zeta(s)|$, we easily obtain that

$$\log S = 3\log|\zeta(\sigma)| + 4\log|\zeta(\sigma + it)| + \log|\zeta(\sigma + 2it)|$$

$$= \sum_{m,\,p} \frac{1}{mp^{m\sigma}}(3 + 4\cos(mt\log p) + \cos(2mt\log p)).$$

However, for every angle α one has that

$$3 + 4\cos\alpha + \cos 2\alpha = 3 + 4\cos\alpha + 2\cos^2\alpha - 1 = 2(\cos\alpha + 1)^2 \geq 0;$$

hence as a sum of positive quantities, $\log S \geq 0$ and $S \geq 1$. If we now assume that $\zeta(1 + it) = 0$ for some real t, we shall reach a contradiction. Indeed considering $\zeta(\sigma + it) = f(\sigma)$ as a function of σ, the assumption reads $f(1) = 0$; hence if $\sigma = 1$ is a zero of order $n(\geq 1)$, then $(\sigma - 1)^n$ is a factor of $f(\sigma)$, that is, of $\zeta(\sigma + it)$, and as $\sigma \to 1^+$, $|\zeta(\sigma + it)| < A(\sigma - 1)^n \leq A(\sigma - 1)$ for some constant A. Next, it follows from Corollary 6.1 that as $\sigma \to 1^+$, $(\sigma - 1)\zeta(\sigma) \to 1$; hence $|\zeta(\sigma)| < 2/(\sigma - 1)$. Finally, from Theorem 6 it follows that $|\zeta(1 + 2it)| = C$, some finite constant.

Consequently, $S < (2/(\sigma - 1))^3(A(\sigma - 1))^4 C = 8A^4C(\sigma - 1)$, and for $\sigma \to 1^+$, $S \to 0$. But this contradicts the inequality $S \geq 1$, which we just proved; hence the assumption $\zeta(1 + it) = 0$ is not tenable, and Theorem 10 is proven.

Corollary 10.1.   *The principal branch of the function $F(s) = \log\zeta(s)$, defined for real $s > 1$ by $F(s) = -\sum_p \log(1 - p^{-s})$ (see Theorem 2) and well defined for arbitrary complex s with $\sigma > 1$ by Corollary 2.3, may be continued also for $s = 1 + it\ (t \neq 0)$.*

PROOF.   By Theorem 5, $\zeta(\sigma + it)$ is analytic for $t \neq 0, \sigma \geq 1$, and $\zeta(1 + it) \neq 0$ by Theorem 10; hence the result follows.

Theorem 11.   *Let C be an arbitrary constant; then there exists a t_0 which depends only on C such that the region $\mathcal{R}$ defined by $1 - C/(\log t) \leq \sigma \leq 2$, and $t \geq t_0$, $|\zeta(s)| = O(\log t)$ holds uniformly in σ.*

The exact meaning of this statement is that, given an arbitrary constant C, we can find t_0 and C_1 (both depending only on C) such that for $1 - C/(\log t) \leq \sigma \leq 2$ and $t \geq t_0$, one has $|\zeta(s)| \leq C_1 \log t$.

REMARK 3.   From the boundedness of $\zeta(s)$ on every bounded closed set not containing $s = 1$, it follows that one may always take $t_0 = 2$ by increasing, if necessary, the constant C_1.

PROOF.   In Lemma 1, if $\sigma > 1$, we may let $k \to \infty$; then the last term vanishes and the integral converges. Also, $\lim_{k\to\infty} S(s; m, k) = \sum_{n=m+1}^{\infty} n^{-s} = \zeta(s) -$

$\sum_{n=1}^{m} n^{-s}$, so that

$$\zeta(s) = \sum_{n=1}^{m} n^{-s} - s \int_{m}^{\infty} \frac{x - [x]}{x^{s+1}} \, dx + \frac{m^{1-s}}{s-1}. \tag{9}$$

This representation remains valid as long as the integral converges, that is, for $\sigma > 0$. In fact in $\mathcal{R}$ for, say $t \geq 2$,

$$|s| \leq 2t, \quad \left| \int_{m}^{\infty} \frac{x - [x]}{x^{s+1}} \, dx \right| < \int_{m}^{\infty} \frac{dx}{x^{\sigma+1}} = \frac{1}{\sigma m^{\sigma}} \quad \text{and} \quad \left| \frac{m^{1-s}}{s-1} \right| \leq \frac{m^{1-\sigma}}{t}.$$

Finally, in $\mathcal{R}$, $|n^{-s}| = n^{-\sigma} = e^{-\sigma \log n} \leq \exp\{-(1 - C/(\log t)) \log n\}$. So far m has been an arbitrary integer. Now if a (large) value of t is given, we select $m[t]$, so that $m \leq t < m + 1$. Then for every $n \leq m$, $|n^{-s}| \leq \exp\{-\log n + C(\log n/\log t)\} \leq n^{-1}e^{C}$ and

$$\left| \sum_{n=1}^{m} n^{-s} \right| \leq \sum_{n=1}^{m} n^{-\sigma} \leq e^{C} \sum_{n=1}^{m} n^{-1} < e^{C}(\log m + 1) \leq e^{C}(\log t + 1).$$

Consequently,

$$|\zeta(s)| \leq \frac{2t}{\sigma m^{\sigma}} + \frac{m^{1-\sigma}}{t} + e^{C}(\log t + 1)$$

$$\leq \frac{2t}{\sigma(t-1)^{\sigma}} + \frac{1}{t^{\sigma}} \left(\frac{t}{m} \right)^{\sigma-1} + e^{C}\log t + e^{C}$$

$$\leq \frac{2}{\sigma} \frac{t^{1-\sigma}}{(1 - t^{-1})^{\sigma}} + \frac{2}{t^{\sigma}} + e^{C} + e^{C}\log t,$$

because $(t/m)^{\sigma-1} \leq 2^{\sigma-1} \leq 2$ for $t \geq 2$, s in $\mathcal{R}$. For $t \geq 2$ and $\sigma \leq 2$, also $1 - t^{-1} \geq \frac{1}{2}$, and $(1 - t^{-1})^{\sigma} \geq \frac{1}{4}$, so that $|\zeta(s)| \leq (8/\sigma)t^{1-\sigma} + 2t^{-\sigma} + e^{C} + e^{C}\log t$. For the given C, we now select t_0 so large that $1 - C/(\log t_0) > \frac{1}{2}$ and $t_0 \geq 2$ both hold. Then if $1 - C/(\log t) \leq \sigma \leq 2$,

$$\frac{8}{\sigma}t^{1-\sigma} + 2t^{-\sigma} \leq 16t^{C/\log t} + 2t^{-1/2} = 16e^{C} + 2t^{-1/2} < 16e^{C} + 2$$

and

$$|\zeta(s)| \leq 2 + 17e^{C} + e^{C}\log t = e^{C}\log t \left(1 + \frac{17 + 2e^{-C}}{\log t} \right).$$

If necessary, we may now increase t_0 once more, to insure that

$$(17 + 2e^{-C})/(\log t_0) < 1;$$

then $|\zeta(s)| \leq 2e^{C}\log t$, and the theorem (actually, slightly more) is proven.

Corollary 11.1. $\zeta(1 + it) = O(\log t)$.

PROOF.   Follows from Theorem 11 because $1 + it \in \mathcal{R}$.

Theorem 12.   *In $\mathcal{R}$, $\zeta'(s) = O(\log^2 t)$ holds uniformly in σ.*

PROOF.   By differentiating (8), $\zeta'(s) = -\sum_{n=1}^{m} n^{-s} \log n$

$$- \int_m^\infty \frac{x - [x]}{x^{s+1}} dx + s \int_m^\infty \frac{(x - [x]) \log x}{x^{s+1}} dx - \frac{m^{1-s}}{(s-1)^2} - \frac{m^{1-s} \log m}{s - 1}.$$

The first sum is estimated as in the proof of Theorem 11 and each term only has the extra factor $\log n \leq \log m \leq \log t$; the same holds for the last term. Next,

$$\left| \int_m^\infty \frac{x - [x]}{x^{s+1}} dx \right| \leq \int_m^\infty \frac{dx}{x^{\sigma+1}} = \frac{1}{\sigma m^\sigma} \leq \frac{1}{\sigma(t - 1)^\sigma} \to 0 \quad \text{for } t \to \infty;$$

also

$$\left| \frac{m^{1-s}}{(s-1)^2} \right| = O(t^{-2} t^{1-\sigma}) = O(t^{-1-\sigma}) \to 0 \quad \text{for } t \to \infty,$$

and only the term $s \int_m^\infty ((x - [x]) \log x) x^{-s-1} dx$ remains to be investigated. One has

$$\left| \int_m^\infty \frac{(x - [x]) \log x}{x^{s+1}} dx \right| \leq \int_m^\infty \frac{\log x}{x^{\sigma+1}} dx = \frac{\log m}{\sigma m^\sigma} \left(\frac{1}{\sigma \log m} + 1 \right) < \frac{2 \log m}{\sigma m^\sigma}$$

again as in Theorem 11, except for the extra factor $\log m \leq \log t$. Hence besides terms that decrease to zero as $t \to \infty$, $\zeta'(s)$ contains only summands that essentially occur already in $\zeta(s)$, with at most an additional factor $\log t$; therefore, all terms are at most of order $O(\log^2 t)$, and the Theorem is proven.

Corollary 12.1. $\zeta'(1 + it) = O(\log^2 t)$.

PROOF.   Evident.

Theorem 13.

$$\frac{\zeta'}{\zeta}(1 + it) = O(\log^9 t).$$

PROOF.   In the proof of Theorem 10 we saw that for $\sigma > 1$,

$$|\zeta^3(\sigma) \zeta^4(\sigma + it) \zeta(\sigma + 2it)| \geq 1;$$

hence

$$|\zeta(\sigma + it)|^{-1} \leq |\zeta(\sigma)|^{3/4} |\zeta(\sigma + 2it)|^{1/4},$$

and using Corollary 6.1 or 6.2 and Theorem 11 we see that there exists some constant D such that, as $t \to \infty$ for fixed $\sigma > 1$ and with $D^{-1} = A$,

$$\frac{1}{|\zeta(\sigma + it)|} < \frac{D\log^{1/4}t}{(\sigma - 1)^{3/4}} \quad \text{or} \quad |\zeta(\sigma + it)| > A\frac{(\sigma - 1)^{3/4}}{\log^{1/4}t}.$$

Also, integrating along a parallel to the real axis, for $\sigma > 1$,

$$\int_1^\sigma \zeta'(x + it)\, dx = \zeta(\sigma + it) - \zeta(1 + it);$$

hence

$$|\zeta(1 + it)| = \left|\zeta(\sigma + it) - \int_1^\sigma \zeta'(x + it)\, dx\right| \geq |\zeta(\sigma + it)| - \left|\int_1^\sigma \zeta'(x + it)\, dx\right|$$

$$> A\frac{(\sigma - 1)^{3/4}}{(\log t)^{1/4}} - \left|\int_1^\sigma \zeta'(x + it)\, dx\right|.$$

By Theorem 12, the last term is less than $B \cdot \log^2 t \int_1^\sigma dx = B(\sigma - 1)\log^2 t$ for some constant B. Hence

$$|\zeta(1 + it)| > A(\sigma - 1)^{3/4}\log^{-1/4}t - B(\sigma - 1)\log^2 t = f(\sigma - 1, t),$$

say. This result is valid for every $\sigma > 1$. We now select $\sigma - 1$ so as to maximize $f(\sigma - 1, t)$. Proceeding as in elementary calculus, it is an easy matter to show that the extremum of $f(\sigma - 1, t)$ is reached for $\sigma = 1 + (3A/4B)^4\log^{-9}t$, and that this is a maximum equal to $(3/B)^3(A/4)^4\log^{-7}t$. Consequently, setting $(3/B)^3(A/4)^4 = C$,

$$|\zeta(1 + it)| > C\log^{-7}t \tag{10}$$

and $|\zeta(1 + it)|^{-1} < (1/C)\log^7 t$. This together with Corollary 12.1 finally leads to $|\zeta'(1 + it)/\zeta(1 + it)| \leq K\log^9 t$, which finishes the proof.

Corollary 13.1. $\zeta'(\sigma + it)/\zeta(\sigma + it) = O(\log^9 t)$ *holds not only for $\sigma \geq 1$ but actually for $\sigma \geq 1 - C_2\log^{-9}t$ with some positive constant C_2.*

PROOF.   Let $\sigma + it \in \mathcal{R}$, $\sigma > 1$; then using Theorem 12,

$$|\zeta(1 + it) - \zeta(\sigma + it)| = \left|\int_\sigma^1 \zeta'(x + it)\, dx\right| \leq C_1\log^2 t \cdot (1 - \sigma)$$

with some finite positive constant C_1. Hence for any σ satisfying $1 - C_2\log^{-9}t \leq \sigma \leq 1$, one obtains, also using (10):

$$|\zeta(\sigma + it)| \geq |\zeta(1 + it)| - C_1(1 - \sigma)\log^2 t > C\log^{-7}t - C_1\log^2 t \cdot C_2\log^{-9}t$$

$$= (C - C_1C_2)\log^{-7}t = K\log^{-7}t.$$

If we select $C_2 < C/C_1$, then $K > 0$, and the Corollary follows like before by use of Theorem 12.

Corollary 13.2. $\zeta(s) \neq 0$ for $\sigma > 1 - C_2 \log^{-9} t$.

PROOF.   Follows almost trivially from Corollary 13.1.

Corollary 13.3.   *The function* $g(s) = \zeta'(s)/\zeta(s) + (s-1)^{-1}$ *is analytic for* $\sigma \geq 1$.

PROOF.   For $\sigma > 1$ this is a consequence of Corollary 2.3, from which follows that $(s-1)\zeta(s)$ is analytic and different from zero for $\sigma > 1$. Hence $f(s) = \log(s-1) + \log \zeta(s)$ has a single-valued branch, real for real $s > 1$, which is analytic and has an analytic derivative; this is precisely $g(s)$. For $\sigma = 1, s \neq 1$ the same conclusion follows from Theorem 10, and for $s = 1$ from Theorem 10, Theorem 5, and Corollary 6.1 (see Problem 8).

Corollary 13.4.   *For* $\sigma \geq 1$, $|\log \zeta(s)| = O(\log^9 t)$.

PROOF.   For $\sigma \geq 2$, $|\zeta(s)| \leq \zeta(2)$ (why?). Also

$$\operatorname{Re} \zeta(s) \geq 1 - \sum_{n=2}^{\infty} n^{-\sigma} \geq 2 - \zeta(2) > 1/3;$$

hence if we plot all values of $\zeta(\sigma + it)$ for $\sigma \geq 2$, all image points are in the right half plane and $|\arg \zeta(\sigma + it)| < \pi/2$. Consequently,

$$|\log \zeta(\sigma + it)| \leq \pi/2 + \log \zeta(2).$$

Also, for $1 - c_2 \log^{-9} t \leq \sigma \leq 2$, by Corollary 13.1,

$$|\log \zeta(\sigma + it) - \log \zeta(2 + it)| = \left| \int_{2+it}^{\sigma+it} \frac{\zeta'}{\zeta}(\sigma + it)\, ds \right|$$

$$\leq \int_{\sigma}^{2} \left| \frac{\zeta'}{\zeta}(\sigma + it) \right| d\sigma$$

$$\leq 2 \max_{\sigma \leq \sigma_1 \leq 2} \left| \frac{\zeta'}{\zeta}(\sigma_1 + it) \right| < K \log^9 t;$$

Corollary 13.4 now follows trivially.

9   COMMENTS ON THE ZETA FUNCTION

We have now obtained a certain amount of information concerning the function $\zeta(s)$. Our study may not have seemed to be too systematic, but it had as its purpose, at least partly, to establish those properties that we want to use in the proof of the PNT. However, I shall have no quarrel with any reader who prefers instead the study of this fascinating zeta function for its own sake (see, e.g., [29] or [10]). Numerous rich rewards in number theory (e.g., improved error term in the PNT), general theory of integration and differentiation (see, e.g. [20]), and other fields of mathematics are in store for any substantial improvement of our still fragmentary knowledge of the behavior of the zeta function. At this point, it is not possible to remain silent on what is probably the most intriguing unsolved problem in the theory of the zeta function and actually in all of number theory—and most likely even one of the most important unsolved problems in contemporary mathematics, namely the famous Riemann hypothesis. This is one among several unproven conjectures, found implicitly or explicitly in Riemann's already-mentioned memoir of 1859. All but one of these conjectures have since been settled (always in the sense expected by Riemann), through the work of Hadamard (1893 and 1896), de la Vallée-Poussin (1896), and von Mangoldt (1895; also 1905). The last, still unsolved problem has to do with the zeros of the zeta function. We easily could prove: (i) if $\sigma > 1$, then $\zeta(s) \neq 0$; (ii) if $\sigma < 0$, then $\zeta(s) = 0$ only for $s = -2n(n = 1, 2, 3, \ldots)$. But the functional equation—probably our most powerful tool so far—does not give us much information on what happens for $0 < \sigma < 1$, in the so-called critical strip. The ease with which we proved that $\zeta(s) \neq 0$ for $\sigma > 1$ (Corollary 2.3) should be contrasted with the rather difficult proof that the inequality $\sigma > 1$ can actually be improved to $\sigma \geq 1$ (Theorem 10). This situation is characteristic of all attempts to penetrate into the critical strip (or even to touch it!) Now certain general considerations show that the equation $\zeta(s) = 0$ *does* have solutions, even infinitely many solutions, other than the even negative integers (which we called "trivial" roots). As we know, these other roots can be found only inside the critical strip. Riemann conjectured (and *this* is the statement known as the Riemann hypothesis) that all these "nontrivial" zeros of the zeta function are at points $s = \frac{1}{2} + it$ of the complex plane, that is, that they all have the real part equal to $\frac{1}{2}$ so that they are located on the "critical line" $\sigma = \frac{1}{2}$. Some progress has been made in this direction as follows: Gram [12], Backlund [4], and Hutchinson [14] computed several of the nontrivial zeros and found them all, as Riemann had expected, on the critical line. This work has also been extended. Lehmer [16], [17] proved that the first few tens of thousands of the nontrivial zeros all have real part equal to $\frac{1}{2}$. Rosser, Yohe, and Schoenfeld [24] raised the number to 3,500,000 zeros, while Brent (see [6] and the papers quoted there) went up to no less than 81,000,000 zeros of $\zeta(s)$, all of which satisfy the requirement Re $\rho = 1/2$ of the Riemann hypothesis.

Changing the point of view somewhat, Hardy [13] proved that there are infinitely many zeros on the critical line. Selberg [25] proved a theorem that in nontechnical terms means essentially that if not all then at least a sizeable fraction of all nontrivial zeros of the Riemann zeta function lie on the critical line. Finally, Levinson [18] proved that at least a third of the nontrivial zeros lie on the critical line. Still the problem of the Riemann hypothesis is open and fascinating and teases the best contemporary minds.

PROBLEMS

1. Give a proof that there are infinitely many primes, using the fact that $\zeta(2) = \pi^2/6$ and is not rational.

2. Use partial summation to prove that for $\sigma > 1$,

$$-\frac{\zeta'(s)}{\zeta(s)} = s\int_1^\infty \psi(x)x^{-s-1}\,dx.$$

   (Hint: By Theorem 9 and Definition 6.10 (3),

$$-\frac{\zeta'(s)}{\zeta(s)} = \sum_{n=1}^\infty \frac{\Lambda(n)}{n^s} = \sum_{n=2}^\infty \frac{\psi(n) - \psi(n-1)}{n^s}.\Bigg)$$

3. (i) Use partial summation to show that for $\sigma > 1$,

$$\frac{1}{\zeta(s)} = \sum_{n=1}^\infty \frac{\mu(n)}{n^s} = s\int_1^\infty \frac{M(x)}{x^{s+1}}\,dx.$$

   (ii) Let A be such that $M(x) = O(x^A)$ for $x \to \infty$; show that the integral converges for $\sigma > A$ and uniformly so for $\sigma \geq A + \varepsilon$ ($\varepsilon > 0$ arbitrarily small).

   (iii) Use parts (i) and (ii) to show that $M(x) = O(x^A) \Rightarrow \zeta(s) \neq 0$ for $\sigma > A$.

4. Prove: For $\sigma \geq 2$, $|\zeta(s)| \leq \zeta(2)$.

5. Prove that $\lim_{k\to\infty}(\sum_{n=1}^k n^{-1} - \log k)$ exists and is finite.

6. Give complete details in the proof of Corollary 6.2. In particular, show how the constants ε' and η that enter the proof have to be determined.

7. Make a sketch showing the "critical strip" $0 \leq \sigma \leq 1$, the "critical line" $\sigma = \frac{1}{2}$, and the zero-free region guaranteed by Corollary 13.2 and Theorem 7.

8. Write out in detail the proof of Corollary 13.3 for $s = 1$.

★9. Set $\chi(s) = \frac{1}{2}s(s-1)\pi^{-s/2}\Gamma(s/2)\zeta(s)$.

   (i) Show that $\chi(s)$ is analytic for every complex s. In other words, show that it has no singular points.
   (Hint: $\zeta(-2n) = 0$ for $n \in \mathbf{Z}^+$; also (see [28] pp. 55 ff), $\lim_{z\to -n}(z + n)\Gamma(z) = (-1)^n/n!$. Study carefully the cases $s = 0$ and $s = 1$.)

   (ii) Prove that $\chi(s)$ satisfies the functional equation $\chi(s) = \chi(1-s)$.
   (Hint: Use Theorem 7.)

   (iii) Use part (i) to find directly $\chi(0)$, $\chi(1)$, and $\chi(2)$; check that $\chi(0) = \chi(1)$ (what is the common value?) as required by part (ii), and find $\chi(-1)$ without further computations.

10. Show that $\lim_{t\to\infty}\sum_{n=1}^\infty t/(n^2 + t^2) = \pi/2$.

BIBLIOGRAPHY

1. L. Ahlfors, *Complex Analysis*, 2nd ed. N.Y.: McGraw-Hill, 1966.

2. R. Apéry, Irrationalité de $\zeta(2)$ et $\zeta(3)$, *Astérisque* **61** (1979) 11–13. See also A. van der Poorten, *Mathem. Intelligencer* **1** (1979) 195–207.

3. T. M. Apostol, *Mathematical Analysis*. Reading, Mass.: Addison-Wesley, 1957.

4. R. J. Backlund, Sur les zéros de la fonction $\zeta(s)$ de Riemann, *Comptes Rendus de l'Acad. Sci.* (Paris) **158** (1914) 1979–1981.

5. M. B. Barban, *Doklady Akademii Nauk UzSSr* **8** (1961) 9–11.

6. R. P. Brent, *Mathematics of Computation* **33** (1979) 1361–1372.

7. V. Brun, Le crible d'Eratosthène et le théorème de Goldbach, *Norske Videnskaps–selskapets Skrifter I*, No. 3. Kristiana (Oslo), 1920.

8. V. Brun, La série $\frac{1}{5} + \frac{1}{7} + \frac{1}{11} + \frac{1}{13} + \cdots$ est convergente ou finie, *Bull. des Sciences Math.* (2), **43** (1919) 100–104, 124–128.

9. Chen, Jing-run, On the representation of a large, even integer as the sum of a prime and the product of at most two primes, *Sci. Sinica* **16** (1973) 157–176.

10. H. M. Edwards, *Riemann's Zeta Function*. New York: Academic Press, 1974.

11. L. Euler, *Institutiones Calculi Differentialis*, Pt. 2, Chapters 5 and 6. St. Petersburg: Acad. Imper. Scient. Petropolitanae, 1755; *Opera Omnia* (1), vol. 10.

12. J. P. Gram, Note sur les zéros de la fonction $\zeta(s)$ de Riemann, *Acta Math.* **27** (1903) 289–304.

13. G. H. Hardy, Sur les zéros de la fonction $\zeta(s)$ de Riemann, *Comptes Rendus de l'Acad. des Sci.* (Paris) **158** (1914) 1012–1014.

14. J. I. Hutchinson, On the zeros of Riemann zeta function, *Transactions of the Amer. Math. Soc.* **27** (1925) 49–60.

15. D. Jackson, *Fourier series and Orthogonal Polynomials*, *Carus Monograph*, No. 6. Menasha, Wisc.: G. Banta, 1941.

16. D. H. Lehmer, On the roots of the Riemann zeta function, *Acta Mathem.* **95** (1956) 291–298.

17. D. H. Lehmer, Extended computation of the Riemann zeta function, *Mathematika* **3** (1956) 102–108.

18. N. Levinson, More than a third of zeros of Riemann's zeta function are on $\sigma = 1/2$, *Advances in Mathematics* **13** (1974) 383–436.

19. E. Meissel, *Mathem. Annalen* **2** (1870) 636–642; **3** (1871) 523–525; **25** (1885) 251–257.

20. M. Mikolás, Differentiation and integration of complex order..., *Acta Mathematica Acad. Sci. Hungar.* **10** (1959) 77–124.

21. H. Rademacher, Beiträge zur Viggo Brunschen Methode in der Zahlentheorie, *Abhandlungen aus dem Math. Seminar der Hamburger Univ.* **3** (1924) 12–30.

22. A. Rényi, On the representation of even integers as sum of a prime and an almost prime, *Izvestia Akad. Nauk SSSR, Ser. Mat.* **12** (1948) 57–78; AMS Translation. Series 2, vol. 19 (1962) 299–321.

23. B. Riemann, *Collected Works of B. Riemann*, edited by H. Weber, 2nd ed. (1892/1902). New York: Dover Publishing, 1953.

24. J. B. Rosser, J. M. Yohe, L. Schoenfeld, *Information Processing, 1968*. Proc. IFIP Congress Edinburgh, 1968, vol. 1, pp. 70–76. Amsterdam: North Holland, 1969.

25. A. Selberg, On the zeros of the Riemann zeta function on the critical line, *Arch. for Math. og Naturv.* **45** (1942) 101–114.

26. A. Selberg, On an elementary method in the theory of primes, *Norske Vid. Selsk. Forh. Trondheim*, (No. 18) **19** (1947) 64–67.

27. A. Selberg, The general sieve method...in prime number theory, *Proceedings of the International Congress of Mathematicians*, Cambridge, Mass., **1** (1950) 286–292.

28. E. C. Titchmarsh, *The Theory of Functions*, 2nd ed. Oxford: Clarendon Press, 1939.

29. E. C. Titchmarsh, *The Theory of the Riemann Zeta Function*. Oxford: Clarendon Press, 1951.

30. Wang Yuan, Several papers on the representation of large integers as a sum of a prime and an almost prime; in particular, *Acta Math. Sinica* **6** (1956) 565–582; *Acta Math. Sinica* **10** (1960) 168–181; *Sciencia Sinica* **11** (1962) 1033–1054 (this is essentially an English translation of the previous paper, plus a most interesting Appendix).

31. D. V. Widder, *Advanced Calculus*, 2nd ed. Englewood Cliffs, N.J.: Prentice-Hall, 1961.

Chapter 9

The Prime Number Theorem

1 INTRODUCTION

As already mentioned in Section 1.4, a considerable part of the theory of functions of a complex variable owes its existence to the efforts made to prove the PNT. It was a resounding success when, at the end of the 19th century, these efforts were finally successful. Since then, sporadic attempts have been made to prove the PNT by purely arithmetic reasonings; in fact, the long and complicated analytic proofs of the simple[†]

Theorem 1. $\pi(x) \sim x/\log x,$

concerning the density of the set of primes, could not be considered satisfactory. However, shortly after Hadamard [7] and de la Vallée-Poussin [13] showed that (i) $\{\zeta(1 + it) \neq 0\}$; and (ii) $\{\zeta(1 + it) \neq 0\} \Rightarrow \{\pi(x) \sim x/(\log x)\}$, it was proven that the converse is also true, namely that $\{\pi(x) \sim x/(\log x)\}$ $\Rightarrow \{\zeta(1 + it) \neq 0\}$. This fact, that the PNT implies the nonvanishing of the zeta function on the line of abscissa $\sigma = 1$, was widely interpreted as meaning that it is futile to try to find an elementary proof of the PNT. Indeed, if $\zeta(1 + it) \neq 0$ were not true, neither would the PNT hold; hence (so the reasoning went) one cannot hope to prove the PNT without using at some stage the fact that $\zeta(1 + it) \neq 0$, and this meant using the theory of functions of a complex variable, as applied to Riemann's zeta function.

It was therefore a great achievement when, in 1949, Selberg [11] and Erdös [5] succeeded in proving the PNT by purely "elementary" arguments. Here "elementary" has to be understood in the technical sense of avoiding the use of complex variables, Fourier analysis, and similar "nonelementary" methods, but it should not be confused with "easy." In fact, the first published proofs were quite difficult. Since then these elementary proofs have been much

---

[†] The reader is reminded that $f(x) \sim g(x)$ means $\lim_{x \to \infty} f(z)/g(x) = 1$.

improved. Several versions have appeared in books (see [8] pp. 359–367, [10] pp. 275–297, and [6] pp. 57–88) in highly polished and very readable form. On the other hand, it has lately become possible to improve these "elementary" proofs from another point of view as well. In their older version, they gave just the asymptotic equality $\pi(x) \sim x/(\log x)$ without any estimate of the so-called error term $E(x) = \pi(x) - x/(\log x)$. At present it is possible to also give estimates for $E(x)$, which approach in accuracy those obtainable by analytic methods (see [2], [3], [9], [14], and [4]). It should be added, however, that these recent versions of "elementary" proofs become again quite sophisticated and that the sharpest results at present are still obtained by analytic methods.

The proof of the PNT that follows has been selected among the numerous available ones mainly because of its transparency. It seems to me that the basic idea of this proof is so simple that it will be grasped without difficulty by every reader who was able to reach the present chapter. Some details might be tedious, but even if some reader should not succeed in unraveling every computation to his own entire satisfaction, he should still have a fairly clear idea of the general scheme of the proof. This scheme is due presumably to Titchmarsh, who published a proof based on it in his justly famous book on the Riemann zeta function [12].

2   SKETCH OF THE PROOF

We already saw in Section 8.4 that for $\sigma > 1$,

$$\frac{\log \zeta(s)}{s} = \int_2^\infty \frac{\pi(x)}{x(x^s - 1)} dx. \tag{1}$$

The function $\log \zeta(s)$ is defined as the principal branch (real for real $s > 1$), and is a well defined, single-valued analytic function for $\sigma \geq 1, s \neq 1$, by Theorem 8.2 and Corollaries 8.23 and 8.9.1. For later use we recall that $\log \zeta(\sigma - it) = \overline{\log \zeta(\sigma + it)}$.

The integral in (1) looks almost like the Mellin transform (see [12] p. 33) $\int_2^\infty \pi(x) x^{-s-1} dx$ of $\pi(x)$; actually, we shall see that it is the Mellin transform not of $\pi(x)$ but of the closely related function

$$f(x) = \sum_{m=1}^\infty \frac{1}{m} \pi(x^{1/m}) = \pi(x) + \tfrac{1}{2}\pi(x^{1/2}) + \tfrac{1}{3}\pi(x^{1/3}) + \cdots .$$

Although this is formally an infinite series, it actually reduces to a finite sum, because if $q = [(\log x)/(\log 2)]$, then for $m > q$, $x^{1/m} < 2$ and $\pi(x^{1/m}) = 0$; hence, all but q terms of the series are zero. The difference between $f(x)$ and

$\pi(x)$ is comparatively small; indeed

$$f(x) - \pi(x) = \sum_{m=2}^{q} \frac{1}{m} \pi(x^{1/m}) \le \sum_{m=2}^{q} \frac{1}{m} x^{1/m} \le (q-1) \cdot \tfrac{1}{2} x^{1/2}$$

$$< \frac{1}{2\log 2} x^{1/2} \log x < x^{1/2} \log x.$$

We also note for later use that

$$f(x) = 0 \quad \text{for } x < 2, \tag{2}$$

$$0 \le f(x) < 2x, \tag{3}$$

$$\pi(x) = f(x) + O(x^{1/2} \log x). \tag{4}$$

Proofs of (2), (3), and (4) are left to the reader.

Once we show that $\int_2^\infty \{\pi(x)/x(x^s - 1)\}\, dx = F(s)$ is the Mellin transform of $f(x)$, i.e., that $F(s) = \int_1^\infty f(x) x^{-s-1}\, dx$, (1) becomes

$$\frac{\log \zeta(s)}{s} = \int_1^\infty f(x) x^{-s-1}\, dx \qquad (\sigma > 1). \tag{1'}$$

This equation may be inverted, that is solved for $f(x)$. Indeed, it is known (see [12] p. 33) that if $F(s) = \int_0^\infty f_0(x) x^{s-1}\, dx$, then $f_0(x) = (2\pi i)^{-1} \lim_{T \to \infty} \int_{\sigma - iT}^{\sigma + iT} F(s) x^{-s}\, ds$. This means, in particular, in the present case

$$f(x) = \frac{1}{2\pi i} \lim_{T \to \infty} \int_{c-iT}^{c+iT} \frac{x^s}{s} \log \zeta(s)\, ds, \tag{5}$$

with any $c > 1$. Assuming that we can explicitly evaluate the integral in (5), our problem is completely solved by (4) with only an additional error term $O(x^{1/2} \log x)$.

One may attempt to compute the integral in (5) as follows: One first may select c to be arbitrarily close to one; then $s \cong 1 + it$ and it seems plausible that the major contribution to the integral comes from values of $t \cong 0$, that is from $s \cong 1$. This suggests setting $h(s) = (s-1)\zeta(s)$, so that $\log \zeta(s) = -\log(s-1) + \log h(s)$, because $\log(s-1)$ is a comparatively simple function and we know from Theorem 8.2, Corollary 8.2.3, Theorem 8.5, and Theorem 8.10 that $h(s)$ is analytic and different from zero for $\sigma \ge 1$, so that there (the principal branch of) $\log h(s)$ is a well-defined single-valued analytic function. Substituting this in (5), we obtain

$$f(x) = \frac{-1}{2\pi i} \lim_{T \to \infty} \int_{c-iT}^{c+iT} \frac{x^s}{s} \log(s-1)\, ds$$

$$+ \frac{1}{2\pi i} \lim_{T \to \infty} \int_{c-iT}^{c+iT} \frac{x^s}{s} \log h(s)\, ds, \qquad (c > 1). \tag{5'}$$

The first integral can be computed explicitly (the operations closely parallel

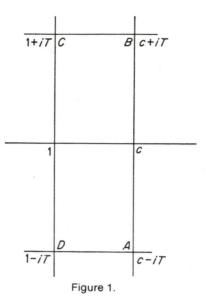

Figure 1.

those we shall perform in Section 4) and we find as the value of the first summand $x/(\log x) + o(x/(\log x))$; however the second integral leads to difficulties. We may start by "moving" the line of integration from $\sigma = c > 1$ to $\sigma = 1$; this is legitimate because inside the rectangular contour of vertices $1 \pm iT, c \pm iT$ the integrand is analytic and tends to zero along the horizontal lines as $T \to \infty$.

By Cauchy's Theorem ([1] p. 510), $\int_{AB} + \int_{BC} + \int_{CD} + \int_{DA} = 0$, hence $\int_{AB} = \int_{AD} + \int_{DC} + \int_{CB}$, and for $T \to \infty$, $\int_{AD} \to 0$, $\int_{CB} \to 0$ so that $\lim_{T \to \infty} \int_{c-iT}^{c+iT} = \lim_{T \to \infty} \int_{1-iT}^{1+iT}$. Consequently, the second summand in (5′) equals

$$\frac{1}{2\pi i} \lim_{T \to \infty} \int_{-T}^{T} \frac{x^{1+it}}{1 + it} \log h(1 + it) \, d(1 + it)$$

$$= \frac{x}{2\pi} \lim_{T \to \infty} \int_{-T}^{T} \frac{\log h(1 + it)}{1 + it} x^{it} \, dt.$$

We could now easily finish the proof by using some form of the Riemann-Lebesgue Theorem (see Lemma 1 in Section 3), if we could show, for instance, that $\lim_{T \to \infty} \int_{-T}^{T} |x^{it}(\log h(1 + it))/(1 + it)| \, dt$ exists and is finite. But actually,

$$\log h(1 + it) = \log \{ it\zeta(1 + it) \} = \log(it) + \log \zeta(1 + it),$$

so that the last integral does *not* approach a finite limit for $T \to \infty$. It would be easy to handle the situation if instead of s we had s^2 in the denominator in (5′), because this would be sufficient to insure the absolute convergence of the integral. Therefore we shall return for a moment to (1') and rewrite it in a slightly modified form that will give us the desired extra factor s in the denominator.

We define the function $g(x) = \int_1^x (f(u)/u)\,du$. One observes that $f(x)$ and $g(x)$ (which may be considered as some sort of average of the values of $f(x)$) have many features in common. So for instance, from $f(x) = 0$ for $x < 2$ it follows that also $g(x) = 0$ for $x < 2$. Also, an elementary consideration shows that if (as we hope) $f(x) \sim x/(\log x)$, then also $g(x) \sim x/(\log x)$. It is less immediate, but we shall prove it, that if $f(x)$ is *positive* and *nondecreasing* (as it actually is here) then the converse also holds and $g(x) \sim x/(\log x)$ can hold only if already $f(x) \sim x/(\log x)$. This is called a *Tauberian argument*. Using the definition of $g(x)$ and integration by parts, the right side of (1') may be transformed as follows:

$$\int_1^\infty \frac{f(x)}{x} \cdot x^{-s}\,dx = \int_1^\infty g'(x)x^{-s}\,dx = g(x)x^{-s}\Big|_1^\infty + s\int_1^\infty g(x)x^{-s-1}\,dx.$$

The integrated part vanishes for $\sigma > 1$. Indeed, $g(1) = 0$ as seen; also by (3), $g(x) \le \int_1^x (2u/u)\,du < 2x$, so that $\lim_{x\to\infty}|g(x)x^{-s}| \le \lim_{x\to\infty} 2x \cdot x^{-\sigma} = 0$ for $\sigma > 1$. Hence (1') may be written equivalently as $\log \zeta(s)/s = s\int_1^\infty g(x)x^{-s-1}\,dx$ or

$$\frac{\log \zeta(s)}{s^2} = \int_1^\infty g(x)x^{-s-1}\,dx \quad \text{for } \sigma > 1. \tag{1''}$$

Just as we solved (1') for $f(x)$, we may now solve (1'') for $g(x)$ and obtain

$$g(x) = \frac{1}{2\pi i} \lim_{T\to\infty} \int_{c-iT}^{c+iT} \frac{x^s}{s^2} \log \zeta(s)\,ds \, (c > 1). \tag{6}$$

The integral in (6) may be now evaluated just as outlined above, with the result that $g(x) = x/(\log x) + o(x/(\log x))$. From this we shall infer that also $f(x) \sim x/(\log x)$; finally it will then follow by (4) that also $\pi(x) \sim x/(\log x)$, which is precisely the statement of the PNT.

3 SOME LEMMAS

Here we shall formulate with precision and prove two statements alluded to before (a version of the Riemann-Lebesgue Theorem and a simple Tauberian theorem); both are weak forms of more general theorems, but are sufficient for our present purpose. We àlso prove for later use the uniform convergence of two series.

Lemma 1 (Riemann-Lebesgue).

(A) *Let $f(t)$ be differentiable on $(0, \infty)$ and be such that*

   (i) *for any η_0, T_0 satisfying $0 < \eta_0 < T_0 < \infty$, $\int_{\eta_0}^{T_0}|f'(t)|\, dt$ exists and is finite; and*

   (ii) $\lim\limits_{\substack{\eta \to 0 \\ T \to \infty}} \int_{\eta}^{T}|f(t)|\, dt$ *exists and is finite.*

   Then $G(y) = \lim\limits_{\substack{\eta \to 0 \\ T \to \infty}} \int_{\eta}^{T} f(t)e^{ity}\, dt$ exists for every real y and $G(y) = o(1)$

   for $y \to \pm\infty$.

(B) *Similarly, let $f(t)$ be differentiable and absolutely integrable over $(-\infty, \infty)$; then $\lim_{y \to \infty}\int_{-\infty}^{\infty} f(t)e^{ity}\, dt = 0$.*

PROOF.   Given $\varepsilon > 0$ arbitrarily small, we determine constants η_0 and T_0 such that for $0 < \eta_1 \le \eta_2 \le \eta_0$ and $T_0 \le T_1 \le T_2 < \infty$, one should have

$$\left| \int_{\eta_1}^{\eta_2} f(t)e^{ity}\, dt \right| \le \int_{\eta_1}^{\eta_2}|f(t)|\, dt < \varepsilon/3$$

and

$$\left| \int_{T_1}^{T_2} f(t)e^{ity}\, dt \right| \le \int_{T_1}^{T_2}|f(t)|\, dt < \varepsilon/3.$$

This is possible because, by assumption, $\lim\limits_{\substack{\eta \to 0 \\ T \to \infty}} \int_{\eta}^{T}|f(t)|\, dt$ exists; this proves the existence of $G(y)$.

Once η_0 and T_0 have been determined, we integrate by parts and find that

$$\left| \int_{\eta_0}^{T_0} f(t)e^{ity}\, dt \right| = \left| \frac{1}{iy} e^{ity}f(t)\Big|_{\eta_0}^{T_0} - \frac{1}{iy} \int_{\eta_0}^{T_0} f'(t)e^{ity}\, dt \right|$$

$$\le \frac{1}{|y|}\left\{ |f(T_0)| + |f(\eta_0)| + \int_{\eta_0}^{T_0}|f'(t)|\, dt \right\}.$$

The bracket being independent of y, we may select $|y|$ so large that $|y|^{-1}\{\cdots\} < \varepsilon/3$; hence

$$|G(y)| = \left| \lim\limits_{\substack{\eta \to 0 \\ T \to \infty}} \int_{\eta}^{T} f(t)e^{ity}\, dt \right|$$

$$\le \lim\limits_{\eta \to 0}\int_{\eta}^{\eta_0}|f(t)|\, dt + \left| \int_{\eta_0}^{T_0} f(t)e^{ity}\, dt \right| + \lim\limits_{T \to \infty}\int_{T_0}^{T}|f(t)|\, dt < \varepsilon,$$

and Lemma 1(A) is proven.

For the proof of Part (B), observe that from $e^{iy} = \cos y + i \sin y$ it follows that it is sufficient to show that $\lim_{y \to \infty} \int_{-\infty}^{\infty} f(t) \cos yt \, dt = \lim_{y \to \infty} \int_{-\infty}^{\infty} f(t) \sin yt \, dt = 0$.

The proof is essentially the same for both integrals, so that it is sufficient to consider only one of them, say the first. By the absolute integrability of $f(t)$, we can find T such that $|\int_T^{\infty} f(t) \cos yt \, dt| \le \int_T^{\infty} |f(t) \cos yt| \, dt \le \int_T^{\infty} |f(t)| \, dt < \varepsilon/3$. Similarly, $|\int_{-\infty}^{-T} f(t) \cos yt \, dt| < \varepsilon/3$. Over the finite interval $(-T, T)$, $|\int_{-T}^{T} f(t) \cos yt \, dt| \le |y^{-1} f(t) \sin yt]_{-T}^{T} - y^{-1} \int_{-T}^{T} f'(t) \sin ty \, dt| \le y^{-1} \{|f(T)| + |f(-T)| + \int_{-T}^{T} |f'(t)| \, dt\} = y^{-1} H(T)$, say. As $H(T)$ is independent of y, we can select y_0 so large that for $|y| \ge y_0$, $|y|^{-1} H(T) < \varepsilon/3$, and this finishes the proof of the lemma.

Lemma 2 (Tauberian).   *Let $f(x)$ be positive and non-decreasing and set $g(x) = \int_1^x (f(u)/u) \, du$; if $g(x) \sim x/(\log x)$, then also $f(x) \sim x/(\log x)$.*

PROOF.   Given any $\varepsilon > 0$, set $y = x(1 + \varepsilon)$. By the assumption on $g(x)$, for sufficiently large x,

$$(1 - \varepsilon^2)(x/(\log x)) < g(x) < (1 + \varepsilon^2)(x/(\log x)),$$

with similar inequalities for $g(y)$. Hence

$$g(y) - g(x) < (1 + \varepsilon^2)\frac{y}{\log y} - (1 - \varepsilon^2)\frac{x}{\log x}$$

$$< \frac{1}{\log x}\{(1 + \varepsilon^2)y - (1 - \varepsilon^2)x\}$$

$$= \frac{x}{\log x}\{(1 + \varepsilon^2)(1 + \varepsilon) - 1 + \varepsilon^2\} = \frac{x}{\log x}\varepsilon(1 + \varepsilon)^2.$$

On the other hand, $f(x)$ does not decrease, so that

$$g(y) - g(x) = \int_x^y \frac{f(u)}{u} \, du \ge f(x) \int_x^y \frac{du}{u} = f(x)\log\frac{y}{x} = f(x)\log(1 + \varepsilon);$$

consequently,

$$f(x) \le \frac{g(y) - g(x)}{\log(1 + \varepsilon)} \le \frac{x}{\log x}\frac{\varepsilon(1 + \varepsilon)^2}{\log(1 + \varepsilon)} \le \frac{x}{\log x}(1 + \varepsilon)^3.$$

In exactly the same way, considering $g(x) - g(z)$ with $z = x(1 - \varepsilon)$, one shows that $f(x) \ge (x/(\log x))(1 - \varepsilon)^3$ for arbitrarily small $\varepsilon > 0$, provided only that x is large enough. Both inequalities together show that $\lim_{x \to \infty}(\log x/x)f(x) = 1$, and the Lemma is proven.

Lemma 3.   *For constant $\sigma \ge 1$ and $2 \le x < \infty$, the series $\sum_{m=1}^{\infty} \pi(x)x^{-ms-1}$ is uniformly convergent.*

PROOF.   In absolute value, the terms of this series are less than $x \cdot x^{-m\sigma-1} = x^{-m\sigma} \leq (2^{-\sigma})^m \leq 2^{-m}$. The series $\sum_{m=1}^{\infty} 2^{-m}$ is convergent and independent of x, and the Lemma is proven.

REMARK 1.   The proof shows that the convergence of the series is uniform with respect to *both* variables x and s; only the uniformity with respect to x will be used.

Lemma 4.   *For constant $\sigma \geq 1$ and $1 \leq x < \infty$, the series*

$$\sum_{m=1}^{\infty} m^{-1}\pi(x^{1/m})x^{-s-1}$$

is uniformly convergent.

PROOF.   We have to show that, given $\varepsilon > 0$ arbitrarily small, we can determine $M = M(\varepsilon)$, but independent of x, such that if $M_2 \geq M_1 \geq M$, then $R = |\sum_{M_1 < m \leq M_2} m^{-1}\pi(x^{1/m})x^{-s-1}|$ satisfies $R < \varepsilon$. To avoid trivial difficulties, we shall assume from the start, without loss of generality, that $\varepsilon \leq 1/2$. Then we shall show that it is sufficient to take $M \geq \varepsilon^{-1}$.

Set $q = [(\log x)/(\log 2)]$; then all terms with $m > q$ vanish. Hence if $M > q$, then $R = 0$; on the other hand, if $M \leq q$, then

$$R \leq \left| \sum_{M_1 < m \leq q} \frac{1}{m} x^{(1/m)-s-1} \right| \leq \sum_{M_1 < m \leq q} \frac{1}{m} x^{(1/m)-\sigma-1}$$

$$\leq \sum_{M_1 < m \leq q} \frac{1}{M_1} x^{(1/M_1)-2}$$

$$= \frac{1}{M_1} x^{(1/M_1)-2}(q - M_1) < \frac{q}{M} x^{(1/M)-2} \leq \frac{\log x}{M \log 2} x^{-3/2}$$

$$= \frac{1}{M \log 2} G(x),$$

with $G(x) = x^{-3/2}\log x$, use having been made of the inequalities $\sigma \geq 1$ and $M_1 \geq M \geq \varepsilon^{-1} \geq 2$.

However, $\max_{1 \leq x < \infty} G(x) = 2/3e$; hence

$$R \leq \frac{2}{3e \log 2} \frac{1}{M} < \frac{1}{M} \leq \varepsilon,$$

because $2/(3e \log 2) < 1$ and $M \geq 1/\varepsilon$, and the Lemma is proven.

REMARK 2.   It follows from this proof that the series converges uniformly in both variables, s and x, within the stated ranges. Only the uniformity of the convergence with respect to x will be needed in what follows.

4   PROOF OF THE PNT

We shall once more go over the complete proof, giving the justifications of those steps that were skipped over in the sketch of the proof (Section 2). Equation (1) was proven in Section 8.5. To obtain (1') we have to present the

PROOF OF $F(s) = \int_1^\infty f(x)x^{-s-1}\,dx = \int_2^\infty \{\pi(x)/x(x^s - 1)\}\,dx$. By its definition, $F(s) = \int_1^\infty (\sum_{m=1}^\infty m^{-1}\pi(x^{1/m})x^{-s-1})\,dx$. By Lemma 4, the series is uniformly convergent; hence (see [1] p. 451) termwise integration is legitimate at least for real $s \geq 1$, when all terms are positive, and

$$F(s) = \sum_{m=1}^\infty \int_1^\infty m^{-1}\pi(x^{1/m})x^{-s-1}\,dx.$$

We now set $x^{1/m} = y$ and obtain

$$\int_1^\infty \frac{1}{m}\pi(x^{1/m})x^{-s-1}\,dx = \int_1^\infty \frac{1}{m}\pi(y)y^{-ms-m}\cdot my^{m-1}\,dy$$

$$= \int_1^\infty \pi(y)y^{-ms-1}\,dy,$$

so that $F(s) = \sum_{m=1}^\infty \int_1^\infty \pi(y)y^{-ms-1}\,dy = \sum_{m=1}^\infty \int_2^\infty \pi(x)x^{-ms-1}\,dx$, because the integrands vanish for $y < 2$, and because it is immaterial what symbol we use for the dummy variable of integration. On the other hand,

$$\int_2^\infty \frac{\pi(x)}{x(x^s - 1)}\,dx = \int_2^\infty \frac{\pi(x)}{x^{s+1}}\frac{dx}{1 - x^{-s}}$$

$$= \int_2^\infty \frac{\pi(x)}{x^{s+1}}\left(\sum_{m=0}^\infty x^{-ms}\right)dx = \int_2^\infty \left(\sum_{m=1}^\infty \pi(x)x^{-ms-1}\right)dx.$$

By Lemma 3, the series is uniformly convergent. Furthermore for real s, all its terms are positive; hence, as already observed, we may integrate termwise. Thus we obtain $\int_2^\infty \{\pi(x)/x(x^s - 1)\}\,dx = \sum_{m=1}^\infty \int_2^\infty \pi(x)x^{-ms-1}\,dx = F(s)$, at least for real $s > 1$. We now observe that both sides of this equality are analytic functions of s, at least for $\sigma > 1$, and that they take on the same values at all points $s > 1$ of the real axis. By the principle of analytic continuation (see [1] p. 519) it now follows that the equality holds for all $\sigma > 1$ and (1') is completely justified.

Next, in order to avoid the difficulties with nonabsolutely convergent integrals, we set $g(x) = \int_1^x (f(u)/u)\,du$ and perform an integration by parts in the right member of (1'), obtaining (1'') as already seen. Equation (1'') may be solved for $g(x)$ by using the inversion formula of [12] (s in [12] (2.15.1) corresponds to $-s$ here) and we obtain (6). In (6) we set $\log \zeta(s) = -\log(s - 1) + \log h(s)$, where $h(s) = (s - 1)\zeta(s)$ is analytic and does not vanish for $\sigma \geq 1$. As already observed, the symbol "log" stands in each case for the principal branch, real for real $s > 1$; this insures that $\log \zeta(s)$ and $-\log(s - 1) + \log h(s)$ have the same imaginary part (namely zero) when $s > 1$; by continuity, this equality persists as long as none of the functions involved vanishes, that is at least for $\sigma \geq 1$, $s \neq 1$. Thus we obtain

$$g(x) = I_1(x) + I_2(x), \tag{6'}$$

with

$$I_1(x) = -\frac{1}{2\pi i} \lim_{T \to \infty} \int_{c-iT}^{c+iT} \frac{x^s}{s^2} \log(s - 1)\, ds$$

and

$$I_2(x) = \frac{1}{2\pi i} \lim_{T \to \infty} \int_{c-iT}^{c+iT} \frac{x^s}{s^2} \log h(s)\, ds$$

valid for any real $c > 1$.

We now show successively that $I_2(x) = o(x/(\log x))$ and that

$$I_1(x) = \frac{x}{\log x} + o\left(\frac{x}{\log x}\right).$$

PROOF THAT $I_2(x) = o(x/(\log x))$.   The integrand $x^s s^{-2} \log h(s)$ is analytic for $\sigma \geq 1$; also, for $|t| \to \infty$,

$$\left| \frac{x^s}{s^2} \log h(s) \right| = \left| \frac{x^{\sigma + it}}{(\sigma + it)^2} \log\{(s - 1)\zeta(s)\} \right|$$

$$\leq \frac{x^\sigma}{|t|^2} (\log |t| + |\log \zeta(s)|)$$

$$< \frac{x^\sigma}{|t|^2} \cdot K \log^9 |t|$$

by Corollary 8.13.4. Using Cauchy's Theorem, the integral around the rectangle of vertices $1 \pm iT$, $c \pm iT$ vanishes, and the two integrals along the horizontal segments contribute together less than

$$2(c - 1) \max_{1 \leq \sigma \leq c} \left| \frac{x^{\sigma + iT}}{(\sigma + iT)^2} \log h(\sigma + iT) \right| \leq 2K(c - 1)\frac{x^c}{T^2} \log^9 T,$$

which is arbitrarily small if T is taken sufficiently large; hence

$$I_2(x) = \frac{1}{2\pi i} \lim_{T \to \infty} \int_{1-iT}^{1+iT} \frac{x^s}{s^2} \log h(s)\, ds$$

$$= \frac{1}{2\pi} \lim_{T \to \infty} \int_{-T}^{T} \frac{x^{1+it}}{(1+it)^2} \log h(1+it)\, dt = \frac{x}{2\pi} I_3(\log x),$$

with

$$I_3(y) = \lim_{T \to \infty} \int_{-T}^{T} \frac{\log h(1+it)}{(1+it)^2} e^{iyt}\, dt.$$

Integrating by parts,

$$\int_{-T}^{T} \frac{\log h(1+it)}{(1+it)^2} e^{iyt}\, dt$$

$$= \frac{e^{iyt} \log h(1+it)}{iy\,(1+it)^2} \Bigg|_{-T}^{T}$$

$$- \frac{1}{iy} \int_{-T}^{T} \frac{\left\{ i(1+it) \dfrac{h'(1+it)}{h(1+it)} - 2i \log h(1+it) \right\} e^{ity}\, dt}{(1+it)^3}.$$

As we already recalled, $|\log h(1+it)| = O(\log^9 t)$; hence taking the limit for $T \to \infty$, the integrated term vanishes. Also, by Theorem 8.13,

$$\frac{h'}{h}(s) = \frac{d}{ds} \log h(s) = \frac{1}{s-1} + \frac{\zeta'}{\zeta}(s) = O(\log^9 t)$$

for $s = 1 + it$ and t sufficiently large, say $t \geq t_0$. However, for $\sigma \geq 1$, $1/(s-1) + \zeta'/\zeta(s)$ is analytic (see Corollary 8.13.3) so that $|1/(s-1) + \zeta'/\zeta(s)|$ is actually bounded by some constant if $s = 1 + it$, $|t| \leq t_0$. Consequently, the numerator in the last integral is $O(t \log^9 t)$, the integral satisfies the conditions of Lemma 1(B), and for $s \to \infty$,

$$\lim_{T \to \infty} \int_{-T}^{T} \cdots\, dt = o(1); \quad \text{hence} \quad \frac{1}{-iy} \lim_{T \to \infty} \int_{-T}^{T} \cdots\, dt = o\left(\frac{1}{y}\right).$$

It follows that $I_3(y) = o(1/y)$, $I_3(\log x) = o(1/(\log x))$, and $I_2(x) = (x/2\pi) I_3(\log x) = o(x/(\log x))$ as claimed.

EVALUATION OF $I_1(x)$. In a certain sense, this is the crux of the proof and yields the principal term. Once more we would like to move the line of integration from $\sigma = c > 1$ to $\sigma = 1$. But now we have to be careful, because $\log(s-1)$ is not analytic at $s = 1$. Hence we consider the contour $ABCDEFG$,

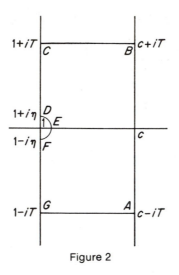

Figure 2

consisting of a rectangle with a small semicircular indentation (DEF in Fig. 2) around $s = 1$. Inside this contour the integrand $s^{-2}x^s\log(s - 1)$ is analytic; hence by Cauchy's Theorem the integral along the contour vanishes:

$$\int_{AB} + \int_{BC} + \int_{CD} + \int_{DEF} + \int_{FG} + \int_{GA} = 0,$$

or

$$\int_{c-iT}^{c+iT} = \int_{c-iT}^{1-iT} + \int_{1-iT}^{1-i\eta} + \int_{FED} + \int_{1+i\eta}^{1+iT} + \int_{1+iT}^{c+iT}.$$

For $T \to \infty$, the integrals $\int_{c-iT}^{1-iT}$ and $\int_{1+iT}^{c+iT}$ along AG and CB, respectively, go to zero as in the evaluation of $I_2(x)$. Therefore,

$$\lim_{T \to \infty} \int_{c-iT}^{c+iT} = \lim_{T \to \infty} \left\{ \int_{1-iT}^{1-i\eta} + \int_{FED} + \int_{1+i\eta}^{1+iT} \right\}. \tag{7}$$

We claim that $J = |\int_{FED}| \to 0$ for $\eta \to 0$. Indeed,

$$J = \left| \int_{-\pi/2}^{\pi/2} \frac{x^{1+\eta e^{i\theta}}}{(1 + \eta e^{i\theta})^2} \log(\eta e^{i\theta}) \cdot \eta i e^{i\theta} \, d\theta \right|$$

$$\leq \frac{x^{1+\eta}}{(1 - \eta)^2} \cdot \eta \int_{-\pi/2}^{\pi/2} |\log(\eta e^{i\theta})| \, d\theta$$

$$< \frac{x^{1+\eta}}{(1 - \eta)^2} \eta \int_{-\pi/2}^{\pi/2} \left(\log \frac{1}{\eta} + |\theta| \right) d\theta = \frac{x^{1+\eta}}{(1 - \eta)^2} \eta \left(\pi \log \frac{1}{\eta} + \frac{\pi^2}{4} \right),$$

which can be made arbitrarily small because $\lim_{\eta \to 0} \eta \log(1/\eta) = 0$. Consequently by (7),

$$\lim_{T \to \infty} \int_{c-iT}^{c+iT} \frac{x^s}{s^2} \log(s-1)\, ds = \lim_{\eta \to 0} \lim_{T \to \infty} \left\{ \int_{1-iT}^{1-i\eta} + \int_{1+i\eta}^{1+iT} \frac{x^s}{s^2} \log(s-1)\, ds \right\}$$

$$= i \lim_{\eta \to 0} \lim_{T \to \infty} \left\{ \int_{-T}^{-\eta} + \int_{\eta}^{T} \frac{x^{1+it}}{(1+it)^2} \log(it)\, dt \right\}$$

and

$$I_1(x) = -\frac{x}{2\pi} I_4(x), \quad \text{with} \quad I_4(x) = \lim_{\eta \to 0} \lim_{T \to \infty} \left\{ \int_{-T}^{-\eta} + \int_{\eta}^{T} \frac{x^{it} \log(it)}{(1+it)^2}\, dt \right\}.$$

The first integral is the complex conjugate of the second. We may show this, for instance, by observing that the integrand $(s^{-2} x^s \log(s-1))$ is real for real $s > 1$, or directly, by replacing t by $-t$ in the first integral. Consequently it is sufficient to compute, say

$$I_5(y) = I_5(y; \eta) = \lim_{T \to \infty} \int_{\eta}^{T} \frac{x^{it} \log(it)}{(1+it)^2}\, dt = \lim_{T \to \infty} \int_{\eta}^{T} \frac{\log(it)}{(1+it)^2} e^{ity}\, dt$$

(where $y = \log x$), then take its complex conjugate, and we obtain

$$I_4(x) = I_4(e^y) = \lim_{\eta \to 0} \left\{ I_5(y) + \overline{I_5(y)} \right\}. \tag{8}$$

It remains to compute $I_5(y)$, and we do it integrating by parts; with $\log i = i\pi/2$, we obtain

$$\int_{\eta}^{T} \frac{\log t + i\pi/2}{(1+it)^2} e^{ity}\, dt$$

$$= \frac{e^{ity}}{iy} \frac{\log t + i\pi/2}{(1+it)^2} \bigg|_{\eta}^{T} - \frac{1}{iy} \int_{\eta}^{T} \frac{t^{-1}(1+it) - 2i(\log t + i\pi/2)}{(1+it)^3} e^{ity}\, dt$$

$$= -\frac{1}{iy} \left\{ \frac{(\log \eta + i\pi/2) e^{i\eta y}}{(1+i\eta)^2} + O\left(\frac{\log T}{T^2} \right) \right.$$

$$+ \int_{\eta}^{T} \frac{e^{ity}}{t(1+it)^3}\, dt + \int_{\eta}^{T} \frac{i + \pi - 2i\log t}{(1+it)^3} e^{ity}\, dt \Bigg\}.$$

Next

$$\int_\eta^T \frac{e^{ity}}{t(1+it)^3}\,dt = \int_\eta^T \frac{e^{ity}}{t}\,dt$$

$$-3\int_\eta^T \frac{i-t}{(1+it)^3}e^{ity}\,dt + i\int_\eta^T \frac{t^2}{(1+it)^3}e^{ity}\,dt.$$

The second integral can be handled by Lemma 1, and it will be easy to show that its contribution to $I_4(e^y)$ is negligible. However Lemma 1 is not immediately applicable to the last integral. Therefore, we first transform it by one more integration by parts, obtaining

$$\int_\eta^T \frac{t^2}{(1+it)^3}e^{ity}\,dt = \frac{t^2}{(1+it)^3}\frac{e^{ity}}{iy}\Bigg|_\eta^T - \frac{1}{iy}\int_\eta^T \frac{2t-it^2}{(1+it)^4}e^{ity}\,dt$$

$$= O\!\left(\frac{1}{Ty}\right) + O\!\left(\frac{\eta^2}{y}\right) - \frac{1}{iy}\int_\eta^T \frac{2t-it^2}{(1+it)^4}e^{ity}\,dt.$$

Putting these terms together and taking the limit for $T \to \infty$, we obtain

$$I_5(y) = -\frac{1}{iy}\left\{\frac{\log\eta + \frac{1}{2}\pi i}{(1+i\eta)^2}e^{i\eta y} + \int_\eta^\infty \frac{e^{ity}}{t}\,dt - 3\int_\eta^\infty \frac{i-t}{(1+it)^3}e^{ity}\,dt\right.$$

$$\left. + \int_\eta^\infty \frac{i+\pi-2i\log t}{(1+it)^3}e^{ity}\,dt - \frac{1}{iy}\int_\eta^\infty \frac{2t-it^2}{(1+it)^4}e^{ity}\,dt + O\!\left(\frac{\eta^2}{y}\right)\right\}.$$

When $\eta \to 0$, the term $O(\eta^2/y)$ vanishes and the last three integrals converge to

$$\int_0^\infty \frac{i-t}{(1+it)^3}e^{ity}\,dt, \quad \int_0^\infty \frac{i+\pi-2i\log t}{(1+it)^3}e^{ity}\,dt, \quad \text{and} \quad \int_0^\infty \frac{2t-it^2}{(1+it)^4}e^{ity}\,dt,$$

respectively; all three satisfy the conditions of Lemma 1 so that for $y \to \infty$ they are all $o(1)$ and contribute to $I_4(e^y)$ terms that are at most $o(1/y)$, $o(1/y)$, and $o(1/y^2)$, respectively.

REMARK 3. Consideration of the second of these integrals should convince the reader that it was essential to formulate Lemma 1 in the generality in which it was done, in particular without restricting $f(t)$ to be bounded on $0 \le t < \infty$.

Adding now to $I_5(y)$ its complex conjugate and taking the limit for $\eta \to 0$, we obtain by (8) successively

$$I_4(e^y) = \frac{1}{y} \lim_{\eta \to 0} \left\{ \left[i(\log \eta + i\pi/2) e^{i\eta y} - i(\log \eta - i\pi/2) e^{-i\eta y} \right] (1 + O(\eta)) \right.$$

$$\left. - \int_\eta^\infty \frac{e^{ity} - e^{-ity}}{it} \, dt \right\} + o\left(\frac{1}{y} \right)$$

$$= \frac{1}{y} \lim_{\eta \to 0} \left\{ (-2 \log \eta \sin \eta y - \pi \cos \eta y)(1 + O(\eta)) \right.$$

$$\left. - 2 \int_\eta^\infty \frac{\sin ty}{t} \, dt \right\} + o\left(\frac{1}{y} \right)$$

$$= -\frac{1}{y} \left\{ \pi + 2 \int_0^\infty \frac{\sin ty}{ty} \, d(ty) \right\} + o\left(\frac{1}{y} \right).$$

The integral is known to be equal to $\pi/2$ (see Problem 12 or [1], Chapter 8, p. 157), so that

$$I_4(e^y) = -\frac{2\pi}{y} + o\left(\frac{1}{y} \right) \quad \text{or} \quad I_4(x) = -\frac{2\pi}{\log x} + o\left(\frac{1}{\log x} \right),$$

and consequently,

$$I_1(x) = -\frac{x}{2\pi} I_4(x) = \frac{x}{\log x} + o\left(\frac{x}{\log x} \right).$$

COMPLETION OF THE PROOF. Going back to (6') we see that we have proven $g(x) = x/(\log x) + o(x/(\log x))$. By Lemma 2 we infer that $f(x) \sim x/(\log x)$. Finally, it follows from (4) that $\pi(x) = f(x) + O(x^{1/2} \log x) = x/(\log x) + o(x/(\log x)) + O(x^{1/2} \log x)$; however, $x^{1/2} \log x = o(x/(\log x))$, so that the last error term may be absorbed into the first, and we have proven that $\pi(x) = x/(\log x) + o(x/(\log x))$, or what is the same, that $\pi(x) \sim x/(\log x)$, the PNT we had set out to prove.

PROBLEMS

1. Prove formally (2), (3), and (4).
2. Justify (5) and (6) by translating appropriately the formulae of [12], quoted just before (5).
3. Compute the first summand in (5') explicitly, show that its value is $x/(\log x) + O(x/(\log^2 x))$. Show that if we know that the second

summand in (5') is $O(x/(\log^2 x))$, it follows beyond the statement of the PNT that $\pi(x) = x/(\log x) + O(x/(\log^2 x))$.

Comment: By working a little harder, we could have shown similarly that $I_1(x) = x/(\log x) + O(x/(\log^2 x))$, but there was no point in doing it, because (i) we were unable to estimate $I_2(x)$ more sharply than $o(x/(\log x))$; and (ii) even if we knew that $g(x) = (x/(\log x)) + O(x/(\log^2 x))$, Lemma 2 does not allow us to infer for $f(x)$ anything better than $f(x) = x/(\log x) + o(x/(\log x))$.

4. Prove that $g(x) = \int_1^x (f(u)/u) \, du$ and $f(x) \sim x/(\log x)$ without any further conditions on $f(x)$ imply that $g(x) \sim x/(\log x)$.

5. Prove the following Tauberian theorem: $g(x) = \int_1^x (f(u)/u) \, du \sim x$ and $f(x) > 0$, nondecreasing $\Rightarrow f(x) \sim x$.

6. Justify the inequality $\varepsilon(1 + \varepsilon)^2/\log(1 + \varepsilon) \le (1 + \varepsilon)^3$ used in the proof of Lemma 2. Within what range of ε does the inequality hold?

7. In the proof of Lemma 2, show that for every $\varepsilon > 0$, $((\log x)/x)f(x) > (1 - \varepsilon)^3$ if x is sufficiently large.

8. In the evaluation of $I_1(x)$, provide a proof that the integrals along the two horizontal segments are less in absolute value than any preassigned $\varepsilon > 0$, if the ordinate T is sufficiently large.

9. Prove: $\max_{1 \le x \le \infty} x^{-3/2} \log x = 2/3e$.

10. Justify among all possible values of $\log i$ the selection $\log i = \pi i/2$ used in the computation of $I_5(y)$.

11. Investigate which of the following limits exist.

(a) $\lim\limits_{T \to \infty} \int_\eta^T \dfrac{e^{ity}}{t} \, dt$;

(b) $\lim\limits_{\eta \to 0} \lim\limits_{T \to \infty} \int_\eta^T \dfrac{e^{ity}}{t} \, dt$;

(c) $\lim\limits_{\eta \to 0} \lim\limits_{T \to \infty} \int_\eta^T \dfrac{e^{ity} - e^{-ity}}{t} \, dt$;

(d) $\lim\limits_{\eta \to 0} \lim\limits_{T \to \infty} \int_\eta^T \dfrac{e^{ity} + e^{-ity}}{t} \, dt$.

12. Use the theorem on residues, as applied to the function $f(z) = e^{iz}/z$, to show that $\int_{-\infty}^{\infty} (\sin z/z) \, dz = \pi$; conclude that $\int_0^{\infty} (\sin z/z) \, dz = \pi/2$. (Hint: Integrate along the real axis, with an indentation (up) at the origin and close contour by a large semicircle in the upper half-plane. Then take imaginary parts and use the fact that $\sin z/z$ is an even function.)

BIBLIOGRAPHY

1. T. M. Apostol, *Mathematical Analysis*. Reading, Mass.: Addison-Wesley, 1957.

2. S. A. Amitsur, Arithmetic linear transformations and abstract prime number theorem, *Canad. J. Math.*, **13**, (1961) 83–109.

3. E. Bombieri, Maggiorazione del resto nel 'Primzahlsatz' col metodo di Erdös-Selberg. *Ist. Lombardo Acad. Sci. Lett. Rend. A.*, **96**, (1962) 343–350.

4. H. G. Diamond and J. Steinig, An elementary proof of the prime number theorem with remainder term, *Inventiones Mathem.* **11** (1970) 199–258.

5. P. Erdös, On a new method in elementary number theory which leads to an elementary proof of the prime number theorem, *Proc. Nat. Ac. Sc.* **35** (1949) 374–384.

6. I. S. Gál, *Lectures on Number Theory*. Minneapolis: Jones Letter Service, 1961.

7. J. Hadamard, Sur la distribution des zéros de la fonction $\zeta(s)$ et ses conséquences arithmétiques, *Bull. Soc. Math. France* **24** (1896) 199–220.

8. G. H. Hardy, and E. M. Wright, *An Introduction to the Theory of Numbers*, 3rd ed. Oxford: The Clarendon Press, 1951.

9. W. B. Jurkat, *Abstracts of Short Communications, International Congress of Mathematicians*, Stockholm, 1962, p. 35.

10. T. Nagell, *Introduction to Number Theory*. New York: Wiley, 1951.

11. A. Selberg, An elementary proof of the prime-number theorem. *Annals of Mathematics* **50** (1949) 305–313.

12. E. C. Titchmarsh, *The Theory of the Riemann Zeta Function*. Oxford: The Clarendon Press, 1951.

13. C. de la Vallée-Poussin, Recherces analytiques sur la théorie des nombres premiers, *Annales de la Soc. Sciences Bruxelles* **20** (1896) 183–256, 281–297.

14. E. Wirsing, Elementare Beweise des Primzahlsatzes mit Restglied I and II, *Journal für die reine u. angw. Mathematik* **211** (1962) 205–214; **214 / 215** (1964) 1–18.

Chapter 10

The Arithmetic of Number Fields

1 INTRODUCTION

In Chapters 8 and 9 the reader had ample opportunity to convince himself
that tools like the theory of functions of a complex variable, or Tauberian
theorems, can be very useful in handling seemingly unrelated problems like the
number $\pi(x)$ of primes up to x. These same theories, while hardly mentioned
in Chapter 7, have been found to be needed in the treatment of certain
partition problems.

It may be puzzling, if not outright disturbing, to the reader that such
apparently farfetched theories seem to be appropriate tools needed in our
attempts to answer simple questions about ordinary integers, questions like the
number of partitions, or the number of primes. Nevertheless, it is hoped that
by now the reader will have accepted this situation as one more of the strange
facts of life.

It should be added that some of the finest mathematicians shared such
apprehensions and felt that the handling of arithmetical questions by analytic
tools is just not "right." They have tried to solve arithmetical problems by
strictly arithmetical means, algebraic problems by algebraic (rather than, say,
topological) means, etc. Occasionally—but by no means always—such en-
deavors of "mathematical purists" have been successful. One shining example
of a success of this kind is the already-mentioned "elementary" proof of the
prime number theorem. There are, however, numerous examples of simple
arithmetic statements for which no elementary (read arithmetical) proofs are
known. Even in some of the cases in which such proofs were found, they are
often complicated and far from transparent, so that the older nonarithmetical
proofs are still at least pedagogically preferable, much to the chagrin of the
purists.

There are still other questions of number theory, for whose treatment other, mainly abstract algebraic, tools have been found useful, or even indispensable. In fact, as already seen in Chapter 1, it was precisely the attempt to solve some of these number-theoretic problems (specifically, some related to certain Diophantine equations, among which Fermat's equation occupies a prominent position) that led in the first place to the discovery of much of abstract algebra, in particular of the *theory of ideals*.

Once the theory of number fields and that of the ideals of their rings of integers was developed, it turned out that in a very natural way some new functions appeared which generalized the Riemann zeta function. In what follows, we shall see some of these functions. It will turn out that with their help it is possible to study problems like the following two:

(i) Are there infinitely many primes in a given arithmetic progression?

(ii) How many such primes (of that arithmetic progression) do not exceed a given bound x?

The theory of ideals will also permit us to study certain types of Diophantine equations and enable us to either find their solutions or prove that there are no solutions.

For some of these questions, algebraic considerations will suffice. For others, one needs both algebraic and analytic tools. One often speaks of *analytic number theory* or *algebraic number theory*, according to the flavor of the tools used to handle the problem. This of course is not necessarily the same as the flavor of the original problem. Also, in some problems considerations of both kinds are needed, and are so thoroughly mixed that many people see little usefulness in the characterization of number-theoretic problems as either analytic or algebraic, or as some other type.

In the following chapters, we shall study some of the problems quoted. We shall also study some of the functions mentioned, in part for their intrinsic interest rather than for applications. In some cases, particularly important or interesting results will be indicated without complete proofs.

In Chapters 10 and 11 knowledge of some concepts of algebra is assumed; in Section 2 of Chapter 10 we shall recall some definitions and theorems. It would be an easy matter to recall *all* needed definitions, and that would make this book more selfcontained. It is, however, hardly sufficient to just read a definition to be able to efficiently handle a really new concept. For that reason, the author takes the point of view that from here on the reader is expected to know the elements of abstract algebra as far as these refer to groups, rings, their isomorphisms and homomorphisms, divisors of zero, integral domains, their fields of fractions, units in rings, vector spaces (finite-dimensional) and their bases, and the elements of the theory of matrices and determinants.

The reader who has not studied the topics mentioned will have no difficulty in finding them in any of a number of excellent textbooks, such as [14], [23], or [5]. The reader who has had even a nodding acquaintance with abstract algebra and who wishes to continue the study of number theory in the following chapters is encouraged to go right ahead and consult some of the textbooks on algebra only if and when the need should arise.

We shall continue to denote the set of ordinary integers (positive, zero, or negative) by **Z**, and its elements by lower-case italic letters (occasionally, however, also by capital italic letters—no confusion will result), rational numbers by capital italic letters, and the set of rationals by **Q**. All other numbers that occur will be complex numbers, i.e., elements of the set **C**; they will be denoted by lower-case Greek letters $\alpha, \beta, \gamma, \ldots$, even if they happen to be real, i.e., already elements of $\mathbf{R} \subset \mathbf{C}$. The subsets of rings that constitute ideals (see Definition 11.1) will be denoted by lower-case German letters, and their classes by capital German letters. We recall that under addition and multiplication (understood without further mention as addition and multiplication in **C**) **Q**, **R**, and **C** are fields while **Z** is a ring, in fact an integral domain.

The particular aspects of the above-mentioned abstract algebraic concepts will be recalled in the concrete setting of number theory in Section 2. Sections 3–6 will discuss in detail one particular number field, its ring of integers, its units, etc. In Sections 7–9 other quadratic fields will be described more briefly and general conclusions about quadratic fields will be drawn in Section 10. Section 11 deals with cyclotomic fields; Section 12 starts out with the consideration of a polynomial that is not irreducible over the rationals, and then briefly (and somewhat superficially) discusses a cubic field. Section 13 discusses integers and units in general algebraic number fields and states Dirichlet's theorem on units.

2 FIELDS AND RINGS OF ALGEBRAIC NUMBERS

Let **S** be an arbitrary set; then the set of polynomials in the variable x with coefficients in **S** is denoted by **S**$[x]$. The set of rational fractions with coefficients in **S** is denoted by **S**(x). In general, if $p(x)$ is a polynomial with coefficients in **S**, we write $p \in \mathbf{S}[x]$ rather than use the more cumbersome notation $p(x) \in \mathbf{S}[x]$. However, if we want to distinguish between the polynomial function $p(x)$ and the numerical value $p(\theta)$ that this function takes when we replace the variable x by the (in general complex) number θ, then we shall use the complete notation.

We recall that $\mathbf{Z}[x]$ is a ring with uniqueness of factorization into *irreducible* polynomials; these are polynomials $p \in \mathbf{Z}[x]$ such that $p = f \cdot g$, $f \in \mathbf{Z}[x]$,

$g \in \mathbb{Z}[x]$ imply that one of the factors is a constant (i.e., an integer). If c is the g.c.d. of the coefficients of $f \in \mathbb{Z}[x]$, f not necessarily irreducible over $\mathbb{Z}$, then c is called the *content* of f. If $c = 1$, f is said to be *primitive*. We shall consider polynomials irreducible over $\mathbb{Z}$ to be also primitive (this does not affect their roots).

Irreducibility is not an absolute property of a polynomial, but is related to a specific field or ring; hence the qualification "over $\mathbb{Z}$" is relevant. In what follows, irreducibility without further indication will be understood over $\mathbb{Q}$.

We recall (see [20] or [23])

Gauss' Lemma. *The product of primitive polynomials is primitive.*

Corollary 1. *If $f \in \mathbb{Z}[x]$ and f splits over $\mathbb{Q}$ (i.e., if $\exists g \in \mathbb{Q}[x]$, $h \in \mathbb{Q}[x] \ni f = g \cdot h$), then f already splits over $\mathbb{Z}$ (i.e., we may select g and h actually in $\mathbb{Z}[x]$).*

Corollary 2 (Eisenstein's Criterion; see [14] or [23]). *Let*

$$f(x) = a_0 x^n + a_1 x^{n-1} + \cdots + a_{n-1} x + a_n \in \mathbb{Z}[x].$$

If there exists a prime p such that $p \nmid a_0$, $p^2 \nmid a_n$, $p \mid a_j$ ($1 \le j \le n$), then f is irreducible over $\mathbb{Q}[x]$.

A polynomial is said to be *monic* if the coefficient of the highest power of x is equal to one. If $f \in \mathbb{Q}[x]$, this is no real restriction, and except for specific mention to the contrary we shall assume that all such polynomials are normalized to be monic. If, however, $f \in \mathbb{Z}[x]$, then $a_0^{-1} f = x^n + a_0^{-1}(a_1 x^{n-1} + \cdots + a_n) \in \mathbb{Q}[x]$, but in general $a_0^{-1} f \notin \mathbb{Z}[x]$, and the condition that f be monic in $\mathbb{Z}[x]$ is a genuine restriction on f.

DEFINITION 1.   Let $f \in \mathbb{Z}[x]$; then any root of the equation $f(x) = 0$ is called an *algebraic number*.

REMARK 1.   By the so-called Fundamental Theorem of Algebra ("For every $f \in \mathbb{C}[x]$, the equation $f(x) = 0$ has at least one root $\theta \in \mathbb{C}$;" see, e.g., [14]), such a root θ of $f(x) = 0$ exists for every $f \in \mathbb{Q}[x]$.

REMARK 2.   There is no loss of generality in Definition 1, in that we ask for $f \in \mathbb{Z}[x]$, rather than $f \in \mathbb{Q}[x]$. Indeed, if $m = $ l.c.m. of the denominators of the coefficients of f, then $g = mf \in \mathbb{Z}[x]$, and g and f have the same zeros.

DEFINITION 2.   Let $f \in \mathbb{Z}[x]$ be monic; then any root θ of the equation $f(x) = 0$ is called an *algebraic integer*.

For a given $f \in \mathbb{Q}[x]$ and θ, such that $f(\theta) = 0$, there exists an irreducible polynomial $p(x)$ such that $p \in \mathbb{Q}[x]$, $p \mid f$, $p(\theta) = 0$. If this polynomial is assumed to be monic, then it is unique; in fact, if θ is an algebraic number, it determines uniquely an irreducible monic polynomial $p \in \mathbf{Q}[x]$, such that $p(\theta) = 0$. As already observed, there is no loss of generality in assuming that $p \in \mathbf{Z}[x]$, but then p in general is not monic. The algebraic number θ is an algebraic integer if and only if $p \in \mathbf{Z}[x]$, with $p(\theta) = 0$, turns out to also be monic. The irreducible polynomial $p \in \mathbf{Z}[x]$ is also the polynomial of lowest degree that has θ as one of its zeros; therefore, it is called the *minimal polynomial* of θ. As p is irreducible, the n roots of $p(x) = 0$ are distinct. We denote them by $\theta = \theta^{(1)}, \theta^{(2)}, \dots, \theta^{(n)}$, and call them the *conjugates* of θ.

DEFINITION 3. $S(\theta) = \sum_{j=1}^{n} \theta^{(j)}$ is called the *trace* of θ; $N(\theta) = \theta^{(1)} \cdot \theta^{(2)} \cdots \theta^{(n)}$ is called the *norm* of θ. It will be denoted either by $N\theta$ or by $N(\theta)$.

We recall the obvious identities $S(\theta) + S(\phi) = S(\theta + \phi)$, $N(\theta) \cdot N(\phi) = N(\theta\phi)$. Also, $N(\theta) = 0 \Leftrightarrow \theta = 0$. If $\partial^0 p$ ($=$ the degree of $p(x)$) is n, then θ is said to be an algebraic number of degree n.

The set of all polynomials in θ with rational coefficients forms a field. From $0 = f(\theta) = \theta^n + \sum_{j=1}^{n} a_j \theta^{n-j}$ it follows that $\theta^n = -\sum_{j=1}^{n} a_j \theta^{n-j}$, next $\theta^{n+1} = -a_1 \theta^n - \sum_{j=2}^{n} a_j \theta^{n-j+1} = a_1 \sum_{j=1}^{n} a_j \theta^{n-j} - \sum_{j=2}^{n} a_j \theta^{n-j+1} = \sum_{j=1}^{n} b_j \theta^{n-j}$, say. Proceeding by induction, it follows that all powers of θ (hence all polynomials in θ with rational coefficients) are represented by polynomials of degree $\leq n - 1$ in θ with rational coefficients. The verification that the field axioms hold for the set $\{f(\theta) | f \in \mathbf{Q}[x], \, p(\theta) = 0\}$ is trivial, except perhaps for the existence of an inverse. The statement (of that existence) itself means that, given $f \in \mathbf{Q}[x]$ and θ algebraic, there exists $g \in \mathbf{Q}[x]$ such that $\{f(\theta)\}^{-1} = g(\theta)$. The proof may be sketched as follows. Let p be the minimal polynomial of θ. By the irreducibility of p, f and p are coprime (as polynomials), and by the analog of the Euclidean algorithm (division of integers replaced by division of polynomials), $\exists s, g \in \mathbf{Q}[x] \ni sp + gf = 1$. If we replace here x by θ, then $p(\theta) = 0$ and we obtain $g(\theta)f(\theta) = 1$, as claimed.

Corollary. $\mathbf{Q}[\theta]$ *and* $\mathbf{Q}(\theta)$ *are the same set.*

DEFINITION 4. The field $\mathbf{K} = \mathbf{Q}(\theta)$ is called an *algebraic number field* and θ is called a *generator* of $\mathbf{K}$.

The algebraic integers of $\mathbf{K}$ form a ring that we shall denote by $\mathbf{I}(\theta)$, or simply by $\mathbf{I}$ if there is no danger of ambiguity. An integer α is a *unit* (i.e., $\alpha \in \mathbf{I}$ and $\alpha^{-1} \in \mathbf{I}$) if and only if $N\alpha = \pm 1$. Integers whose ratio is a unit are called *associates*. For each algebraic θ of degree n, $\mathbf{I}(\theta)$ has an *integral basis*, i.e., there exists a set of n elements $\{\omega_1, \omega_2, \dots, \omega_n\}$, $\omega_j \in \mathbf{I}$ ($j = 1, 2, \dots, n$), such that $\sum_{j=1}^{n} a_j \omega_j$ yield (in fact, uniquely) each element of $\mathbf{I}$ when the a_js ($j = 1, 2, \dots, n$) range over all rational integers. If θ is of degree n, then $\mathbf{K}$ is said to be of degree n over $\mathbf{Q}$, or in symbols, $[\mathbf{K} : \mathbf{Q}] = n$. In fact, $\mathbf{K}$ is an n-dimensional vector space over $\mathbf{Q}$, with a basis $1, \theta, \theta^2, \dots, \theta^{n-1}$. Every element of $\mathbf{K}$ is of the form $\alpha = f(\theta)$, $f \in \mathbf{Q}[x]$, $\partial^0 f \leq n - 1$, and this representation is unique. If we replace θ successively by its conjugates, we obtain the n *field conjugates* $\alpha^{(j)} = f(\theta^{(j)})$ of α. These may all be distinct, and in that case α itself is an algebraic number of degree n and generates $\mathbf{K}$ (i.e., $\mathbf{K} = \mathbf{Q}(\theta) = \mathbf{Q}(\alpha)$), or the n values of $f(\theta^{(j)})$ may fall into t identical sets of m distinct values each, $m < n$, so that $n = m \cdot t$. In that case α is an algebraic number of degree m and $\mathbf{Q}(\alpha) \subsetneqq \mathbf{Q}(\theta)$.

The solutions $\theta = e^{(2\pi i m)/n}$ $(m = 0, 1, \ldots, n - 1)$ of $x^n - 1 = 0$ are called nth *roots of unity*. If $(m, n) = 1$, then that particular nth root of unity is not a root of unity of any lower order than n and is said to be a *primitive nth root of unity*. Except for the trivial root $\theta = \theta_0 = 1$, the others (and in particular the primitive nth roots of unity) are solutions of the *cyclotomic equation*

$$\frac{x^n - 1}{x - 1} = x^{n-1} + x^{n-2} + \cdots + x + 1 = 0.$$

If $n = p$, a prime, then the *cyclotomic polynomial* $x^{p-1} + x^{p-2} + \cdots + x + 1$ is irreducible (see Section 11).

3   THE QUADRATIC FIELD $\mathbf{Q}(\sqrt{2})$

Let us consider the polynomial $p(x) = x^2 - 2$. It has rational integral coefficients; hence $p \in \mathbf{Z}[x]$. In fact, as $z \in \mathbf{Q} \subset \mathbf{R} \subset \mathbf{C}$, we also have $p \in \mathbf{Q}[x]$, $p \in \mathbf{R}[x]$, and $p \in \mathbf{C}[x]$. These four similar-looking assertions convey, however, somewhat different information. Indeed, $x^2 - 2 = (x - \sqrt{2})(x + \sqrt{2})$, with $\pm\sqrt{2} \in \mathbf{R}$, but $\pm\sqrt{2} \notin \mathbf{Q}$. It follows that $p(x)$, as a polynomial over either the real or the complex field, is not irreducible but splits into two linear factors. On the other hand, $p(x)$ does not split over $\mathbf{Z}$. Indeed, if $x^2 - 2 = (ax + b)(cx + d) = acx^2 + (ad + bc)x + bd$, with $a, b, c, d \in \mathbf{Z}$, we have $a = c = \pm 1$, $ad + bc = 0$, $bd = 2$. Hence $0 = ad + bc = a(d + b) = \pm(d + b)$, so that b and d are the two roots of the equation $X^2 - 0 \cdot X - 2 = 0$ (i.e., and not by chance, of $p(X) = 0$), and (perhaps up to a change of signs) $b = -d = \sqrt{2}$, so that $b, d \notin \mathbf{Q}$, contrary to $b, d \in \mathbf{Z}$, and the conclusion that p is irreducible over $\mathbf{Z}$ follows. By Gauss' Lemma it now follows that p is also irreducible over $\mathbf{Q}$. The present result illustrates the already-mentioned fact that irreducibility is not an absolute property of a polynomial but depends on the underlying field.

If we denote by θ either $\sqrt{2}$, or $-\sqrt{2}$, then the set $A + B\theta$ $(A, B \in \mathbf{Q})$ is a field (under ordinary addition and multiplication in $\mathbf{C}$). This is denoted by $\mathbf{Q}(\theta)$ or, more specifically in our case, by $\mathbf{Q}(\sqrt{2})$. It is obvious that the set $\mathbf{Q}(\theta)$ is exactly the same regardless of which one of the zeros of the irreducible polynomial $p(x) = x^2 - 2$ we denote by θ. The two distinct zeros are *algebraic conjugates* of each other. Observe, however, that they are *not* complex conjugates of each other; as real numbers, each is its own complex conjugate. The verification of all field axioms is trivial (use has to be made of $\theta^2 = 2$), except perhaps for the existence of a multiplicative inverse. The latter can be seen in a variety of ways. The simplest is, perhaps, for $A, B \in \mathbf{Q}$, not both zero, to set $(A + B\theta)^{-1} = x + y\theta$ and to show that the rationals $x = A/(A^2 - 2B^2)$, $y = -B/(A^2 - 2B^2)$ satisfy the equation identically. A remark concerning the non-vanishing of the denominator may, however, be in order. If we assume otherwise, then $2 = (A/B)^2$, so that $\sqrt{2} = A/B \in \mathbf{Q}$, which is false (why?). Reasoning differently and without any computation, we may observe that $A^2 - 2B^2 = N(A + B\theta) = 0 \Rightarrow A + B\theta = 0 \Rightarrow A = B = 0$, which is precisely the excluded case.

A more formal way to obtain the inverse, and one that works equally well for irreducible polynomials $p \in \mathbf{Q}[x]$ of arbitrary degrees, is the following.

The condition that $x + y\theta$ be the multiplicative inverse of $A + B\theta$ can be written as

$$1 = (A + B\theta)(x + y\theta) = Ax + Ay\theta + Bx\theta + By\theta^2$$
$$= Ax + 2By + \theta(Ay + Bx). \tag{1}$$

If x and y are rationals, then the first-degree equation (1) in x and y can hold with irrational θ only if the coefficient of θ vanishes. In that case, (1) reduces to $Ax + 2By = 1$. Hence if (1) admits rational solutions x, y, these satisfy the system

$$Bx + Ay = 0$$
$$Ax + 2By = 1. \tag{2}$$

The system (2) has unique solutions, provided that the determinant

$$\begin{vmatrix} B & A \\ A & 2B \end{vmatrix} = 2B^2 - A^2 \neq 0.$$

If that is the case, the solutions will indeed be rational, being obtained by rational operations from the rationals A, B, and 1. Said determinant, however, is precisely $-N(A + B\theta)$, and hence is different from zero unless $A = B = 0$ (excluded case); the proof is complete. The actual solutions for x and y are of course those already seen.

Let us denote by $\mathbf{K} = \mathbf{Q}(\theta)$ the field of rational fractions in θ which is, as seen, identical to the field of polynomials of first degree in θ. Each element $\alpha \in \mathbf{K}$ is of the form $\alpha = A + B\theta$, $A, B \in \mathbf{Q}$. It has (as already observed) the norm $N(\alpha) = A^2 - 2B^2$, and its trace is $S(\alpha) = 2A$.

There is still another way to obtain the same field. (By "same" field of course, we mean, here and in what follows, a field isomorphic to the given one.) Let us start with the set $\mathbf{Q}[x]$ of all polynomials with rational coefficients. Let

$f \in \mathbf{Q}[x]$ be one such polynomial. We may set $f(x) = p(x)q_1(x) + r(x)$, $q_1, r \in \mathbf{Q}[x]$, and, as $\partial^0 p = 2$, $r(x) = A + Bx$. We now form equivalence classes of polynomials by letting f and g belong to the same class, if, after division by $p(x)$, their remainders are the same. This means that if g belongs to the class of f and $g(x) = p(x)q_2(x) + s(x)$, then $s(x) = r(x)$. We may write this symbolically as

$$ f \sim g \Leftrightarrow f = pq_1 + r, g = pq_2 + r. $$

We may also think of f and g as belonging to the same class if, upon ignoring an appropriate multiple (by a polynomial in $\mathbf{Q}[x]$) of p, the polynomials f and g become equal. This again means that $f - g = pq$, $q \in \mathbf{Q}[x]$. It is easy to verify that this relation $f \sim g$ is indeed an equivalence relation, and that the classes are equivalence classes. We denote the class of f by $\{f\}$. Clearly, $f \sim r$, so that the class of f also contains $r(x) = Ax + B$. This polynomial is often selected as the representative of its class and $\{f\} = \{r\}$. These classes of polynomials are in fact residue classes modulo the polynomial p, in a way entirely analogous to the residue classes of integers modulo an integer m in Chapter 2. Addition of classes is obvious: If $f \sim A_1 + B_1 x$ and $g \sim A_2 + B_2 x$, then $f + g \sim A_1 + A_2 + (B_1 + B_2)x$. As for multiplication,

$$ f \cdot g \sim A_1 A_2 + (A_1 B_2 + A_2 B_1)x + B_1 B_2 x^2 $$
$$ = A_1 A_2 + 2B_1 B_2 + (A_1 B_2 + A_2 B_1)x + (x^2 - 2)B_1 B_2 $$
$$ \sim A_1 A_2 + 2B_1 B_2 + (A_1 B_2 + A_2 B_1)x. $$

If $f \sim A + Bx$, then the class inverse to f is represented by $C + Dx$ if $(A + Bx)(C + Dx) \sim 1$. By previous results, that means that $AD + CB = 0$ and $AC + 2BD = 1$. These two equations in C and D are precisely the system (2) and we know that it has unique solutions $C = A/(A^2 - 2B^2)$, $D = -B/(A^2 - 2B^2)$. Under this identification, it is clear that $\mathbf{L} = \mathbf{Q}[x]/\{p(x)\}$, the set of polynomials with rational coefficients considered modulo $p(x)$, is a field. It is instructive to write out explicitly a correspondence that defines a $1:1$ map between $\mathbf{K}$ and $\mathbf{L}$ and to verify that this is in fact an isomorphism $\mathbf{K} \simeq \mathbf{L}$.

There is even still another way to look at the field $\mathbf{K}$ (or $\mathbf{L}$). All its elements are of the form $A + B\theta$ (or $A + Bx$, respectively) and may be identified with the ordered pair of rationals (A, B). These pairs are added by the rule $(A, B) + (C, D) = (A + B, C + D)$ and are multiplied by a "scalar" C (i.e., by an element of $\mathbf{Q}$), by the rule $C(A, B) = (AC, BC)$. It is easy to verify that these elements of $\mathbf{K}$ satisfy all postulates of a two-dimensional vector space over $\mathbf{Q}$, spanned by the elements 1 and θ. We formalize some of the results obtained so far in

Theorem 1. *The set* $\mathbf{K} = \{A + B\theta | \theta$ *a zero of* $x^2 - 2\} = \mathbf{Q}(\sqrt{2})$ *is a field under ordinary addition and multiplication. It is isomorphic to* $\mathbf{L} = \mathbf{Q}[x]/\{x^2 - 2\}$.

4 THE RING I($\sqrt{2}$) OF THE INTEGERS OF K = Q($\sqrt{2}$) AND ITS UNITS

As elements of $\mathbf{K} = \mathbf{Q}(\theta)$ $(\theta^2 = 2)$, the integers of $\mathbf{I} = \mathbf{I}(\theta)$ have to be of the form $\alpha = A + B\theta$, $A, B \in \mathbf{Q}$, $\theta^2 = 2$. In particular, $\alpha^2 = A^2 + 2AB\theta + B^2\theta^2 = A^2 + 2B^2 + 2AB\theta$. However, $B\theta = \alpha - A$, so that $\alpha^2 = A^2 + 2B^2 + 2A(\alpha - A)$ and α satisfies the equation

$$\alpha^2 - 2A\alpha + A^2 - 2B^2 = 0,$$

which may also be written as

$$\alpha^2 - S(\alpha) \cdot \alpha + N(\alpha) = 0.$$

If α is to be an integer, we know that the coefficients of this monic equation have to be rational integers; hence $S(\alpha) = 2A = a$, $N(\alpha) = A^2 - 2B^2 = b$, $a, b \in \mathbf{Z}$. Let us assume first that a is odd, $a = 2m + 1$, say. Then $A = m + 1/2$, $b = A^2 - 2B^2 = m^2 + m + 1/4 - 2B^2$. As $b \in \mathbf{Z}$, one has $2B^2 - 1/4 = c \in \mathbf{Z}$, so that $8B^2 - 1 = 4c \equiv 0 \pmod 4$. This, however, is not possible. Indeed, if $B = s/k$ in reduced form, s, $k \in \mathbf{Z}$, then $k = 1$ or 2 (otherwise $8B^2 \notin \mathbf{Z}$). In either case, $8B^2$ is even, so that $8B^2 - 1 \equiv 0$ does not hold even modulo 2. It follows that a has to be even, $a = 2m$, say, so that $A = m$ and $2B^2 = A^2 - b = m^2 - b$. This shows that B has to be an integer (otherwise $2B^2 \notin \mathbf{Z}$), say c, and then α satisfies the equation

$$\alpha^2 - 2m\alpha + m^2 - 2c^2 = 0, \tag{3}$$

with the solutions $\alpha = m \pm c\sqrt{2}$. It follows that a necessary condition for α to be an integer in $\mathbf{Q}(\sqrt{2})$ is to be of this form with $m, c \in \mathbf{Z}$. Conversely, if α is of this form it is an element of $\mathbf{Q}(\sqrt{2})$ and also satisfies (3); hence it is an algebraic integer of $\mathbf{Q}(\sqrt{2})$ and the necessary condition is also sufficient. We formalize these results in

Theorem 2. *The ring* $\mathbf{I} = \mathbf{I}(\sqrt{2})$ *of integers in* $\mathbf{K} = \mathbf{Q}(\sqrt{2})$ *consists of the set* $\{m + c\sqrt{2} | m, c \in \mathbf{Z}\}$.

For later use, let us also observe the following: If $\theta^2 = 2$, then the set $\{m + c\sqrt{2}\}$ runs through *all* integers in $\mathbf{I}$ when m and c run independently through all rational integers. For that reason, $\{1, \theta\}$ are said to form an *integral basis* for **I**. The determinant

$$D = \begin{vmatrix} 1 & \theta \\ 1 & -\theta \end{vmatrix} = \begin{vmatrix} 1 & \sqrt{2} \\ 1 & -\sqrt{2} \end{vmatrix} = -2\sqrt{2}$$

has the property that its square is rational (even a rational integer). Indeed, $\Delta = D^2 = 8 = 4\theta^2$, and Δ is called the field *discriminant* of **K**.

The *units* of **I** constitute the subset of integers of norm equal to ± 1. As the norm of $\alpha = m + c\sqrt{2}$ is $N(\alpha) = m^2 - 2c^2$, α is a unit precisely when one of the following two equations holds:

$$c^2 - 2m^2 = 1 \tag{4}$$

or

$$c^2 - 2m^2 = -1. \tag{4'}$$

These equations are a particular case of the so called *Pell equation*. (This is a misnomer! Pell had hardly anything to do with this equation. The name is presumably due to an error of Euler, but the error stuck, and we are stuck with the name! See [12], [8], or [1] for more on the history of this equation.) The Pell equation reads, in general,

$$x^2 - dy^2 = a. \tag{5}$$

If a is a square, say $a = b^2$, then (5) always has at least the solutions $x = \pm b$; otherwise, (5) may have no solutions. If d is a square, the solution is rather trivial. Indeed, if $d = g^2$, then (5) becomes $(x + gy)(x - gy) = a$. Let $d_1 d_2 = a$ be any factorization of a, with $d_1 \equiv d_2 \pmod{2g}$. To each such factorization we have a unique solution of the system

$$x + gy = d_1$$
$$x - gy = d_2,$$

namely $x = (d_1 + d_2)/2$, $y = (d_1 - d_2)/2g$; both are integers and are the only solutions of (5), with $d = g^2$. (Why did we require $d_1 \equiv d_2 \pmod{2g}$?)

If d is not a perfect square, the general problem of the solution of (5) is much more interesting and far more difficult. At present, however, we are interested only in the particular cases $d = 2$ and $a = \pm 1$. Both equations have obvious solutions: $c = \pm 1$, $m = 0$ for (4) and $c = \pm 1$, $m = \pm 1$ for (4'). These, however, are not the only solutions. Before we look for any other solutions, let us make the following two remarks:

(i) The obvious solutions $c = \pm 1$, $m = 0$, of (4) correspond to the units $\pm 1 + 0 \cdot \theta = \pm 1$ of **Z** itself.

(ii) The four *distinct* mentioned solutions of (4'), namely $\pm 1 \pm \theta = \pm 1 \pm \sqrt{2}$ are not *unrelated*. Indeed, $-1 - \sqrt{2} = -(1 + \sqrt{2})$, $-1 + \sqrt{2} = -(1 - \sqrt{2})$, and $-(1 - \sqrt{2}) = 1/(1 + \sqrt{2})$.

More generally, with α, $-\alpha$ is also a unit so that it is sufficient to find all positive solutions of (4) and (4'). Similarly, if $0 < \alpha \le 1$, then α^{-1} is also a unit and $\alpha^{-1} \ge 1$. It follows that it is sufficient to determine the set $U_1 = \{\alpha$ a unit, $\alpha \ge 1\}$. The set $U = \{\alpha, -\alpha, \alpha^{-1}, -\alpha^{-1} | \alpha \in U_1\}$ will then contain all units.

In order to find some of the other solutions of (4), (4'), let us recall that (4') is the explicit form of the condition $N(\alpha) = -1$, or $N(c + m\sqrt{2}) = -1$. If $\beta = \alpha^k$, then $N(\beta) = \{N(\alpha)\}^k = (-1)^k$, and β is also a unit. Specifically, if $\beta = \alpha^k = c_k + m_k\sqrt{2}$ and k is even, then c_k, m_k occur as solutions of (4); if k is odd, then they occur as solutions of (4'). We may verify, e.g., that for $k = 0$, one has $\beta = 1$; this is one of the solutions of (4) (namely $c = 1$, $m = 0$).

Among the six solutions of (4), (4') obtained so far (namely $\pm 1, \pm 1 \pm \sqrt{2}$), $\alpha = 1 + \theta = 1 + \sqrt{2}$ is the only one with $\alpha > 1$. It immediately leads to an infinite sequence of solutions $(1 + \theta)^{2k}$ for (4) and to another infinite sequence of solutions $(1 + \theta)^{2k-1}$ for (4'). So, e.g., $(1 + \theta)^0 = 1$ (solution of (4)), $(1 + \theta)^1 = 1 + \theta$ (solution of (4')), $(1 + \theta)^2 = 1 + 2\theta + \theta^2 = 3 + 2\theta$ (we check that this corresponds to a solution of (4), because $(3 + 2\sqrt{2})(3 - 2\sqrt{2}) = 9 - 8 = 1$), $(1 + \theta)^3 = 1 + 3\theta + 3\theta^2 + \theta^3 = 7 + 5\theta$ (solution of (4'), because $(7 + 5\sqrt{2})(7 - 5\sqrt{2}) = -1$), etc.

For negative values of k we obtain solutions $\alpha = c + m\sqrt{2}$ of (4), (4') with $0 < \alpha < 1$ because $(1 + \theta)^{-1} = (1 + \sqrt{2})^{-1} < 1$, and so also $(1 + \sqrt{2})^{-k} < 1$.

The question of course immediately arises whether there are any other solutions in addition to those obtained so far. Fortunately, the answer is "no!" As we already know it is sufficient to verify that there are no other units $\alpha = m + c\sqrt{2} > 1$. If there were any, then for such units α either $1 < m + c\sqrt{2} < 1 + \sqrt{2}$, or for some $k, 0 < k \in \mathbf{Z}$, $(1 + \sqrt{2})^k < m + c\sqrt{2} < (1 + \sqrt{2})^{k+1}$. In the first case, $m^2 - 2c^2 = \pm 1$, so that $-1 < m - c\sqrt{2} < 1$ and, adding, $0 < 2m < 2 + \sqrt{2}$, that is $0 < m < 1 + \sqrt{2}/2$, so that $m = 1$. This reduces the inequalities for c to $0 < c\sqrt{2} < \sqrt{2}$, which is impossible for $c \in \mathbf{Z}$. The second case reduces to the first through the observation that it implies the existence of a unit $\beta = (m + c\sqrt{2})(1 + \sqrt{2})^{-k}$ that satisfies $1 < \beta < 1 + \sqrt{2}$, shown to be impossible. We have also shown that $1 + \sqrt{2}$ is the smallest unit of $\mathbf{Q}(\sqrt{2})$ larger than one. As this unit generates all other units, it is called the *fundamental unit* of the ring (or of the corresponding field).

5   FACTORIZATION IN $\mathbf{K} = \mathbf{Q}(\sqrt{2})$

Before we can discuss factorization in $\mathbf{K}$ we have to clarify what we mean by a prime number. In Chapter 1, we defined the integer $p > 1$ to be a prime by the property that $a, b, p \in \mathbf{Z}$, $a > 1$, $b > 0$, $p = a \cdot b \Rightarrow b = 1$. In Theorem 3.4 we then showed that the integers that are primes by this definition are precisely those which also have the property that $p|a \cdot b$, $p \nmid a \Rightarrow p|b$. From that point on, we could (and did when convenient) use this second property as an equivalent definition of primality. The reader will remember that the proof of Theorem 3.4, while not difficult, was surprisingly nontrivial. Now we shall better understand why. Indeed the property of being a prime according to the first definition no longer defines the same set of integers as the second definition, in all numbers fields. In order to avoid confusion, we shall reserve the term "prime" for those integers that qualify under the second definition, and we shall call those which have the property described by the first definition, more properly *indecomposables*.

What are now the indecomposable elements of $\mathbf{I} = \mathbf{I}(\sqrt{2})$? A simple criterion is that $\alpha = a + b\theta$ is indecomposable (for every θ, not only for $\theta = \sqrt{2}$) if $N\alpha = p$, a rational prime. Indeed if $\alpha = \beta\gamma$, then $N\alpha = N\beta \cdot N\gamma = p$ with $N\beta$, $N\gamma \in \mathbf{Z}$, and this implies that either $N\beta = 1$ or $N\gamma = 1$, so that in every factorization of α in $\mathbf{K}$, one of the factors is a unit, and that is what we call an indecomposable element.

Unfortunately, the converse is false. There are indecomposable elements $\alpha \in \mathbf{I}$ for which $N\alpha$ is not a rational prime. In the present case, with $\theta = \sqrt{2}$, this happens, e.g., with $\alpha = 3(= 3 + 0 \cdot \theta \in \mathbf{I})$, for which $N\alpha = 9$. (Why is $\alpha = 3$ an indecomposable element? See Problem 7.)

Before we ask any new questions, let us observe that $\mathbf{K} = \mathbf{Q}(\theta)$ is precisely the field of fraction of the ring $\mathbf{I}(\theta)$. Indeed given $\alpha = A + B\theta \in \mathbf{K}$, let d be the least common multiple of the denominators of A and B, so that $A = a/d$, $B = b/d$; then $\alpha = (a + b\theta)/d$, with $a + b\theta \in \mathbf{I}$, $d \in \mathbf{I}$. Conversely, consider $(a + b\theta)/(c + d\theta)$, $a, b, c, d \in \mathbf{Z}$, so that $a + b\theta \in \mathbf{I}$, $c + d\theta \in \mathbf{I}$. We know that $(c + d\theta)^{-1} = A + B\theta$, $A \in \mathbf{Q}$, $B \in \mathbf{Q}$; hence $(a + b\theta)/(c + d\theta) = (a + b\theta)(A + B\theta) = aA + \theta(aB + bA) + Bb\theta^2 = aA + 2bB + \theta(aB + bA) = C + D\theta$, with $C, D \in \mathbf{Q}$ and $\alpha = C + D\theta \in \mathbf{Q}(\theta)$. At one point we have used $\theta^2 = 2$, valid only in the presently studied field; in the general case of a quadratic field, we more generally replace θ^2 with an expression of the form $u + v\theta$, $u, v \in \mathbf{Q}$, with the same result.

We now ask what can be said about the decomposition of the integers in $\mathbf{I}(\theta)$ into indecomposable elements. To answer this question, it is useful to look back at Chapter 3. While the discussion there concerned only the ring $\mathbf{Z}$ of integers of the field $\mathbf{Q}$ (which is the field of fractions of $\mathbf{Z}$), none of the special

properties of $\mathbf{Z}$ were used except the following: There exists a map $\mathbf{Z} \xrightarrow{\phi} \mathbf{R}^+$ from the ring of integers into the non-negative reals, namely $\phi(a) = |a|$, with the property that if $a, b \in \mathbf{Z}$, $b \neq 0$, then there exist $q, r \in \mathbf{Z}$ such that $a = bq + r$ and $\phi(r) < \phi(b)$. The reader is invited to verify this statement, which is not all that obvious from the formulation given in Chapter 3. The method of effectively finding q and r was called the Euclidean algorithm (see Problem 3 in Chapter 3). The existence of the Euclidean algorithm led immediately (see same Problem 3) to the existence of a greatest common divisor for any two elements of the ring of integers. If an element is defined as indecomposable by the property that $\alpha = \beta\gamma \Rightarrow$ either β or γ is a unit, then the existence of a greatest common divisor leads (see Corollary 3.6.3, Theorem 3.4, Corollary 3.4.1) to the proof of Theorem 3.3, i.e., to the uniqueness of factorization into indecomposable elements. Once more, no specific properties of the ring $\mathbf{Z}$ or of its field of fractions $\mathbf{Q}$ were used, beyond the existence of the Euclidean algorithm. For that reason, it is sufficient to verify the existence of a Euclidean algorithm in $\mathbf{I}(\sqrt{2})$ (or for that matter in any $\mathbf{I}(\theta)$) to be certain that the factorization of its integers into their indecomposable elements is unique up to units and the order of the factors.

Before we attempt to define such an algorithm, let us remember that the units in $\mathbf{I}(\sqrt{2})$ are by no means just ± 1, as in $\mathbf{Z}$, but are infinitely many in number—some very large and some very small in absolute value—so that even if we succeed in proving the existence of a Euclidean algorithm in $\mathbf{I}(\sqrt{2})$, the arithmetic on $\mathbf{I}(\sqrt{2})$ is likely to be rather different from that in $\mathbf{Z}$.

6   THE EUCLIDEAN ALGORITHM IN $\mathbf{Q}(\sqrt{2})$

To construct a Euclidean algorithm in $\mathbf{Q}(\sqrt{2})$, we try to mimic the procedure used in $\mathbf{Q}$. There we had the mapping $\phi(a) = |a|$. We already saw several instances that suggest that the norm may replace the absolute value as a measure of size. With this analogy in mind, we are led to say that we have a Euclidean algorithm in $\mathbf{I}(\theta)$ (even in general, not only for $\theta = \pm\sqrt{2}$) if for every $\alpha, \beta \in \mathbf{I}(\theta)$ there exist $\kappa, \rho_1 \in \mathbf{I}$ such that $\alpha = \beta\kappa + \rho_1$ with $|N\rho_1| < |N\beta|$. If we divide by β, then α/β and $\rho = \rho_1/\beta$ become elements of $\mathbf{Q}(\theta)$, while κ stays in $\mathbf{I}$. The previous tentative formulation of a Euclidean algorithm is now equivalent to the statement that for every $\alpha/\beta \in \mathbf{Q}(\theta)$ we can find an integer $\kappa \in \mathbf{I}$ such that $|N(\rho/\beta)| = |N(\alpha/\beta - \kappa)| < 1$. To summarize our reasonings so far: The essential uniqueness of factorization in $\mathbf{I}(\theta)$ will follow for those θ for which we can prove the analog of the following lemma.

Lemma 1.   *Let $\theta^2 - 2 = 0$; then for every $\alpha = A + B\theta$ and $\beta = C + D\theta$, $A, B, C, D \in \mathbf{Q}, \neq 0$, there exists $\kappa \in \mathbf{I}(\theta)$ and $\rho = (\alpha/\beta) - \kappa \in \mathbf{Q}(\theta)$ such that $|N\rho| < 1$.*

PROOF.   Since $\mathbf{Q}(\theta)$ is the field of fractions of $\mathbf{I}(\theta)$, $\alpha/\beta \in \mathbf{Q}(\theta)$; hence $\alpha/\beta = E + F\theta, E, F \in \mathbf{Q}$. Let $\kappa = e + f\theta$, with e and f the rational integers closest to E and F, respectively, so that $|E - e| \leq 1/2$, $|F - f| \leq 1/2$. It follows that $|N\rho| = |N(\alpha/\beta) - \kappa)| = |N(E - e + (F - f)\theta)| = |(E - e)^2 - 2(F - f)^2| \leq \max\{2^{-2}, 2 \cdot 2^{-2}\} = 1/2 < 1$. This finishes the proof of

Theorem 3. *For $\theta^2 = 2$, the field $\mathbf{Q}(\theta)$ is Euclidean (i.e., it has a Euclidean algorithm), so that its ring of integers $\mathbf{I}(\theta)$ has the property of (essential) uniqueness of factorization.*

REMARK 3.   In the proof of Lemma 1 we obtained $|N\rho| < 1/2$, while all we needed was $|N\rho| < 1$. This suggests that one could perhaps prove a theorem much stronger than Theorem 3, and that is indeed the case. By using refinements of the above method it has been shown that $\mathbf{Q}(\sqrt{m})$ is Euclidean for exactly five negative values on m, namely $m = -1, -2, -3, -7, -11$, and for 16 positive values, namely $2, 3, 5, 6, 7, 11, 13, 17, 19, 21, 29, 33, 37, 41, 57, 73$, and for no others (see [12] and [7]).

REMARK 4.   In fields with uniqueness of factorization, the concepts of "prime" and "indecomposable" coincide. (Why? See Problem 12.)

7   THE FIELD $\mathbf{Q}(\sqrt{-5})$

Let us consider the polynomial $p(x) = x^2 + 5$. We observe that p is irreducible over $\mathbf{Z}$, $\mathbf{Q}$, and also $\mathbf{R}$, but not over $\mathbf{C}$; indeed $p(x) = (x - i\sqrt{5})$ $(x + i\sqrt{5})$, $i\sqrt{5} \in \mathbf{C}$, and $p(x)$ vanishes if we substitute for x one of the complex numbers $\theta = \pm\sqrt{-5} = \pm i\sqrt{5}$ $(i = \sqrt{-1})$. In either case, $\theta^2 = -5$, and proceeding as in the case $\theta = \sqrt{2}$, we reach the following conclusions:

The set $\mathbf{K} = \mathbf{Q}(\theta) = \{A + B\theta | A, B \in \mathbf{Q}, \theta^2 = -5\}$ is a field under ordinary addition and multiplication. It is isomorphic to the field $\mathbf{L} = \mathbf{Q}[x]/\{p(x)\}$, where now $p(x) = x^2 + 5$. This field is a vector space over $\mathbf{Q}$, spanned by $\{1, \theta | \theta^2 = -5\}$. Its discriminant is

$$\Delta = \begin{vmatrix} 1 & \theta \\ 1 & -\theta \end{vmatrix}^2 = \begin{vmatrix} 1 & i\sqrt{5} \\ 1 & -i\sqrt{5} \end{vmatrix}^2 = (-2i\sqrt{5}) = -20.$$

We observe that, as in the case $\theta^2 = 2$, $\Delta = 4\theta^2$.

The algebraic conjugate of θ now coincides with the complex conjugate $-\theta$, which had not been the case for $p(x) = x^2 - 2$. Each element of $\mathbf{K}$ is of the form $\alpha = A + B\theta$ and has the trace $S(\alpha) = A + B\theta + (A - B\theta) = 2A$ (just as in $\mathbf{Q}(\sqrt{2})$) and the norm $N\alpha = (A + B\theta)(A - B\theta) = A^2 + 5B^2$. There is a difference between this norm and the norm $A^2 - 2B^2$ in $\mathbf{Q}(\sqrt{2})$, in that some algebraic numbers of $\mathbf{Q}(\sqrt{2})$ have positive norms and others have negative norms. In the present case, $A^2 + 5B^2$ is never negative.

The integers $\alpha = A + B\theta$ of $\mathbf{Q}(\sqrt{-5})$ again have to satisfy the equation $x^2 - S(\alpha)x + N(\alpha) = 0$, which now, written explicitly, reads

$$x^2 - 2Ax + A^2 + 5B^2 = 0.$$

For α to be an integer, we need $2A = a$, $A^2 + 5B^2 = b$, $a, b \in \mathbf{Z}$. If a is odd, then $5B^2 = k - 1/4$, $k \in \mathbf{Z}$, so that $20B^2 \equiv 3 \pmod 4$, which is impossible for any rational B. Hence $a = 2m$, even, $A = m$, $b = m^2 + 5B^2$, so that B has to be an integer, say c (no denominator except 5 could be cancelled by the coefficient; but if $B = r/5$, $(r, 5) = 1$, then $5B^2 = r^2/5 \notin \mathbf{Z}$), and $b = m^2 + 5c^2$. The integer α now satisfies the equation

$$x^2 - 2mx + m^2 + 5c^2 = 0,$$

so that $\alpha = m \pm \sqrt{-5c^2} = m \pm c\sqrt{-5}$, and $\mathbf{I} = \mathbf{I}(\sqrt{-5})$ consists of the subset of $\mathbf{Q}(\sqrt{-5})$, with both coefficients rational integers, i.e., $\mathbf{I}(\sqrt{-5}) = \{a + b\sqrt{-5} \mid a, b \in \mathbf{Z}\}$.

The units $\alpha = a + b\sqrt{-5}$ of $\mathbf{I}$ also have to satisfy one of the two equations $N\alpha = a^2 + 5b^2 = \pm 1$. Obviously there are no solutions with -1, and if we consider only the solutions with $+1$, there are only the two solutions $a = \pm 1$, the units of $\mathbf{Z}$. This fact seems to suggest a much simpler arithmetic structure of $\mathbf{Q}(\sqrt{-5})$ than that of $\mathbf{Q}(\sqrt{2})$. This hope, however, turns out to be shortlived. Indeed we now have to turn to the indecomposable elements of $\mathbf{I}$ and may wish to attempt to prove the existence of a Euclidean algorithm to be certain of the uniqueness of factorization in $\mathbf{I}$. As in all algebraic number fields, the following proposition holds.

Proposition.   *If $N\alpha = p$, a rational prime, then α is indecomposable in $\mathbf{I}$.*

Not every rational prime can be the norm of an element of $\mathbf{I}$. Indeed, $N\alpha = a^2 + 5b^2$. If a and b are of the same parity, then $a^2 + 5b^2$ is even; however, 2 is the only even prime and $2 = a^2 + 5b^2$ has no solutions in integers. If exactly one of a, b is even, then $a^2 + 5b^2 \equiv 1 \pmod 4$. Hence, only the primes $p \equiv 1 \pmod 4$ can be norms of indecomposable integers of $\mathbf{I} = \mathbf{I}(\sqrt{-5})$. The converse is not true (e.g., $13 \neq a^2 + 5b^2$; however, $29 = 3^2 + 5 \cdot 2^2$). In any case, for any $\alpha \in \mathbf{I}$, $N\alpha \neq 3$, $N\alpha \neq 7$, etc. Also, $\alpha = 3 \ (= 3 + 5 \cdot 0^2) \in \mathbf{I}$ is an indecomposable element. Indeed, if $3 = \beta \cdot \gamma$, then $N(3) = 9 = N\beta \cdot N\gamma$, and if neither β nor γ is a unit, then $N\beta = N\gamma = 3$, which we saw is impossible. It follows by similar reasoning that all rational primes $p \equiv 3 \pmod 4$ are indecomposable in $\mathbf{I}$.

As already observed, the converse of the proposition does not hold. As an example consider $\alpha = 1 + 2\sqrt{-5} \in \mathbf{I}$,

$$N\alpha = (1 + 2\sqrt{-5})(1 - 2\sqrt{-5}) = 21 = 3 \cdot 7 \neq p. \tag{6}$$

Nevertheless, $1 + 2\sqrt{-5}$ (and similarly, $1 - 2\sqrt{-5}$) is indecomposable in $\mathbf{I}$. Indeed, if $\alpha = 1 + 2\sqrt{-5} = \beta \cdot \gamma$, then $N\alpha = N\beta \cdot N\gamma = 21 = 3 \cdot 7$. As 3 and 7 cannot be norms, it follows that either $N\beta = 1$ or $N\gamma = 1$, and α is an indecomposable element.

It is worthwhile to take a second look at (6). We see that 21 has two distinct factorizations into indecomposable elements; it equals both $3 \cdot 7$ and $(1 + 2\sqrt{-5})(1 - 2\sqrt{-5})$. This shows that it would be futile to try to find a Euclidean algorithm in $\mathbf{I}(\sqrt{-5})$. If one did exist, it would imply the uniqueness of factorization in $\mathbf{I}$, and we just saw that factorization is not unique in $\mathbf{I}$. The situation is very different from that in $\mathbf{Q}(\sqrt{2})$, where we had a Euclidean algorithm and hence uniqueness of factorization in $\mathbf{I}(\sqrt{2})$.

Perhaps it will help the understanding of the situation somewhat if we consider the following facts:

(i) the integers 3 and $1 + 2\sqrt{-5}$ of $\mathbf{I}$ have no nontrivial common factor in $\mathbf{I}$; nevertheless,

(ii) with proper determination of the square roots,

$$3 = \sqrt{2 + \sqrt{-5}} \sqrt{2 - \sqrt{-5}} \quad \text{and}$$

$$1 + 2\sqrt{-5} = \sqrt{2 + \sqrt{-5}} \sqrt{-2 + 3\sqrt{-5}} . \tag{7}$$

Indeed, the first equality is clear and the second follows from $(1 + 2\sqrt{-5})^2 = 1 + 4\sqrt{-5} - 20 = -19 + 4\sqrt{-5} = (2 + \sqrt{-5})(-2 + 3\sqrt{-5})$. Hence it seems that 3 and $1 + 2\sqrt{-5}$ do have, after all, the nontrivial divisor $\sqrt{2 + \sqrt{-5}}$ in common. This factor, and in fact all factors that occur in (7), are integers, as the reader can easily check. So, e.g., $\sqrt{2 + \sqrt{-5}}$ satisfies the equation

$$x^4 - 4x^2 + 9 = 0; \tag{8}$$

the reader is invited to verify the integrality of the other factors.

(iii) In the same way, $7 = \sqrt{(2 + 3\sqrt{-5})(2 - 3\sqrt{-5})}$, and (6) may be written as

$$21 = \sqrt{2 + \sqrt{-5}} \sqrt{2 - \sqrt{-5}} \sqrt{-2 + 3\sqrt{-5}} \sqrt{-2 - 3\sqrt{-5}}$$

$$= \sqrt{2 + \sqrt{-5}} \sqrt{-2 + 3\sqrt{-5}} \sqrt{2 - \sqrt{-5}} \sqrt{-2 - 3\sqrt{-5}} . \tag{9}$$

Equation (9), while somewhat complicated, is much more satisfying than (6) because we see exactly the same factors on both sides of the equal sign, only in changed order. It may remind us of an equality like $24 = 6 \cdot 4 = 8 \cdot 3$, written by use of primes as $24 = 2 \cdot 3 \cdot 2 \cdot 2 = 2 \cdot 2 \cdot 2 \cdot 3$, where only the order of the factors changed.

The problem we are facing is, however, that the factors in (9) do not belong to $\mathbf{I} = \mathbf{I}(\theta)$ $(\theta^2 = 5)$. In fact, as we see from (8) and from the similar equations satisfied by the other factors of (9), these belong to fields of degree 4. It is precisely this fact, that we don't dispose of them in $\mathbf{I}$, that leads to the paradoxical situation of (6), much like the corresponding one encountered with the "Hilbert Primes" in Section 3.3. Even the remedy appears somewhat the same. Indeed, if we denote by $a = (3, 1 + 2\sqrt{-5})$ the greatest common divisor of these two integers of $\mathbf{I}$, and similarly for the other couples $\mathfrak{b} = (3, 1 - 2\sqrt{-5})$, $\mathfrak{c} = (7, 1 + 2\sqrt{-5})$, and $\mathfrak{d} = (7, 1 - 2\sqrt{-5})$, we can write equation (6) as

$$21 = a \cdot \mathfrak{b} \cdot \mathfrak{c} \cdot \mathfrak{d} = a \cdot \mathfrak{c} \cdot \mathfrak{b} \cdot \mathfrak{d}.$$

But how do we overcome the difficulty that these algebraic integers, like $\sqrt{2} + \sqrt{-5}$ and the others, are simply not available in $\mathbf{I}$? As already mentioned, two methods have been tried. In the first, one attempts to enlarge the original field. In our example, one would adjoin to $\mathbf{Q}$, besides $\sqrt{-5}$ (which leads to our present $\mathbf{Q}(\sqrt{-5})$), also, say $\sqrt{2} + \sqrt{-5}$. It may be shown that there exists an algebraic number (even an integer), say ϕ, such that $\mathbf{Q}(\phi)$ contains both $\sqrt{-5}$ and $\sqrt{2} + \sqrt{-5}$ (can you indicate ϕ?). In the new field (enlarged so as to contain all the factors in (9)), we have re-established the uniqueness of factorization of 21, and perhaps with luck, that of all integers of our old $\mathbf{I}$. However, the new field is so large that it may well contain new integers (not present in $\mathbf{I}$), for which the old difficulty of nonuniqueness of factorization reappears. One may then attempt to enlarge the field still more, by adjoining to it all the g.c.d. of all integers in the new field, and so on. There is at least an outside chance that this process will eventually come to a halt. Unfortunately, when this problem (connected to what is known in the literature as the problem of the class field tower) was solved, it became clear that there are fields for which this procedure does not end (see [11], [21]).

The other approach to solving the problem of nonuniqueness of factorization uses the fact that while the g.c.d. of two integers, say α and β, of $\mathbf{I}$ may not belong to $\mathbf{I}$, some of its multiples do. The set of those multiples certainly contains at least the set $\{\lambda\alpha + \mu\beta\}$, where λ and μ range over all integers of $\mathbf{I}$. We denote this set by (α, β). This leads to no confusion with the notation for the g.c.d. itself. Indeed, should it happen that this g.c.d. does in fact belong to $\mathbf{I}$, say $\delta = (\alpha, \beta)$, then the set $\{\lambda\alpha + \mu\beta\}$ coincides with the set $\{\rho\delta | \rho \in \mathbf{I}\}$, just as was the case in $\mathbf{Z}$. In any case, however, the set $a = \{\lambda\alpha + \mu\beta | \lambda, \mu \in \mathbf{I}\}$ is a vector space over $\mathbf{I}$. We already met with such sets over $\mathbf{Z}$ and have called them *modules* (see Definition 3.5). We observe that if $\gamma \in a$, then for every $\rho \in \mathbf{I}$, also $\rho\gamma \in a$. Sets like a are called *ideals*. Their precise definition and properties will be found in Chapter 11. Here we only anticipate the remark that ideals of the type mentioned above $a = \{\rho\delta | \rho \in \mathbf{I}\}$ with a single generator are called *principal ideals*. Every integer of $\mathbf{I}$ is in a $1 : 1$ correspondence with the principal ideal of its multiples. To distinguish such principal ideals from their generators, we shall write (α) for the principal ideal of the multiplies of $\alpha \in \mathbf{I}$.

By introducing ideals we shall dispose of the machinery to be developed in Chapter 11, and in particular (see Theorem 11.20) we shall succeed in re-establishing uniqueness of factorization in number fields, if not for all integers, at least for their ideals.

8 THE FIELD $\mathbf{Q}(\sqrt{5})$

The polynomial $p(x) = x^2 - 5$, like $x^2 - 2$, leads to an algebraic number field $\mathbf{K} = \mathbf{Q}(\theta)$, where now, however, $\theta = \sqrt{5}$ rather than $\sqrt{2}$. The elements of the field are $\alpha = A + B\theta$ $(\theta^2 = 5)$, the trace is $S(\alpha) = 2A$ as in the previous case, the norm is $N(\alpha) = A^2 - 5B^2$, and many of the results are the same for both fields; however, there are also differences. As before, $\alpha \in \mathbf{I}\,(= \mathbf{I}(\theta),\ \theta^2 = 5)$ is an integer if it is one of the solutions of the equation

$$x^2 - S(\alpha)x + N(\alpha) = 0,$$

with $S(\alpha) = a$, $N(\alpha) = b$, $a, b \in \mathbf{Z}$. In the present case this equation, written out explicitly, reads

$$x^2 - 2Ax + A^2 - 5B^2 = 0.$$

If $2A = a$ is odd, say $a = 2m + 1$, then $A = m + 1/2$, $b = A^2 - 5B^2 = m^2 + m + 1/4 - 5B^2$, whence $5B^2 = m^2 + m - b + 1/4$. This equality is not possible with B either an integer or an irreducible rational with denominator other than 2; but if we set $B = (2k + 1)/2 = k + 1/2$, then $5B^2 = 5(k^2 + k + 1/4) = 5(k^2 + k) + 1 + 1/4$, and we obtain that $b = m^2 + m - 5(k^2 + k) - 1$, an integer. If on the other hand a is even, say $a = 2m$, $A = m$ and $b = A^2 - 5B^2 = m^2 - 5B^2$, which shows that B also has to be an integer. The result of these considerations is:

Theorem 4.   *The integers of $\mathbf{Q}(\sqrt{5})$ are of the form $(a + b\sqrt{5})/2$, where a and b are rational integers of the same parity (either both even, or both odd).*

It is clear that 1 and $\theta = \sqrt{5}$ are no longer an integral basis for $\mathbf{I}$, because they yield only the integers of the form $c + d\sqrt{5}$ (corresponding only to a and b even in Theorem 4). However, $\{1, (1 + \sqrt{5})/2\}$ do form an integral basis. Indeed, $a + b(1 + \sqrt{5})/2 = (2a + b + b\sqrt{5})/2$, and $2a + b, b$ form a couple of integers of the same parity, and all such couples are obtained as a and b range independently over all rational integers. The number $\alpha = (1 + \sqrt{5})/2$ is of course an integer (it is the case of a and b both odd, in Theorem 4) and satisfies the equation $x^2 - x - 1 = 0$. Its (algebraic) conjugate is the "other" root of the same equation, i.e., $\alpha^{(2)} = (1 - \sqrt{5})/2$; hence

$$D = \begin{vmatrix} 1 & \alpha^{(1)} \\ 1 & \alpha^{(2)} \end{vmatrix} = \begin{vmatrix} 1 & (1 + \sqrt{5})/2 \\ 1 & (1 - \sqrt{5})/2 \end{vmatrix} = -5,$$

so that the discriminant becomes $\Delta = D^2 = 5$. In this case therefore, $\Delta = \theta^2$, rather than $4\theta^2$ as in the previous two cases.

Finally, let us recall that the field $\mathbf{Q}(\sqrt{5})$ is one of only 16 real quadratic fields that have a Euclidean algorithm; consequently, factorization in its ring of integers is unique.

9 THE GAUSSIAN FIELD **K** = **Q**($\sqrt{-1}$)

This is one of the first algebraic number fields studied. As its name suggests, its arithmetic has been clarified by Gauss (before 1800; see [10]). By following the procedures used for $\mathbf{Q}(\sqrt{d})$ in the previous cases, we find that the discriminant $\Delta = D^2 = 4d = -4$, and the integers are of the form $a + bi$, $a, b \in \mathbf{Z}, \sqrt{-1} = i$. The norm $N(a + bi) = (a + bi)(a - bi) = a^2 + b^2 > 0$. The units have $N(a + bi) = 1$, so that either $a = \pm 1, b = 0$ or $a = 0, b = \pm 1$, and this leads to the four units $\pm 1, \pm i$. It is a field with uniqueness of factorization because, as mentioned in Section 6, it even has a Euclidean algorithm (see Problem 13). Consequently (see Remark 4), the concepts of "prime" and "indecomposable" coincide.

What are these prime elements of $\mathbf{K} = \mathbf{Q}(\sqrt{-1})$? Let $\pi = a + bi$ be such a prime. Then $N\pi = \pi\bar{\pi} \in \mathbf{Z}$; hence π divides some rational integer, and consequently at least one rational prime, say p. If $\pi | p$, then $\pi \nmid q$ for any other rational prime q. Indeed, $(p, q) = 1$, so that $mp + nq = 1$ for some $m, n \in \mathbf{Z}$ and $\pi | p, \pi | q \Rightarrow \pi | 1$, which is false because primes are not units. If $\pi | p$, then $p = \pi\lambda, \lambda \in \mathbf{I}\ (= \mathbf{I}(i))$. From $p = \pi\lambda$ it follows that $Np = p^2 = N\pi \cdot N\lambda$. There are now only two possibilities. Either

(i) $N\pi = p^2, N\lambda = 1$, so that λ is a unit and π is an associate of p, that is, that p itself is a prime in **K**, or else

(ii) $N\pi = N\lambda = p$. In this case $\pi = a + bi\ (a \cdot b \neq 0)$, and also λ is a prime in **K**, say $\lambda = c + di$, with $N\pi = a^2 + b^2 = N\lambda = c^2 + d^2 = p$. Moreover, $\pi\lambda = (a + ib)(c + id) = ac - bd + i(ad + bc) = p$, so that $ad + bc = 0, ac - bd = p$.

Considered as a system of two equations in c and d, with determinant

$$\begin{vmatrix} a & b \\ a & -b \end{vmatrix} = -2ab \neq 0,$$

these equations yield $c = a, d = -b$, so that $\lambda = a - ib = \bar{\pi}$, the conjugate (both algebraic and complex) of π. We now have to establish when if ever π and $\bar{\pi}$ are associates. If $\rho = \pi/\bar{\pi} = (a + ib)/(a - ib) = (a^2 - b^2)/(a^2 + b^2) + i(2ab/(a^2 + b^2))$, then $\rho = \pm 1$ requires $a \cdot b = 0$, which is false, while $\rho = \pm i$ requires $a = \pm b$, so that $\pi = a(1 \pm i)$. In that case, $N\pi = p = 2a^2$, possible only for $p = 2$, $a = \pm 1$. Hence for $p = 2$, $2 = -i(1 + i)^2$ and $\pi = 1 + i$ is associated to $\bar{\pi} = 1 - i$ (indeed $(-i)\pi = \bar{\pi}$), so that 2 is essentially a square, $2 = (-i)\pi^2$.

In case of odd p, as seen, either p itself is a prime in $\mathbf{K}$ or p *splits* into two nonassociated primes conjugate to each other. If $p \equiv 3 \pmod{4}$, the first case occurs. Indeed $N\pi = a^2 + b^2 = p \equiv 3 \pmod{4}$ is not possible because $a^2 + b^2 \equiv 0$, 1, or 2 $\pmod{4}$. On the other hand, if $p \equiv 1 \pmod{4}$, then $p = a^2 + b^2$ always has a solution and $p = (a + ib)(a - ib) = \pi \cdot \bar{\pi}$. Indeed $p \equiv 1 \pmod{4}$ $\Rightarrow (-1/p) = +1$ (see Theorem 5.2(6)), so that $p | a^2 + 1$ for some $a \in \mathbf{Z}$; hence if p is a prime, $p | (a + i)(a - i)$, and for either $+$ or $-$, $p | a \pm i$. This however is false, as neither of $(a \pm i)/p$ is a Gaussian integer. We conclude that if $p \equiv 1 \pmod{4}$, the Diophantine equation

$$a^2 + b^2 = p \tag{10}$$

always has a solution. As already seen, this solution is essentially unique. Obviously, any of the four associates of π could be selected instead of π, and it is convenient to say that (a, b), $(-a, b)$, $(a, -b)$, $(-a, -b)$, (b, a), $(-b, a)$, $(b, -a)$, and $(-b, -a)$ are eight solutions of (10).

Let us formalize the results obtained concerning the Gaussian field $\mathbf{Q}(\sqrt{-1})$ in:

Theorem 5. *The field* $\mathbf{K} = \mathbf{Q}(\sqrt{-1})$ *is a Euclidean field, isomorphic to* $\mathbf{Q}[x]/\{(x^2 + 1)\}$. *Its discriminant is* $\Delta = -4$. *It has the property of uniqueness of factorization of integers into indecomposable (i.e., prime) elements, and its integers are of the form* $a + ib$ *with* $a, b \in \mathbf{Z}$. *The norm of* $\alpha = a + ib$ *is* $N\alpha = a^2 + b^2$. *There are four units in* $\mathbf{K}$, *namely* ± 1, $\pm i$. *The rational primes behave in* $\mathbf{K}$ *as follows:*

(i) *If* $p \equiv 1 \pmod{4}$, *so that* $(\Delta/p) = 1$, *then* p *splits into two nonassociated but conjugate primes of* $\mathbf{K}$ *and* $p = \pi \cdot \bar{\pi}$.

(ii) *If* $p \equiv 3 \pmod{4}$, *so that* $(\Delta/p) = -1$, *then* p *remains a prime in* $\mathbf{K}$.

(iii) *If* $p = 2$, *so that* $(\Delta/p) = 0$, *then* p *is associated to the square of a prime in* $\mathbf{K}$, *namely* $2 = (-i)\pi^2$, $\pi = 1 + i$.

Consequently, the primes of $\mathbf{K}$ *are, up to associates,*

(i) *the factors* $\pi = a + ib$ *of the rational primes* $p \equiv 1 \pmod{4}$;

(ii) *the rational primes* $p \equiv 3 \pmod{4}$; *and*

(iii) *the prime divisor* $\pi = 1 + i$ *of the rational prime* $p = 2$.

For later use we also record as a theorem the following result.

Theorem 6. *The Diophantine equation*

$$x^2 + y^2 = p$$

has no solution if $p \equiv 3 \pmod 4$; it has the four solutions $x = \pm 1, y = \pm 1$ if $p = 2$ and has essentially a single solution if $p \equiv 1 \pmod 4$. In this case we have eight solutions, by counting solutions that differ by the order or sign of one of the variables x, y as different solutions.

The proof of Theorem 6 is contained in preceding reasonings, to which one may add the remark that for $p \equiv 1 \pmod 4$, $x \neq y$ (they must even be of opposite parity), so that the eight mentioned solutions can never reduce to four, as in the case of $p = 2$.

10 QUADRATIC FIELDS

After the study of these four particular cases of quadratic fields, the reader should have no difficulty in proving by himself the following general theorem.

Theorem 7. *Let d be a squarefree integer, positive or negative. Then $\mathbf{K} = \mathbf{Q}(\sqrt{d}) = \{A + B\sqrt{d} | A, B \in \mathbf{Q}\} \subset \mathbf{C}$ is a field under ordinary addition and multiplication. If $\mathbf{L} = \mathbf{Q}[x]/\{p(x)\}$ where $p(x) = x^2 - d$, then these two fields are isomorphic, $\mathbf{K} \simeq \mathbf{L}$. If $d \equiv 1 \pmod 4$, then $\{1, (1 + \sqrt{d})/2\}$ is an integral basis for the ring of integers $\mathbf{I} = \mathbf{I}(\sqrt{d})$, and the discriminant is*

$$\Delta = D^2 = \begin{vmatrix} 1 & (1 + \sqrt{d})/2 \\ 1 & (1 - \sqrt{d})/2 \end{vmatrix}^2 = (-\sqrt{d})^2 = d.$$

If $d \equiv 2 \pmod 4$ or $d \equiv 3 \pmod 4$, then $\{1, \sqrt{d}\}$ is an integral basis for $\mathbf{I}$ and the discriminant is

$$\Delta = D^2 = \begin{vmatrix} 1 & \sqrt{d} \\ 1 & -\sqrt{d} \end{vmatrix}^2 = 4d.$$

If $d < 0$, all units of $\mathbf{I}$ are also roots of unity. Specifically, if $d = -1$, the units are $\pm 1, \pm i$; if $d = -3$, $\mathbf{I}$ contains the six units $\pm 1, (\pm 1 \pm i\sqrt{3})/2$, with all combinations of signs; for all other $d < 0$, $\mathbf{I}$ has only the two units ± 1.

If $d > 0$, the positive units are given by the powers of the fundamental unit; this is obtained as a solution of Pell's equation $x^2 - dy^2 = \pm 1$ if $d \equiv 2, 3 \pmod 4$, and of $x^2 - dy^2 = \pm 4$ if $d \equiv 1 \pmod 4$.

The odd rational primes split in $\mathbf{K}$ as follows:

(i) *If $(\Delta/p) = +1$, then p splits into two nonassociated prime factors.*

(ii) *If $(\Delta/p) = -1$, then p remains a prime factor in $\mathbf{K}$.*

(iii) *If $(\Delta/p) = 0$ (i.e., if $p|\Delta$), then p splits into two essentially identical (i.e., associated) factors*

For $p = 2$ essentially the same rule holds, but (Δ/p) has to be understood as a quadratic residue or non-residue modulo 8.

The "factors" are integers of **K** (i.e., elements of **I**), or equivalently, principal ideals, if **K** is a field with uniqueness of factorization; otherwise, they have to be understood as (not necessarily principal) ideal factors.

Proofs of Theorem 7 can be found, e.g., in [12] or in [20]; the reader is, however, encouraged to prove it to the extent he finds possible before consulting either of these texts.

We saw in Section 6 that there are five imaginary quadratic fields and 16 real quadratic fields with Euclidean algorithms, hence with uniqueness of factorization into indecomposable elements. In addition, the fields $\mathbf{Q}(\sqrt{m})$ with $m = -19, -43, -67, -163$ are also fields with uniqueness of factorization, although they have no Euclidean algorithm. The complete set of such fields with positive m is not known, but it is conjectured to be infinite.

11   CYCLOTOMIC FIELDS

Let us consider the polynomial $z^n - 1$ $(0 < n \in \mathbf{Z})$; it is not irreducible because $z^2 - 1 = (z - 1)(z^{n-1} + z^{n-2} + \cdots + z + 1)$. Clearly if ζ is a solution of $z^n - 1 = 0$, then $\zeta^n = e^{2\pi i k}$ for some integer k, and $\zeta = e^{2\pi i k/n} = \cos 2\pi k/n + i \sin 2\pi k/n$, $k = 0, 1, \ldots, n - 1$. These n complex numbers all have absolute value one and are all distinct. Indeed, if we plot them on a graph, so that $x = \cos 2\pi k/n$, $y = \sin 2\pi k/n$, we obtain n points that subdivide the unit circle into n equal arcs, each of $2\pi/n$ radians, starting with $x = 1$, $y = 0$ on the real axis. For this reason, the polynomial (with the factor $z - 1$ removed) is called the nth *cyclotomic polynomial*. We recall (see Section 6.9) that the numbers $\zeta = e^{2\pi i k/n}$ are called nth roots of unity. If $k = dk_0$, $n = dn_0$ with $(k_0, n_0) = 1$, $d > 1$, then $\zeta = e^{2\pi i k_0/n_0}$ is already an n_0th root of unity with $n_0 < n$. If $(k, n) = 1$, then ζ is not a root of unity of any lower order and is called a *primitive nth root of unity*. If we adjoin a primitive nth root of unity ζ to **Q**, we obtain the *cyclotomic field* $\mathbf{Q}(\zeta)$. This has particular relevance for the treatment of many Diophantine equations, and particularly Fermat's equation. In what follows, we shall be interested mainly in the case $n = p$ with p a rational prime. In that case, as already mentioned in Section 2, the cyclotomic polynomial $p_n(z) = z^{p-1} + z^{p-2} + \cdots + z + 1$ is irreducible over **Q**.

PROOF.   Set $z = y + 1$; then $p_n(z) = p_n(y + 1) = q_n(y)$, say, and it is clear that p_n and q_n are either both irreducible or both reducible. In fact there is a

$1:1$ correspondence between their irreducible factors. We now observe that $z^p - 1 = (y + 1)^p - 1 = y^p + py^{p-1} + \cdots + py = y(y^{p-1} + pyh(y) + p)$, and removing the factor $z - 1 = y$, $p_n(z) = q_n(y) = y^{p-1} + pyh(y) + p$, where $h \in \mathbf{Z}[y]$. By Eisenstein's Criterion (see Corollary 2, Section 2), $q_n(y)$ is irreducible over $\mathbf{Q}$ and hence so is $p_n(z)$, as claimed.

As ζ^k ($k = 0, 1, \ldots, p - 2$) are the $p - 1$ zeros of $p_n(z)$, it follows that these powers are conjugates of ζ. From the irreducibility of p_n it now follows that if ζ is any of the pth roots of unity ($\zeta \neq 1$), then the powers $1, \zeta, \zeta^2, \ldots, \zeta^{p-2}$ are linearly independent over $\mathbf{Q}$, while $\zeta^{p-1} = -(1 + \zeta + \zeta^2 + \cdots + \zeta^{p-2})$; hence the field $\mathbf{Q}(\zeta)$ consists of the sums $A_0 + A_1\zeta + \cdots + A_{p-2}\zeta^{p-2}$, where the A_js ($j = 0, 1, \ldots, p - 2$) run independently through all rationals. It follows that $\mathbf{Q}(\zeta)$ is a $(p - 1)$-dimensional vector space over $\mathbf{Q}$ with a basis $(1, \zeta, \ldots, \zeta^{p-2}\}$. The proof that this set forms a field under addition and multiplication follows the same pattern as in the cases already discussed and is left to the reader (see Problem 15).

What are the integers in $\mathbf{Q}(\zeta)$? In the quadratic fields we found as integers either the elements $a + b\theta$ ($\theta =$ generator of the field and an integer) or the elements $(a + b\theta)/2$ (a, b integers of the same parity). In fields of higher degrees, the situation is still more complicated. It comes therefore as a pleasant surprise that all integers of $\mathbf{Q}(\zeta)$, $\zeta = e^{2\pi i/p}$, are given by $a_0 + a_1\zeta + a_2\zeta^2 + \cdots + a_{p-2}\zeta^{p-2}$, where the a_js ($j = 0, 1, \ldots, p - 2$) run independently through all rational integers. Hence $1, \zeta, \zeta^2, \ldots, \zeta^{p-2}$ form an integral basis for the ring $\mathbf{I} = \mathbf{I}(\zeta)$ of integers of $\mathbf{Q}(\zeta)$. The proof of this fact will not be given here (see [12] or [20]). The converse, that all such polynomials in ζ represent integers, is trivial; indeed the integers of $\mathbf{I}$ form a ring, hence are closed under addition and multiplication, and all a_js as well as ζ belong to $\mathbf{I}$.

By generalizing from the quadratic case, we now define the discriminant Δ by $\Delta = D^2$ where D is the determinant that has as first row an integral basis and in successive rows the (algebraic) conjugates of these basis elements. The value of Δ does not depend on the choice of the basis (see Problem 16). In the present case, with basis elements the successive powers of ζ, we have

$$D = \begin{vmatrix} 1 & \zeta & \zeta^2 & \cdots & \zeta^{p-2} \\ 1 & \zeta^2 & \zeta^4 & \cdots & \zeta^{2(p-2)} \\ \vdots & & & & \\ 1 & \zeta^{p-1} & \zeta^{2(p-1)} & \cdots & \zeta^{(p-2)(p-1)} \end{vmatrix} = \begin{vmatrix} \zeta & \zeta^2 & \cdots & \zeta^{p-1} \\ \zeta^2 & \zeta^4 & \cdots & \zeta^{2(p-1)} \\ \vdots & & & \\ \zeta^{p-1} & \zeta^{2(p-1)} & \cdots & \zeta^{(p-1)^2} \end{vmatrix}$$

The first equality defines D; the second follows by observing that the second determinant is equal to the first multiplied by $\zeta^{1+2+\cdots+(p-1)} = \zeta^{p(p-1)/2} = 1$. The second determinant is a Vandermonde determinant and equals

$\prod_{1 \leq \nu < \mu \leq p-1}(\zeta^{\nu} - \zeta^{\mu})$; hence we obtain for the discriminant the value

$$\Delta = D^2 = \prod_{1 \leq \nu < \mu \leq p-1} (\zeta^{\nu} - \zeta^{\mu})^2 = \prod_{\substack{\nu \neq \mu \\ 1 \leq \nu, \mu \leq p-1}} (\zeta^{\nu} - \zeta^{\mu}).$$

One may show (see Problem 17) that $\Delta = (-1)^{(p-1)/2} p^{p-2}$. We shall not discuss in detail the units and indecomposable elements of $I(\zeta)$ ($\zeta = (e^{2\pi i/p})$), but shall content ourselves with the following remarks.

ζ itself is a root of unity, hence a unit, and so are all its powers, which are also, as already observed, its conjugates. Hence $N\zeta = \zeta^{1+2+\cdots+(p-1)} = \zeta^{p(p-1)/2} = 1$. Among the units we find, e.g., $1 + \zeta$ (and consequently all the powers, positive and negative, of $1 + \zeta$) and, more generally, $\varepsilon = (1 - \zeta^s)/(1 - \zeta)$ for $(s, p) = 1$. The problem of finding independent sets of generators of the full group of units is still not solved in all its generality (see, however, [22], [2], [15]).

The integer $\lambda = 1 - \zeta$ is an indecomposable element. Indeed, $N\lambda = (1 - \zeta)(1 - \zeta^2) \cdots (1 - \zeta^{p-1})$. To estimate this product, we observe that the cyclotomic polynomial has the zeros ζ^j ($j = 1, 2, \ldots, p - 1$); hence $(z^p - 1)/(z - 1) = 1 + z + \cdots + z^{p-1} = \prod_{j=1}^{p-1}(z - \zeta^j)$. If we here set $z = 1$, we obtain $p = N\lambda$. We already know that this condition (while not necessary) is sufficient to insure that λ is indecomposable.

There is, however, a most important question concerning these cyclotomic fields $Q(\zeta)$, $\zeta = e^{2\pi i/p}$ that has to be clarified before any arithmetic can be attempted in these fields. Are the rings of integers of these fields domains with uniqueness of factorization? Euler assumed (in 1753) that this was the case at least for $p = 3$, and even a century later Lamé believed this to be the case for all primes. Legendre [16] proved the uniqueness of factorization for $p = 3$ and for $p = 5$, and Dirichlet [9] also had such a proof. The theorem of the uniqueness of factorization in $I(\zeta)$ was proved successively for other primes, $p = 7, 11, 13, \ldots$. However in spite of all their efforts, the German and French mathematicians of the middle of the 19th century did not succeed in abstracting from the many particular cases in which they had been successful some general method that would yield the uniqueness of factorization for all cyclotomic fields $Q(\zeta)$, $\zeta = e^{2\pi i/n}$ or at least for the cases $n = p$, a rational prime. One may well imagine their frustration when they found proofs for $p = 3, 5, 7, 11, 13, 17$ and even for $p = 19$ (for historic reviews see [8], [17]), each proof slightly different from the others, but the general theorem still eluded them. This mystery was soon elucidated. Indeed, Cauchy [6] showed that for the very next prime, $p = 23$, the factorization in the ring of integers $I(e^{2\pi i/23})$ fails to be unique. Kummer had already shown a similar result, but his work, published in a rather obscure place, had not become known until somewhat later. Further work, which culminated with Masley [18], [19], showed that $p = 19$ was in fact the last prime for which $\mathbf{I}(e^{2\pi i/p})$ enjoys the property of uniqueness of factorization. It is quite likely that Kummer had discovered the nonuniqueness of factorization in the general cyclotomic fields at about the same time as Cauchy (or earlier). It was in fact at least in part this recognition (the other incentive was his attempt to generalize the quadratic reciprocity law) that led him to the definition of ideals, which permits us to re-establish a certain uniqueness of factorization.

12   OTHER ALGEBRAIC NUMBER FIELDS

Let us consider the polynomial $f(x) = x^4 + 2x^3 - x^2 - x + 2$; it has rational integer coefficients, so that $f \in \mathbf{Z}[x] \subset \mathbf{Q}[x]$. However contrary to our previous examples, f is not irreducible over $\mathbf{Q}$ (or equivalently, by Gauss' Lemma, over $\mathbf{Z}$). Indeed $f(x) = (x^3 - x + 1)(x + 2)$. The last factor has the rational integral zero $x = -2$, which already belongs to $\mathbf{Q}$; hence adjoining it to $\mathbf{Q}$ does not enlarge the field. The first factor is irredudible over $\mathbf{Q}$. (Prove this; see Problem 18.) If $p(x) = x^3 - x + 1$, then $p(-1) = 1, p(-2) = -5$, so that there exists a real $\theta, -2 < \theta < -1$, for which $p(\theta) = 0$. We may use either Newton's method or any other convenient one to determine that $\theta \cong -1.325\dots$. If we factor $x - \theta$ out of $p(x)$, we obtain as quotient a quadratic polynomial. This can be solved by the usual formula and yields the other two zeros, $\theta^{(2)} \cong .6625\dots + .5628\dots i$ and $\theta^{(3)} \cong .6625\dots - .5628\dots i$. We recall that in general, if $p(x) \in \mathbf{Q}[x]$ is an irreducible polynomial of degree n, its complex zeros, if any, occur in pairs of complex conjugates. If there are r_1 real zeros and r_2 pairs of complex conjugate ones, then $n = r_1 + 2r_2$ (either r_1 or r_2 may of course be zero). For later use we record that the number $r = r_1 + r_2 - 1$ is an important parameter of the field $\mathbf{Q}(\theta)$ ($p(\theta) = 0$, $p \in \mathbf{Q}[z]$, p irreducible).

Returning now to our specific irreducible polynomial $p(x) = x^3 - x + 1$, let θ be one of its zeros. We do not have to specify which one; in fact any will do and we denote it by θ. Then $\theta^3 = \theta - 1, \theta^4 = \theta^2 - \theta, \theta^5 = \theta^3 - \theta^2 = \theta - 1$ $-\theta^2 = -\theta^2 + \theta - 1, \theta^6 = -\theta^3 + \theta^2 - \theta = -\theta + 1 + \theta^2 - \theta = \theta^2 - 2\theta + 1 = (\theta - 1)^2$ (naturally!), etc. It follows inductively that for any $n \in \mathbf{Z}$, $\theta^n = A + B\theta + C\theta^2$. Consequently, any polynomial in θ with rational coefficients reduces to the form $A + B\theta + C\theta^2$. The set $\{A + B\theta + C\theta^2 | A, B, C \in \mathbf{Q}, \theta^3 - \theta + 1 = 0\}$ forms a field denoted by $\mathbf{Q}(\theta)$. As in the previous cases, the verifications are trivial, except for the existence of a multiplicative inverse. This we proceed to do in detail.

Let us assume that $A, B, C \in \mathbf{Q}$ and that A, B, C do not all three vanish, and set $(A + B\theta + C\theta^2)^{-1} = x + y\theta + z\theta^2$. We want to show that there exists a unique solution in rationals x, y, z. If we multiply both sides by $A + B\theta + C\theta^2$ and replace θ^3 and θ^4 by their values found above, we find that θ satisfies the equation

$$Ax - Cy - Bz - 1 + (Bx + (A + C)y + (B - C)z)\theta$$
$$+ (Cx + By + (A + C)z)\theta^2 = 0. \tag{11}$$

If x, y, z are rational and not all coefficients in (11) vanish, then it would follow that θ satisfies a quadratic equation with rational coefficients. This, however, is not possible, because θ is a root of an irreducible cubic equation. (How exactly does this follow? See Problem 19.) Hence all coefficients of (11) vanish, so that x, y, and z satisfy the linear system

$$Ax - Cy - Bz = 1$$
$$Bx + (A + C)y + (B - C)z = 0 \tag{12}$$
$$Cx + By + (A + C)z = 0.$$

All operations needed to solve for x, y, z are rational and are applied to A, B, C, and 1, also all rational; hence if the determinant of (12) does not vanish, we shall have shown that every nonzero element of $\mathbf{Q}(\theta)$ has a unique inverse in $\mathbf{Q}(\theta)$, as claimed. This also completes the verification that $\mathbf{Q}(\theta)$ is a field. The determinant of (12), however, turns out (see Problems 20 and 21) to be precisely the norm $N(A + B\theta + C\theta^2)$ and can vanish only if $A = B = C = 0$, i.e., in the one excluded case. The proof is complete.

13   INTEGERS AND UNITS IN ALGEBRAIC NUMBER FIELDS

Having reached this point, in the previous cases we asked questions about the integers and units in those fields. If, e.g., $p \in \mathbf{Q}[x]$ is irreducible, $\partial^0 p = 3$ and $p(\theta) = 0$, we would like to know for $\alpha = A + B\theta + C\theta^2$ under what additional conditions on A, B, C (A, B, $C \in \mathbf{Q}$) α is an integer, and in particular a unit. These conditions are much more complicated in fields of higher degrees; they are, however, not beyond attack. So, e.g., we know that α is an integer if and only if it satisfies an equation of the form

$$x^3 - S(\alpha)x^2 + P(\alpha)x - N(\alpha) = 0,$$

where $S(\alpha) = \alpha + \alpha' + \alpha''$, $P(\alpha) = \alpha\alpha' + \alpha'\alpha'' + \alpha\alpha''$, $N(\alpha) = \alpha\alpha'\alpha''$ (α, α', α'' the three roots of $p(x) = 0$) and $S(\alpha) = a$, $P(\alpha) = b$, $N(\alpha) = c$ are all rational integers. Now these symmetric functions can be expressed as polynomials in A, B, and C. In the case studied, with $p(x) = x^3 - x + 1$, it is easy to compute $S(\alpha) = 3A + 2C$; we also know $N(\alpha)$ (namely the determinant of (12)), and it is cumbersome but not difficult to also express $P(\alpha)$ as a polynomial in A, B, and C (see Problem 22). It then follows that α is an integer precisely when there exist $a, b, c \in \mathbf{Z}$, such that $S(\alpha) = a$, $P(\alpha) = b$, and $N(\alpha) = c$ hold. The translation of these implicit conditions on A, B, and C into explicit ones generally does not lead to any simple statements.

As for units, these have to satisfy the same conditions as all integers and also the additional one $c = \pm 1$.

We know that, for $p(x) = x^3 - x + 1$, each solution θ of $p(x) = 0$ is a unit (because $N(\theta) = -1$) and so are all its (positive and negative) integral powers. However the determination of all units of a cubic field is a rather difficult problem (see, e.g., [3], [4]).

Still more difficult is the determination of the indecomposable elements and the existence or nonexistence of a Euclidean algorithm. No really general theorems are known. The algebraic number fields are being studied one by one. There are only 334 fields with known Euclidean algorithms and among them 109 cubic fields with such algorithms. Whether there are any other fields, cubic or of higher degrees, with Euclidean algorithm is unknown but unlikely. On the other hand, there exist many algebraic number fields without a Euclidean algorithm but still with uniqueness of factorization in their rings of integers. Still, for many algebraic number fields, and in particular for the so-important cyclotomic fields $\mathbf{Q}(e^{2\pi i/p})$ with $p > 19$, we know that the factorization is not unique and the use of ideals becomes indispensable. For its general interest, we here add an important theorem concerning units in algebraic number fields. We shall not use this theorem anywhere in this book and do not give its proof. The interested reader may find it, e.g., in [13].

We say that a unit ε of some field $\mathbf{Q}(\theta)$ is expressible by or dependent on the units $\varepsilon_1, \varepsilon_2, \ldots, \varepsilon_s$ of that field if $\varepsilon^a = \varepsilon_1^{a_1}\varepsilon_2^{a_2} \cdots \varepsilon_s^{a_s}$ holds with some rational integers $a, a_1, a_2, \ldots, a_s$, not all zero. The unit ε is said to be independent of $\varepsilon_1, \varepsilon_2, \ldots, \varepsilon_s$ if any such relation implies that $a = a_1 = a_2 = \cdots = a_s = 0$. This statement is equivalent to the following:

DEFINITION 5. Let $a_1, a_2, \ldots, a_s$ be rational integers; if a relation $\sum_{j=1}^{s} a_j \log \varepsilon_j = 0$ among the s units ε_j $(j = 1, 2, \ldots, s)$ of an algebraic number field implies that all coefficients a_j vanish, then the s units are said to be independent. If such a relation holds for integers a_j not all zero, the s units are said not to be independent.

DEFINITION 6. Let θ be an algebraic number of degree n and set $\mathbf{K} = \mathbf{Q}(\theta)$; also let $\theta = \theta^{(1)}, \theta^{(2)}, \ldots, \theta^{(n)}$ be the conjugates of θ. The fields $\mathbf{K}^{(j)} = \mathbf{Q}(\theta^{(j)})$ $(j = 1, 2, \ldots, n)$ are called the fields conjugate to $\mathbf{K}$.

With this terminology we now state:

Theorem 8 (Dirichlet). *The units of an algebraic number field* $\mathbf{K}$ *form an abelian group under multiplication. If* $\mathbf{K}$ *has* r_1 *real conjugate fields and* $2r_2$

complex conjugate ones, then the group of units has $r + 1 \, (= r_1 + r_2)$ generators. Of these, r are of infinite order and one is a root of unity. Equivalently, every unit ε of **K** *is of the form*

$$\varepsilon = \zeta^a \eta_1^{a_1} \cdots \eta_r^{a_r},$$

where $\eta_1,\ldots,\eta_r$ are r independent units, $a_1, a_2,\ldots,a_r$ are rational integers, and a may take one of the t values $0,1,\ldots,t-1$. Here t is the number of roots of unity in **K**.

The units $\eta_j \ (j = 1, 2,\ldots,r)$ are called *fundamental units*.

For future use we also give here the definition of the *regulator* of the field **K**. Let us denote by $\mathbf{K}^{(1)}, \mathbf{K}^{(2)},\ldots,\mathbf{K}^{(r_1)}$ the r_1 real conjugate fields of **K** and by $\mathbf{K}^{(r_1+1)},\ldots,\mathbf{K}^{(r_1+r_2+1)},\ldots,\mathbf{K}^{(r_1+2r_2)}$ the $2r_2$ complex conjugate fields, with $\mathbf{K}^{(r_1+s)}$ the complex conjugate of $\mathbf{K}^{(r_1+r_2+s)}$. For each superscript $j \ (j = 1, 2,\ldots,n, \ n = r_1 + 2r_2)$, let $c_j = 1$ if $1 \le j \le r_1$, $c_j = 2$ if $r_1 < j \le n$. With these notations we have:

DEFINITION 7. The *regulator* R of the field **K** is the determinant

$$R = \begin{vmatrix} c_1 \log |\eta_1^{(1)}| & c_1 \log |\eta_2^{(1)}| & \cdots & c_1 \log |\eta_r^{(1)}| \\ c_2 \log |\eta_1^{(2)}| & c_2 \log |\eta_2^{(2)}| & \cdots & c_2 \log |\eta_r^{(2)}| \\ \vdots & & & \\ c_r \log |\eta_1^{(r)}| & c_r \log |\eta_2^{(r)}| & \cdots & c_r \log |\eta_r^{(r)}| \end{vmatrix},$$

where $\eta_j^{(k)}$ is the conjugate of the fundamental unit η_j in the field $\mathbf{K}^{(k)}$.

PROBLEMS

1. Prove that $\sqrt{2}$ is not rational; more generally, show that unless m is a perfect square $\sqrt{m}$ is not rational.
2. Let $p \in \mathbf{Q}[x]$ be irreducible, of second degree; let θ be any of its zeros, and set $\mathbf{K} = \mathbf{Q}(\theta)$. Also, set $\mathbf{L} = \mathbf{Q}[x]/\{p(x)\}$. Write out explicitly a $1:1$ correspondence between the elements of **K** and those of **L** and prove that this correspondence is an isomorphism. Can you generalize this result by dropping the requirement $\partial^0 p = 2$?
3. Show that if $\mathbf{K} = \mathbf{Q}(\theta)$ and $[\mathbf{K}:\mathbf{Q}] = n$, then **K** is an n-dimensional vector space over **Q** with $\{1, \theta, \theta^2,\ldots,\theta^{n-1}\}$ as a basis.
4. In the case of Pell's equation $x^2 - dy^2 = a$ with $d = g^2$ a perfect square, only factorizations of a of the form $a = d_1 d_2, d_1 \equiv d_2 \pmod{2g}$ were considered. Why?

5. Solve the following Pell equations, or prove that they have no solutions.
   (a) $x^2 - y^2 = 1$
   (b) $x^2 - 4y^2 = 2$
   (c) $x^2 - 25y^2 = 36$
   (d) $x^2 - 25y^2 = 56$.

6. Determine all units in the field $\mathbf{K} = \mathbf{Q}(\sqrt{3}\,)$.

7. Show that all rational primes $p \equiv 3 \pmod 8$ remain primes (i.e., indecomposable) in the field $\mathbf{K} = \mathbf{Q}(\sqrt{2}\,)$.

8. Show that all rational primes $p \equiv 3 \pmod 4$ are indecomposable in the field $\mathbf{K} = \mathbf{Q}(\sqrt{-5}\,)$.

9. Establish the equations satisfied by the four algebraic numbers $\sqrt{2} \pm \sqrt{-5}\,, \sqrt{-2} \pm 3\sqrt{-5}$ and show that all four numbers are algebraic integers of degree 4. (Caution: For the last result one has to verify the irreducibility of the minimal polynomials of those numbers.)

10. (a) Find an algebraic number ϕ such that $\sqrt{-5} \in \mathbf{Q}(\phi)$ and $\sqrt{2 + \sqrt{-5}} \in \mathbf{Q}(\phi)$.

    (b) Find an algebraic number ψ such that all factors of (9) belong to $\mathbf{Q}(\psi)$.

11. Find the minimal equation over $\mathbf{Q}$ satisfied by the algebraic number ψ, obtained in Problem 10.

12. Show that in fields with uniqueness of factorization the two logically distinct properties

    (i) $\pi = \alpha \cdot \beta,\ \pi, \alpha, \beta \in \mathbf{I},\ N\pi > 1 \Rightarrow |N\alpha| = 1 \text{ or } |N\beta| = 1$; and
    (ii) $\pi | \alpha \cdot \beta \Rightarrow \pi | \alpha \text{ or } \pi | \beta$

    define the same set of integers of $\mathbf{I}$.

13. Prove that the Gaussian field $\mathbf{K} = \mathbf{Q}(i)$ is Euclidean. (Hint: use the scheme of Section 6.)

14. Prove Theorem 7. (No harm if you cannot prove it completely.)

15. Prove that the set $\{\, f \in \mathbf{Q}[\zeta] | \zeta^p = 1 \,\}$ is a subfield of $\mathbf{C}$.

16. Prove that the discriminant of an algebraic number field does not depend on the choice of the basis. (Hint: represent the elements of one basis in the other basis and consider the value of the determinant of its coefficients.)

17. Evaluate the discriminant Δ of the cyclotomic field $\mathbf{Q}(\zeta)$, $\zeta = e^{2\pi i/p}$.

18. Prove that the polynomial $x^3 - x + 1$ is irreducible over $\mathbf{Q}$.

19. Show that if θ is the zero of an irreducible cubic polynomial, then it cannot satisfy a quadratic equation with rational coefficients, not all zero.

20. Compute the norm $N\alpha = N(A + B\theta + C\theta^2)$, where θ is a zero of $x^3 - x + 1$.

21. Compute the determinant of system (12).

22. Let θ be any of the zeros of the polynomial $x^3 - x + 1$ and let $\alpha = A + B\theta + C\theta^2 \in \mathbf{Q}[\theta]$; also let α' and α'' be the field conjugates of α. Compute $P(\alpha) = \alpha\alpha' + \alpha'\alpha'' + \alpha''\alpha$ as a polynomial in A, B, and C.

BIBLIOGRAPHY

1. W. W. Adams and L. J. Goldstein, *Introduction to Number Theory*, Englewood Cliffs, N.J.: Prentice Hall, 1976.

2. A. B. Ayoub, *On the Fundamental Units of Prime Cyclotomic Fields*. Unpublished dissertation, Temple University, 1980.

3. P. Barrucand and H. Cohn, *Journal of Number Theory*, **2**, 1970, pp. 7–21.

4. L. Bernstein and H. Hasse, *Pacific Journal of Math.*, **30**, 1969, pp. 293–365.

5. G. Birkhoff and S. MacLane, *A Survey of Modern Algebra*, New York: Macmillan, 1959.

6. A. L. Cauchy, *Comptes Rendus de l'Acad. des Sciences* (Paris), **24**, (1947) pp. 578–584.

7. H. Chatland and H. Davenport, *Canadian Journal of Math.*, **2**, 1950, pp. 289–296.

8. L. E. Dickson, *History of the Theory of Numbers*, New York: Chelsea Publishing Co., 1952.

9. G. L. Dirichlet, Mémoire read at the Académie Royale des Sciences (Institut de France) on July 11, 1825; published in the *Journal f. d. reine. u. angew. Mathem.*, **3**, 1828, pp. 354–357 Werke, vol 1, pp. 21–46.

10. C. F. Gauss, *Disquisitiones Arith.*, (Translated from the second Latin edition, 1870, by A. Clark). New Haven: Yale University Press, 1966.

11. E. S. Golod and I. R. Shafarevich, *Izvestiya Akad. Nauk SSSR Ser. Mat.*, **28**, 1964, pp. 261–272.

12. G. H. Hardy and E. M. Wright, *An Introduction to the Theory of Numbers*, 4th ed., Oxford: The Clarendon Press, 1960.

13. E. Hecke, *Theorie der algebraischen Zahlen*, Leipzig: Akad. Verlagsgesellschaft, 1923; New York: Chelsea Publishing Co., 1948.

14. I. N. Herstein, *Topics in Algebra*, Waltham, Mass.: Ginn & Co., 1964.

15. E. E. Kummer, *Collected Papers*, *Vol. I*. Edited by André Weil. New York: Springer Verlag, 1975.

16. A. M. Legendre, *Essai sur la Théorie des Nombres*, Paris: Duprat, 1808. (Legendre gives credit for many results to Sophie Germaine).

17. H. W. Lenstra, Mathematical Intelligencer, **2**, 1979, pp. 6–15, 73–77, 99–103.

18. J. M. Masley, *Inventiones Mathematica*, **28**, 1975, pp. 243–244; *Compositio Mathematica*, **33**, 1976, pp. 179–186; see also dissertation, Princeton University.

19. J. M. Masley and H. L. Montgomery, *Journal f. d. reine u. angew. Mathem.*, **286/287**, 1976, pp. 248–256.

20. R. Pollard, *The Theory of Algebraic Numbers* (Carus Monograph No. 9). New York: J. Wiley & Sons, 1950.

21. P. Roquette, *Proceedings of the International Conference on Algebraic Number Theory*, London Math. Soc. (NATO Advanced Study Institute) and International Mathem. Union. Edited by J. W. S. Cassel and A. Fröhlich. London: Academic Press. Washington, D.C.: Thompson Book Co., 1967, pp. 231–249.

22. C. L. Siegel, *Lectures on advanced, analytic number theory*. Bombay: Tata Institute of Fundam. Research 1965.

23. B. L. van der Waerden, *Modern Algebra* (2 vol.), New York, F. Ungar, 1950.

Chapter 11

Ideal Theory

1 INTRODUCTION

As seen in Chapter 10, there are infinitely many algebraic number fields in which the uniqueness of factorization of integers fails (it is not known whether there are infinitely many fields with uniqueness of factorization). As already mentioned, of the two approaches to remedy the situation, the introduction of ideals has proved extremely successful while the continued extension of fields led to a certain disappointment (infinite class field towers). For this reason, we here interrupt the study of algebraic number fields, to develop the theory of ideals in rings of algebraic integers.

2 DEFINITIONS AND ELEMENTARY PROPERTIES OF IDEALS

DEFINITION 1. Let $\mathbf{I}$ be a ring of algebraic integers in some number field $\mathbf{K}$ of degree n over $\mathbf{Q}$, and let $\alpha_j \in \mathbf{I}(1 \leq j \leq k)$. The set $\mathfrak{a} = \{\lambda_1\alpha_1 + \cdots + \lambda_k\alpha_k\}$ obtained when the λ_js range independently over $\mathbf{I}$ is said to be an *ideal*. $\alpha_1, \alpha_2, \ldots, \alpha_k$ are called the *generators* of $\mathfrak{a}$, in symbols $\mathfrak{a} = (\alpha_1, \alpha_2, \ldots, \alpha_k)$. Two ideals are identical, $\mathfrak{a} = \mathfrak{b}$, if they consist of the same integers.

REMARK 1. The set $\{0\}$ consisting of the only element zero is an ideal, according to Definition 1. It will be denoted by $\mathfrak{o}$, but unless specific mention is made to the contrary, we shall always assume tacitly that any ideal $\mathfrak{a}$ under consideration is not $\mathfrak{o}$. Also the whole ring $\mathbf{I}$ is clearly an ideal. For uniformity of notation we shall denote it by $\mathfrak{i}$ when we want to consider it as an ideal.

REMARK 2. If $\alpha_j = a_j \in \mathbf{Z}$, then $\mathfrak{a} = \{kd\}$, where $d = (\alpha_1, \ldots, \alpha_k)$. Hence although conceptually considered as the g.c.d., $(\alpha_1, \ldots, \alpha_k) = d$ is a number and not a set, there will be no real danger of confusion (and there will be some important advantages) if we use the same symbol $(\alpha_1, \alpha_2, \ldots, \alpha_k)$ to stand also for the ideal generated by $\alpha_1, \alpha_2, \ldots, \alpha_k$.

Theorem 1.   *If* $\alpha \in \mathfrak{a}, \beta \in \mathfrak{a}, \lambda \in \mathbf{I}$ *then* $\alpha \pm \beta \in \mathfrak{a}, \lambda\alpha \in \mathfrak{a}$.

PROOF.   Left to the reader.

REMARK 3.   The generators of a given ideal are by no means unique. In particular, one may suppress among them or add to them any integers, α_0, say, which is not "linearly independent" of the others, that is any integer which can be represented as a sum $\alpha_0 = \Sigma_i\lambda_i\alpha_i$ with α_i generators of $\mathfrak{a}$, $\lambda_i \in \mathbf{I}$. Also, without changing the ideal one may add or subtract from any generator, products of any other generator by integers of the ring.

DEFINITION 2.   An ideal $\mathfrak{a} = (\alpha)$ that can be generated by a single generator is called a *principal ideal*.

Theorem 2.   *Two principal ideals* $\mathfrak{a} = (\alpha)$ *and* $\mathfrak{b} = (\beta)$ *are equal if and only if the generators* α *and* β *are associates, i.e., if their ratio is a unit.*

PROOF. $\mathfrak{a} = \mathfrak{b} \Leftrightarrow \alpha|\beta$ and $\beta|\alpha$.

THEOREM 3.   *In* $\mathbf{Z}$ *all ideals are principal.*

PROOF.   Left to the reader.

REMARK 4.   The ideal (1) is identical with the ring $\mathbf{I}$ or, by Remark 1, $(1) = \mathfrak{i}$; conversely, it follows from Theorem 2 that if $(\alpha) = \mathfrak{i}$, then α is a unit.

DEFINITION 3.   Given the ideals $\mathfrak{a} = (\alpha_1, \alpha_2, \ldots, \alpha_m)$ and $\mathfrak{b} = (\beta_1, \beta_2, \ldots, \beta_k)$ we call the *product* of $\mathfrak{a}$ and $\mathfrak{b}$ the ideal $\mathfrak{c}$ generated by all products $\alpha_i\beta_j$.

REMARK 5. $\mathfrak{c}$ depends only on $\mathfrak{a}$ and $\mathfrak{b}$, not on the particular set of generators. (Why?)

Theorem 4.   *The multiplication of ideals is commutative and associative.*

PROOF.   This follows from the corresponding properties of the multiplication of integers.

NOTATION.   We shall denote products of ideals by themselves as powers. For instance, $\mathfrak{a} \cdot \mathfrak{a} = \mathfrak{a}^2$, $\mathfrak{a} \cdots \mathfrak{a} = \mathfrak{a}^m$, $\mathfrak{a}^0 = 1 = \mathfrak{i} = \mathbf{I}$.

DEFINITION 4.   If $\mathfrak{a} = \mathfrak{b} \cdot \mathfrak{c}$, we say that $\mathfrak{a}$ has the *factors* $\mathfrak{b}$ and $\mathfrak{c}$ and that $\mathfrak{b}$ and $\mathfrak{c}$ *divide* $\mathfrak{a}$; in symbols $\mathfrak{b}|\mathfrak{a}$, $\mathfrak{c}|\mathfrak{a}$.

The statements of the following theorem are immediate consequences of the definitions, and their proofs are left to the reader.

Theorem 5. *If* $\mathfrak{a} = (\alpha)$ *and* $\mathfrak{b} = (\beta)$ *are principal ideals, then*

(1) $\mathfrak{b}|\mathfrak{a} \Leftrightarrow \beta|\alpha$; *for all ideals* $\mathfrak{a}, \mathfrak{b}, \mathfrak{c}, \mathfrak{d}$ *one has*

(2) $\mathfrak{a}|\mathfrak{b}, \mathfrak{b}|\mathfrak{c} \Rightarrow \mathfrak{a}|\mathfrak{c}$;

(3) $\mathfrak{a}|\mathfrak{b} \Rightarrow \mathfrak{a}\mathfrak{d}|\mathfrak{b}\mathfrak{d}$;

(4) $\mathfrak{i}|\mathfrak{a}$;

(5) $\mathfrak{a}|\mathfrak{a}$.

DEFINITION 5. We say that $\alpha(\in \mathbf{I})$ is divisible by an ideal $\mathfrak{a}$ if and only if the principal ideal (α) is divisible by $\mathfrak{a}$, that is if and only if $\mathfrak{a}|(\alpha)$.

Theorem 6. $\mathfrak{a}|\mathfrak{b} \Rightarrow \mathfrak{a} \supset \mathfrak{b}$ (*that is, each integer of* $\mathfrak{b}$ *belongs to* $\mathfrak{a}$).

PROOF. Follows from Definition 3.

Corollary 6.1. $\mathfrak{a}|\mathfrak{i} \Rightarrow \mathfrak{a} = \mathfrak{i}$.

PROOF. Left to the reader.

DEFINITION 6. An ideal $\mathfrak{p} \neq \mathfrak{i}$ is said to be a prime ideal if it has no other factors except $\mathfrak{p}$ itself and $\mathfrak{i}$.

DEFINITION 7. A set $\alpha_1,\ldots,\alpha_k \in \mathfrak{a}$ is said to form a *basis* for $\mathfrak{a}$ if every integer $\alpha \in \mathfrak{a}$ has exactly one representation of the form $\alpha = a_1\alpha_1 + \cdots + a_k\alpha_k$, with $a_j \in \mathbf{Z}(1 \le j \le k)$.

Theorem 7. *Every ideal* $\mathfrak{a}(\neq \mathfrak{o})$ *has a basis.*

PROOFS. May be found in [4] pp. 78–79 or [1] pp. 116–117, but will not be given here because the theorem itself, quoted for its intrinsic interest, will not be used.

Theorem 8. *There exist only finitely many ideals containing a given integer* $a \in \mathbf{Z}$.

PROOF. Let $\mathfrak{a} = (\alpha_1, \alpha_2,\ldots,\alpha_m)$ contain $a \in \mathbf{Z}$. Then by Remark 3, $\mathfrak{a} = (\alpha_1,\ldots,\alpha_m, a)$. If $(\omega_1,\ldots,\omega_n)$ is an integral basis, then $\alpha_j = a_{j1}\omega_1 + \cdots + a_{jn}\omega_n(a_{jk} \in \mathbf{Z})$. Clearly, there exist $q_{jk} \in \mathbf{Z}$, $r_{jk} \in \mathbf{Z}$ such that $a_{jk} = aq_{jk} + r_{jk}$ with $0 \le r_{jk} < a$. Hence if we set $\sum_{k=1}^{n} r_{jk}\omega_k = \beta_j$ and use Remark 3 once more, we obtain $\mathfrak{a} = (\beta_1, \beta_2,\ldots,\beta_m, a)$. Here a is fixed and the β_j each may take only finitely many values, namely those of the sums $\sum_{k=1}^{n} r_{jk}\omega_k$ with $0 \le r_{jk} \le a - 1$. Also, by Remark 3 we know that we need to keep only independent elements as generators, and it follows that there can exist at most n such elements if the degree of the field is n. Hence $m \le n$ and the total number of ideals $\mathfrak{a} = (\beta_1, \beta_2,\ldots,\beta_m, a)$ that may contain $a \in \mathbf{Z}$ is indeed finite.

3   DIVISIBILITY PROPERTIES OF IDEALS

Theorem 9.   *An ideal $\mathfrak{a}$ has only a finite number of factors.*

PROOF.   First we recall that by Theorem 6, $\mathfrak{b}|\mathfrak{a} \Leftrightarrow \mathfrak{b} \supset \mathfrak{a}$. Next, by Theorem 1, $\alpha \in \mathfrak{a}, \lambda \in \mathbf{I} \Rightarrow \lambda\alpha \in \mathfrak{a}$. Taking in particular λ as product of all conjugates of α (Why is λ an integer? Why is λ in $\mathbf{K}$?), $\lambda\alpha = N\alpha \in \mathfrak{a}$; hence if $\mathfrak{b}|\mathfrak{a}$, then $N\alpha = a(\in \mathbf{Z})$ belongs also to $\mathfrak{b}$. However by Theorem 8, the number of ideals containing a given $a(\in \mathbf{Z})$ is finite and the Theorem is proven.

Theorem 10.   *Given an ideal $\mathfrak{a}$, there exists an ideal $\mathfrak{b}$ such that $\mathfrak{a} \cdot \mathfrak{b} = (a)$, $a \in \mathbf{Z}^+$, that is, such that the product is a principal ideal generated by a positive rational integer.*

In view of the great importance of this theorem, it is recommended that the reader make the necessary effort to understand thoroughly the not very simple proof.

PROOF OF THEOREM 10.   Let $\mathfrak{a} = (\alpha_0, \alpha_1, \ldots, \alpha_m)$ and consider the polynomial $f(x) = \alpha_0 x^m + \alpha_1 x^{m-1} + \cdots + \alpha_{m-1}x + \alpha_m$; by Remark 3 we may assume that $\alpha_0 \neq 0$. Each element α of $\mathbf{K}$ is of the form $\alpha = p(\theta)$, where θ generates $\mathbf{K}$ of degree n, and $p(x) \in \mathbf{Z}[x]$ is a polynomial of degree $\leq n - 1$. The conjugates of α are by definition $\alpha^{(j)} = p(\theta^{(j)})$, where $\theta^{(j)}$ ($j = 1, 2, \ldots, n$) are the conjugates of θ. In particular, let $\alpha_i = p_i(\theta)$ ($i = 0, 1, \ldots, m$). In the polynomial $g(x) = \prod_{j=1}^{n}\{ p_0(\theta^{(j)})x^m + \cdots + p_m(\theta^{(j)})\} = \sum_{j=0}^{N} c_j x^{N-j}$ ($N = m \cdot n$) the coefficients c_j are symmetric polynomials with integral rational coefficients of $\theta = \theta^{(1)}, \theta^{(2)}, \ldots, \theta^{(n)}$; hence (see [4]), they are polynomials with rational integral coefficients in the coefficients $A_1, A_2, \ldots, A_n$ of the irreducible equation satisfied by θ. These A_j being rational, so are the coefficients c_j. On the other hand, they are products of integers, as follows from the product representation of $g(x)$. Consequently the c_js are rational integers, and $g \in \mathbf{Z}[x]$. Also, $g(x) = f(x)h(x)$, so that $h(x) = g(x)/f(x)$ is a rational function with coefficients in $\mathbf{K}$. However, again from the product representation of $g(x)$, it is clear that $h(x)$ is a polynomial in x with integer coefficients; consequently $h \in \mathbf{I}[x]$, say, $h(x) = \beta_0 x^k + \cdots + \beta_{k-1}x + \beta_k$, ($\beta_j \in \mathbf{I}$, $\beta_0 \neq 0$). If we define $\mathfrak{b} = (\beta_0, \beta_1, \ldots, \beta_k)$ we claim that $\mathfrak{a} \cdot \mathfrak{b} = (c)$, where c is the content of $g(x)$. Indeed by its definition, $c = (c_0, c_1, \ldots, c_N)$. It is a consequence of Gauss' Lemma (see Section 10.2) that $c|c_j$ (for all j) $\Rightarrow c|\alpha_i\beta_j$ (all i, j) $\Rightarrow (c) \supset (\alpha_0\beta_0, \alpha_0\beta_1, \ldots, \alpha_i\beta_j, \ldots) = \mathfrak{a} \cdot \mathfrak{b}$. Also $\{c_j/c\}$ is a set of rational integers whose g.c.d. is one. Therefore (recall Problem 9 in Chapter 3) there exist integers $a_j \in \mathbf{Z}$ such that

$$a_0 \frac{c_0}{c} + a_1 \frac{c_1}{c} + \cdots + a_j \frac{c_j}{c} + \cdots + a_N \frac{c_N}{c} = 1$$

or

$$a_0 c_0 + \cdots + a_j c_j + \cdots + a_N c_N = c.$$

Replacing the c_js by their values $\sum_k \alpha_k \beta_{j-k}$,

$$c = \sum_{j=0}^{N} a_j \sum_k \alpha_k \beta_{j-k} \Rightarrow c \in \mathfrak{a} \cdot \mathfrak{b} \Rightarrow (c) \subset \mathfrak{a} \cdot \mathfrak{b}.$$

These two opposite inclusions prove that $\mathfrak{a}\mathfrak{b} = (c)$, as claimed.

Theorem 11.   *If (γ) is a principal ideal and $(\gamma)\mathfrak{a} = (\gamma)\mathfrak{b}$, then $\mathfrak{a} = \mathfrak{b}$.*

PROOF.   By hypothesis, for every $\alpha \in \mathfrak{a}$, $\gamma\alpha$ equals an integer of the form $\gamma\beta$, $\beta \in \mathfrak{b}$; or, to each integer $\alpha \in \mathfrak{a}$ there corresponds an integer $\beta \in \mathfrak{b}$ such that $\alpha = \beta$. This shows that $\mathfrak{a} \subset \mathfrak{b}$. By symmetry, $\mathfrak{b} \subset \mathfrak{a}$, whence $\mathfrak{a} = \mathfrak{b}$ follows immediately.

Theorem 12 (Cancellation Law).   *If $\mathfrak{a}$, $\mathfrak{b}$, $\mathfrak{c}$ are ideals of $\mathbf{K}$, then $\mathfrak{c}\mathfrak{a} = \mathfrak{c}\mathfrak{b} \Rightarrow \mathfrak{a} = \mathfrak{b}$.*

PROOF.   By Theorem 10 we can find an ideal $\mathfrak{d} \Rightarrow \mathfrak{d}\mathfrak{c} = (c)$. Hence $\mathfrak{c}\mathfrak{a} = \mathfrak{c}\mathfrak{b} \Rightarrow \mathfrak{d}\mathfrak{c}\mathfrak{a} = \mathfrak{d}\mathfrak{c}\mathfrak{b} \Rightarrow (c)\mathfrak{a} = (c)\mathfrak{b} \Rightarrow \mathfrak{a} = \mathfrak{b}$ by Theorem 11.

REMARK 6.   One should remember that in the statement and proof of Theorem 12 (like in other statements and proofs) it is tacitly assumed that $\mathfrak{c} \neq \mathfrak{o}$.

Theorem 13. $\mathfrak{a} \supset \mathfrak{b} \Rightarrow \mathfrak{a} | \mathfrak{b}$.

REMARK 7.   Theorem 13 is the converse of Theorem 6.

PROOF OF THEOREM 13. $\mathfrak{a} \supset \mathfrak{b} \Rightarrow \mathfrak{c}\mathfrak{a} \supset \mathfrak{c}\mathfrak{b}$ for any $\mathfrak{c}$. In particular, if $\mathfrak{c}$ has been chosen so that $\mathfrak{c}\mathfrak{a} = (d)$, then each element of $\mathfrak{c}\mathfrak{b}$ is a multiple of d, and therefore $\mathfrak{c}\mathfrak{b}$ is of the form $(d)(\gamma_1, \gamma_2, \ldots, \gamma_r) = \mathfrak{c}\mathfrak{a}(\gamma_1, \gamma_2, \ldots, \gamma_r)$. By Theorem 12 cancellation is permitted, and

$$\mathfrak{c}\mathfrak{b} = \mathfrak{c}\mathfrak{a}(\gamma_1, \ldots, \gamma_r) \Rightarrow \mathfrak{b} = \mathfrak{a}(\gamma_1, \ldots, \gamma_r) \Rightarrow \mathfrak{a} | \mathfrak{b}.$$

REMARK 8.   In more general situations it is convenient to make a distinction between the concepts of maximal ideal, prime ideal, and irreducible ideal; but these objects coincide in the present setting. The three conceptually distinct properties which an ideal may have—

  (i) To have no factors except i and itself (irreducibility),
 (ii) To divide at least one factor if it divides a product of two ideals (primality),
(iii) Not to be contained properly in any other ideal except i (maximality)

—go together for ideals in algebraic number fields. An ideal has either all three or none of them. This is the reason why we did not formally define irreducible, prime, and maximal ideals, but only prime ideals (Definition 6), which we characterized by property (i). On account of Theorems 6 and 13 it is clear that properties (i) and (iii) are equivalent. One can also easily prove directly (iii) ⇔ (ii), but we shall not do it because the implication (i) or (iii) ⇒ (ii) will come out anyway, as a corollary, and the implication (ii) ⇒ (i) or (iii) is left to the reader.

4   UNIQUENESS OF FACTORIZATION OF IDEALS INTO PRIME IDEALS

Theorem 14.   *If* $\mathfrak{a}|\mathfrak{b}$, $\mathfrak{a} \neq \mathfrak{b}$, *then* $\mathfrak{a}$ *has fewer factors than* $\mathfrak{b}$.

PROOF. $\mathfrak{b} = \mathfrak{a}\mathfrak{c}$, $\mathfrak{c} \neq \mathfrak{i}$;   therefore every factor of $\mathfrak{a}$ is also a factor of $\mathfrak{b}$. However $\mathfrak{b}$ has at least one factor which is not a factor of $\mathfrak{a}$, namely $\mathfrak{b}$ itself.

Theorem 15.   *Every ideal* $\mathfrak{a}$ *can be factored into prime ideals.*

PROOF.   If in every factorization $\mathfrak{a} = \mathfrak{b}\mathfrak{c}$ either $\mathfrak{b} = \mathfrak{i}$ or $\mathfrak{c} = \mathfrak{i}$, thèn $\mathfrak{a}$ is a prime ideal and the theorem holds. Otherwise, by Theorem 9 $\mathfrak{a}$ contains only a finite number of factors; by Theorem 14, each of the ideals $\mathfrak{b}$ and $\mathfrak{c}$ contains fewer factors than $\mathfrak{a}$. Hence repeating the reasoning on them, in a finite number of steps we reach ideals with at most one factor, that is prime ideals.

Theorem 16.   *If p and q are distinct rational primes and the prime ideal* $\mathfrak{p}|(p)$, *then* $\mathfrak{p} \nmid (q)$.

PROOF.   By Corollary 3.6.3, $(p, q) = 1 \Rightarrow \exists m, n \ni mp + nq = 1$ or $(p, q) = (1)$. (This illustrates the double meaning of the symbol (p, q), which, however, cannot lead to any ambiguity!) If $\mathfrak{p}|(p)$ and $\mathfrak{p}|(q)$, then every multiple of p and every multiple of q belongs to $\mathfrak{p}$. But $\mathfrak{p}$ is an ideal; hence by Theorem 1 it contains in particular $mp + nq = 1$ so that $\mathfrak{p} = (1)$, which is contrary to the definition of a prime ideal.

Corollary 16.1.   *In every field of algebraic numbers there exist infinitely many primes ideals.*

PROOF.   Left to the reader.

DEFINITION 8.   Let $\mathfrak{a}$ and $\mathfrak{b}$ be ideals in **K**. If there exists in **K** an ideal $\mathfrak{d}$ such that:

   (i) $\mathfrak{d}|\mathfrak{a}$, $\mathfrak{d}|\mathfrak{b}$; and
   (ii) $\mathfrak{c}|\mathfrak{a}$, $\mathfrak{c}|\mathfrak{b} \Rightarrow \mathfrak{c}|\mathfrak{d}$,

then $\mathfrak{d}$ is called the *greatest common ideal divisor* (g.c.d.) of $\mathfrak{a}$ and $\mathfrak{b}$.

Theorem 17. *Any two ideals* $\mathfrak{a}$ *and* $\mathfrak{b}$ *of* **K** *have a g.c.d.* $\mathfrak{d}$; *in symbols* $\mathfrak{d} = (\mathfrak{a}, \mathfrak{b})$.

PROOF.   If $\mathfrak{a} = (\alpha_1, \alpha_2, \ldots, \alpha_r)$ and $\mathfrak{b} = (\beta_1, \beta_2, \ldots, \beta_s)$, then one verifies that $\mathfrak{d} = (\alpha_1, \ldots, \alpha_r, \beta_1, \ldots, \beta_s)$ has the required properties.

Corollary 17.1. *The elements of* $(\mathfrak{a}, \mathfrak{b})$ *are of the form* $\alpha + \beta$ *with* $\alpha \in \mathfrak{a}$, $\beta \in \mathfrak{b}$.

PROOF.   The proof follows from the fact that $\mathfrak{d} = (\alpha_1, \ldots, \alpha_r, \beta_1, \ldots, \beta_s)$ (see Proof of Theorem 17).

DEFINITION 9.   If $(\mathfrak{a}, \mathfrak{b}) = \mathfrak{i}$, then $\mathfrak{a}$ and $\mathfrak{b}$ are called *coprime*.

Corollary 17.2. $(\mathfrak{a}, \mathfrak{b}) = \mathfrak{i} \Rightarrow \exists \alpha \in \mathfrak{a}, \beta \in \mathfrak{b} \ni \alpha + \beta = 1$.

PROOF. $1 \in \mathfrak{i}$; hence the result follows from Corollary 17.1.

Theorem 18.   *If* $\mathfrak{p}$ *is a prime ideal, then* $\mathfrak{p}|\mathfrak{a}\mathfrak{b}, \mathfrak{p} \nmid \mathfrak{a} \Rightarrow \mathfrak{p}|\mathfrak{b}$.

PROOF. $\mathfrak{p} \nmid \mathfrak{a} \Rightarrow (\mathfrak{p}, \mathfrak{a}) = \mathfrak{i} \Rightarrow \exists \alpha \in \mathfrak{a}, \pi \in \mathfrak{p} \ni \alpha + \pi = 1$ (because of Corollary 17.2). Therefore for every $\beta \in \mathfrak{b}, \beta\alpha + \beta\pi = \beta$. By assumption, $\mathfrak{p}|\mathfrak{a}\mathfrak{b}$ so that $\beta\alpha \in \mathfrak{p}$; also, $\beta\pi \in \mathfrak{p}$. Consequently for every $\beta \in \mathfrak{b}, \beta \in \mathfrak{p}$ also holds, so that $\mathfrak{b} \subset \mathfrak{p}$ or $\mathfrak{p}|\mathfrak{b}$, by Theorem 13.

Corollary 18.1. $\mathfrak{p}|(\alpha)(\beta), \mathfrak{p} \nmid (\alpha) \Rightarrow \mathfrak{p}|(\beta)$.

PROOF.   Take $\mathfrak{a} = (\alpha)$ and $\mathfrak{b} = (\beta)$ in Theorem 18.

REMARK 9.   This proves the implication (i) or (iii) $\Rightarrow$ (ii) alluded to in Remark 8.

Theorem 19.   *If* $\mathfrak{p}$ *is a prime ideal and* $\mathfrak{p}|\mathfrak{a}_1 \ldots \mathfrak{a}_r$, *then* $\mathfrak{p}$ *divides at least one ideal* $\mathfrak{a}_j$ $(1 \le j \le r)$.

PROOF.   By induction on r starting with $r = 2$, which is Theorem 18.

The reader will already have sensed a certain parallelism between the last few theorems and some results of Chapter 3 leading to the unique factorization in **Z**. This is indeed the case and we are actually ready to draw the principal conclusion, namely:

Theorem 20.   *The ideals of* **K** *factor into prime ideals and this factorization is unique except for order.*

PROOF.   The same as that of Theorem 3.3, using Theorem 19 instead of Corollary 3.4.1.

5   IDEAL CLASSES AND THE CLASS NUMBER

DEFINITION 10.   Two integers α and β of **I** are said to be *congruent modulo an ideal* $\mathfrak{a}$, in symbols $\alpha \equiv \beta$ (mod $\mathfrak{a}$), if $\mathfrak{a} | \alpha - \beta$ or equivalently (because of Definition 5), if $\mathfrak{a} | (\alpha - \beta)$.

Theorem 21.   *Congruence modulo an ideal is an equivalence relation.*

PROOF.   Left to the reader.

Theorem 22. $\alpha \equiv \beta$ (mod $\mathfrak{a}$) *and* $\mathfrak{c} | \mathfrak{a} \Rightarrow \alpha \equiv \beta$ (mod $\mathfrak{c}$).

PROOF.   Left to the reader.

Corollary 22.1.   *Let* $\alpha, \beta, \gamma, \delta, \lambda \in$ **I** *and let* $\mathfrak{a}$ *be an ideal in* **K**; *then* $\alpha \equiv \beta$ (mod $\mathfrak{a}$), $\gamma \equiv \delta$ (mod $\mathfrak{a}$) $\Rightarrow \alpha \pm \gamma \equiv \beta \pm \delta$ (mod $\mathfrak{a}$), $\alpha\gamma \equiv \beta\delta$ (mod $\mathfrak{a}$), *and* $\lambda\alpha \equiv \lambda\beta$ (mod $\mathfrak{a}$).

PROOF.   Left to the reader.

Theorem 23.   *The set of residue classes modulo an ideal is finite.*

PROOF OF THEOREM 23.   Let $\mathfrak{a}$ be an ideal in **K**. By Theorem 10 we can find an ideal $\mathfrak{b}$ so that $\mathfrak{a} \cdot \mathfrak{b} = (c), c > 0$. Let $\omega_1, \omega_2, \ldots, \omega_n$ be an integral basis; then every $\alpha \in$ **I** has a representation $\alpha = a_1\omega_1 + \cdots + a_n\omega_n$ ($a_j \in$ **Z**). If we first consider the number of residue classes modulo c, it is clear that each coefficient a_j may belong to only c residue classes; consequently, α can belong to at most (actually, exactly!) c^n residue classes (mod c). This is also the number of residue classes (mod (c)). By Theorem 22 it now follows that there are at most c^n residue classes mod $\mathfrak{a}$.

DEFINITION 11.   The number of residue classes of integers of **I** modulo an ideal $\mathfrak{a}$ is called the *norm* of $\mathfrak{a}$, in symbols $N(\mathfrak{a})$ or $N\mathfrak{a}$. In the case of principal ideals we write $N((\alpha))$ to avoid confusion with $N\alpha = N(\alpha)$.

The reader may wonder whether this terminology could not lead to confusion. Indeed, we already have a definition for the norm $N(\alpha)$ of an integer $\alpha \in$ **I**. Now we just gave a new definition for the norm of the corresponding principal ideal, $N((\alpha))$. As a matter of fact, the two are closely related by:

Theorem 24.   *If* $\alpha \in$ **I**, *then* $N((\alpha)) = |N(\alpha)|$.

The proof will not be given, and this theorem, quoted mainly for completeness, will be used only once (in the proof of Lemma 3). Proofs may be found, for instance, in [4] p. 107 or [1] p. 119; see also Problem 17.

REMARK 10. $N\mathfrak{a} = 1 \Leftrightarrow \mathfrak{a} = \mathfrak{i}$, because $N\mathfrak{a} = 1$ means that all integers are congruent to each other and hence to $0 (\in \mathfrak{a})$ so that all belong to $\mathfrak{a}$.

Theorem 25.   *If* $\mathfrak{a}$ *is an ideal and* $\mathfrak{p}$ *a prime ideal of* **K**, *then* $N\mathfrak{a} \cdot N\mathfrak{p} = N(\mathfrak{a}\mathfrak{p})$.

PROOF. $\mathfrak{p} \neq i \Rightarrow \mathfrak{a}\mathfrak{p} \neq \mathfrak{a}$; hence $\exists \alpha \in \mathbf{I} \ni \mathfrak{a}|\alpha,\ \mathfrak{a}\mathfrak{p} \nmid \alpha$, or equivalently (see Definition 5), $\mathfrak{a}|(\alpha),\ \mathfrak{a}\mathfrak{p} \nmid (\alpha)$. We also note that if $(\alpha) = \mathfrak{a} \cdot \mathfrak{b}$, then $\mathfrak{p} \nmid \mathfrak{b}$; otherwise $\mathfrak{a}\mathfrak{p}|\mathfrak{a}\mathfrak{b} \Rightarrow \mathfrak{a}\mathfrak{p}|(\alpha)$, which is false.

Consider now the $N\mathfrak{a}$ residues classes mod $\mathfrak{a}$ and let $\alpha_1, \alpha_2, \ldots, \alpha_{N\mathfrak{a}}$ be integers of $\mathbf{I}$ incongruent mod $\mathfrak{a}$; similarly select $\pi_1, \pi_2, \ldots, \pi_{N\mathfrak{p}}$ as representatives of the $N\mathfrak{p}$ distinct residue classes mod $\mathfrak{p}$. We claim that the $N\mathfrak{a} \cdot N\mathfrak{p}$ integers $\alpha\pi_j + \alpha_k$ $(1 \leq j \leq N\mathfrak{p}, 1 \leq k \leq N\mathfrak{a})$ have the following two properties:

(i) No two of them are congruent mod $\mathfrak{a}\mathfrak{p}$.

(ii) Every integer $\gamma \in \mathbf{I}$ is congruent to one of them mod $\mathfrak{a}\mathfrak{p}$.

This will finish the proof that there exist exactly $N\mathfrak{a} \cdot N\mathfrak{p}$ residue classes mod $\mathfrak{a}\mathfrak{p}$, as asserted by the theorem.

PROOF if (i). Assume $\alpha\pi_{j_1} + \alpha_{k_1} \neq \alpha\pi_{j_2} + \alpha_{k_2}$, with $\alpha\pi_{j_1} + \alpha_{k_1} \equiv \alpha\pi_{j_2} + \alpha_{k_2}$ (mod $\mathfrak{a}\mathfrak{p}$); by Theorem 22, $\alpha\pi_{j_1} + \alpha_{k_1} \equiv \alpha\pi_{j_2} + \alpha_{k_2}$ (mod $\mathfrak{a}$) or (remember: $\mathfrak{a}|\alpha$) $\alpha_{k_1} \equiv \alpha_{k_2}$ (mod $\mathfrak{a}$), so that $k_1 = k_2$, and the original congruence reduces to $\alpha\pi_{j_1} \equiv \alpha\pi_{j_2}$ (mod $\mathfrak{a}\mathfrak{p}$), that is, $\mathfrak{a}\mathfrak{p}|\mathfrak{a}\mathfrak{b}\,(\pi_{j_1} - \pi_{j_2})$, or $\mathfrak{p}|\mathfrak{b}\,(\pi_{j_1} - \pi_{j_2})$. But $\mathfrak{p} \nmid \mathfrak{b}$; hence $\mathfrak{p}|(\pi_{j_1} - \pi_{j_2})$ which is possible only if $j_1 = j_2$ and the two integers were not distinct, contrary to our assumption.

PROOF of (ii). If $\gamma \in \mathbf{I}$, then it is congruent to some α_j (mod $\mathfrak{a}$), that is $\gamma = \alpha' + \alpha_j, \alpha' \in \mathfrak{a}$. From $\mathfrak{p} \nmid \mathfrak{b}$ follows that $((\alpha), \mathfrak{p}\mathfrak{a}) = (\mathfrak{b}\mathfrak{a}, \mathfrak{p}\mathfrak{a}) = \mathfrak{a}$, so that by Corollary 17.1, $\alpha' \in \mathfrak{a} \Rightarrow \alpha' = \lambda\alpha + \mu(\lambda \in \mathbf{I}, \mu \in \mathfrak{p}\mathfrak{a})$. By the definition of the π_js, $\exists k(1 \leq k \leq N\mathfrak{p}) \ni \lambda \equiv \pi_k$ (mod $\mathfrak{p}$), that is $\lambda = \pi' + \pi_k, \pi' \in \mathfrak{p}$. Consequently,

$$\gamma = \lambda\alpha + \mu + \alpha_j = (\pi' + \pi_k)\alpha + \mu + \alpha_j = (\pi'\alpha + \mu) + \pi_k\alpha + \alpha_j.$$

However $\pi'\alpha \in \mathfrak{p}\mathfrak{a}, \mu \in \mathfrak{p}\mathfrak{a}$, so that $\gamma \equiv \pi_k\alpha + \alpha_j$ (mod $\mathfrak{a}\mathfrak{p}$) as claimed, and the proof is complete.

Theorem 26. *Let $\mathfrak{a}, \mathfrak{b}, \mathfrak{c}$ be ideals in $\mathbf{K}$; then $\mathfrak{a} \cdot \mathfrak{b} = \mathfrak{c} \Rightarrow N\mathfrak{a} \cdot N\mathfrak{b} = N\mathfrak{c}$.*

PROOF. The proof is by induction on the number of factors of $\mathfrak{b}$. If $\mathfrak{b} = \mathfrak{p}$, $N\mathfrak{a} \cdot N\mathfrak{p} = N(\mathfrak{a}\mathfrak{p})$ by Theorem 25; if Theorem 26 is known to hold for $\mathfrak{b}$ containing $k - 1$ prime ideal factors, then it also holds for $\mathfrak{b} = \mathfrak{p}_1 \ldots \mathfrak{p}_{k-1}\mathfrak{p}$ as follows: Let $\mathfrak{b} = \mathfrak{b}_1\mathfrak{p}$; then $N(\mathfrak{a} \cdot \mathfrak{b}) = N(\mathfrak{a} \cdot \mathfrak{b}_1 \cdot \mathfrak{p}) = N(\mathfrak{a}\mathfrak{b}_1) \cdot N\mathfrak{p}$ by Theorem 25. Also, by the induction assumption, $N(\mathfrak{a}\mathfrak{b}_1) = N\mathfrak{a} \cdot N\mathfrak{b}_1$; hence $N(\mathfrak{a} \cdot \mathfrak{b}) = N\mathfrak{a} \cdot N\mathfrak{b}_1 \cdot N\mathfrak{p} = N\mathfrak{a} \cdot N(\mathfrak{b}_1 \cdot \mathfrak{p})$ by Theorem 25. Replacing $\mathfrak{b}_1 \cdot \mathfrak{p}$ by $\mathfrak{b}$, the theorem is proven.

Theorem 27. *If $\mathfrak{a}$ is an ideal of $\mathbf{K}$, then $\mathfrak{a}|N\mathfrak{a}$ (i.e., $\mathfrak{a}|(N\mathfrak{a})$).*

PROOF. If $\alpha_1, \alpha_2, \ldots, \alpha_{N\mathfrak{a}}$ are a maximal set of integers incongruent mod $\mathfrak{a}$ (so that there is exactly one out of every residue class), then the same property also belongs to the set $\alpha_1 + 1, \alpha_2 + 1, \ldots, \alpha_{N\mathfrak{a}} + 1$; consequently $\alpha_1 + \alpha_2 + \cdots + \alpha_{N\mathfrak{a}} \equiv (\alpha_1 + 1) + \cdots + (\alpha_{N\mathfrak{a}} + 1) \pmod{\mathfrak{a}}$ or simplifying, $0 \equiv 1 + 1 + \cdots + 1 = N\mathfrak{a} \pmod{\mathfrak{a}}$.

Theorem 28. *For every $m \in \mathbf{Z}^+$, there exist only finitely many ideals $\mathfrak{a}$ such that $N\mathfrak{a} = m$.*

PROOF. By Theorem 27, $\mathfrak{a} | m$; by Theorem 8 there exist only finitely many such ideals $\mathfrak{a}$ for any given $m \in \mathbf{Z}^+$.

DEFINITION 12. Two ideals $\mathfrak{a}$ and $\mathfrak{b}$ of $\mathbf{K}$ are said to be *equivalent*, in symbols $\mathfrak{a} \sim \mathfrak{b}$, if there exist algebraic integers $\alpha, \beta \in \mathbf{I}$ such that $(\alpha)\mathfrak{a} = (\beta)\mathfrak{b}$.

Theorem 29. *The equivalence of ideals is an equivalence relation.*

PROOF.

  (i) $\mathfrak{a} \sim \mathfrak{a}$ because $(1)\mathfrak{a} = (1)\mathfrak{a}$.
  (ii) $\mathfrak{a} \sim \mathfrak{b}$ means $(\alpha)\mathfrak{a} = (\beta)\mathfrak{b}$ for some $\alpha, \beta \in \mathbf{I}$, so that $(\beta)\mathfrak{b} = (\alpha)\mathfrak{a}$ and $\mathfrak{b} \sim \mathfrak{a}$.
  (iii) $\mathfrak{a} \sim \mathfrak{b}, \mathfrak{b} \sim \mathfrak{c} \Leftrightarrow (\alpha)\mathfrak{a} = (\beta_1)\mathfrak{b}, (\beta_2)\mathfrak{b} = (\gamma)\mathfrak{c} \Leftrightarrow (\beta_2)(\alpha)\mathfrak{a} = (\beta_2)(\beta_1)\mathfrak{b} = (\beta_1)(\gamma)\mathfrak{c} \Leftrightarrow (\beta_2\alpha)\mathfrak{a} = (\beta_1\gamma)\mathfrak{c}$ with $\alpha, \beta_1, \beta_2, \gamma \in \mathbf{I}$ and $\mathfrak{a} \sim \mathfrak{c}$.

DEFINITION 13. The classes induced by the equivalence relation $\sim$ among ideals are called ideal classes.

REMARK 11. The principal ideals are equivalent to each other (and to $\mathfrak{i} = (1)$); hence they form one of the classes, sometimes called the *principal class*.

We come now to one of the fundamental results of the theory of ideals, namely:

Theorem 30. *In every field $\mathbf{K}$ of algebraic numbers, the number of ideal classes is finite.*

The proof of Theorem 30 requires several lemmas.

Lemma 1. *Let $\mathfrak{a}, \mathfrak{b}, \mathfrak{c}, \mathfrak{d}$ be ideals in $\mathbf{K}$; then $\mathfrak{a} \sim \mathfrak{b}, \mathfrak{c} \sim \mathfrak{d} \Rightarrow \mathfrak{a}\mathfrak{c} \sim \mathfrak{b}\mathfrak{d}$.*

PROOF. By assumption $\exists \alpha, \beta, \gamma, \delta \in \mathbf{I} \ni (\alpha)\mathfrak{a} = (\beta)\mathfrak{b}, (\gamma)\mathfrak{c} = (\delta)\mathfrak{d}$; consequently $(\alpha\gamma)\mathfrak{a}\mathfrak{c} = (\beta\delta)\mathfrak{b}\mathfrak{d}$, so that $\mathfrak{a}\mathfrak{c} \sim \mathfrak{b}\mathfrak{d}$.

Lemma 2. *For every field $\mathbf{K}$ of algebraic numbers there exists a positive integer $m = m(\mathbf{K})$ with the property that in every ideal $\mathfrak{a}$ of $\mathbf{K}$ there exists an integer $\alpha \in \mathfrak{a}$ such that $|N\alpha| \leq m \cdot N\mathfrak{a}$.*

PROOF. Let $\omega_1, \ldots, \omega_n$ be an integral basis for $\mathbf{K}$. Each ω_j is (uniquely) represented by a polynomial in the generator θ of $\mathbf{K}$, $\omega_j = g_j(\theta)$. If $\theta =$

$\theta^{(1)}, \theta^{(2)}, \ldots, \theta^{(n)}$ are the conjugates of θ, then the conjugates of ω_j are $\omega_j^{(k)} = g_j(\theta^{(k)})$. Let $M = \prod_{k=1}^{n}\{\sum_{j=1}^{n}|\omega_j^{(k)}|\}$; then M has the required property and we may take $m = [M] + 1$. Indeed, for every ideal $\mathfrak{a}$ we can determine an integer r such that $r^n \le N\mathfrak{a} < (r + 1)^n$. Next consider the set of integers of **I** represented by $a_1\omega_1 + a_2\omega_2 + \cdots + a_n\omega_n$ with $0 \le a_j \le r$. Each a_j may take $r + 1$ distinct values; hence, we obtain $(r + 1)^n$ different integers of **I**. But these cannot all be incongruent mod $\mathfrak{a}$ because there exist only $N\mathfrak{a} < (r + 1)^n$ residue classes mod $\mathfrak{a}$. Hence among these integers there are at least two, say $\alpha = a_1\omega_1 + \cdots + a_n\omega_n$ and $\beta = b_1\omega_1 + \cdots + b_n\omega_n$, such that $\alpha \ne \beta$ but $\alpha \equiv \beta \pmod{\mathfrak{a}}$, and with $0 \le a_j \le r$, $0 \le b_j \le r$. Then $0 \ne \gamma = \alpha - \beta \equiv 0 \pmod{\mathfrak{a}}$ so that $\gamma \in \mathfrak{a}$. Also $|N\gamma| = |\prod_{k=1}^{n}\gamma^{(k)}|$; however,

$$|\gamma^{(k)}| = \left|\sum_{j=1}^{n}(a_j - b_j)\omega_j^{(k)}\right| \le \sum_{j=1}^{n}r|\omega_j^{(k)}| = r\sum_{j=1}^{n}|\omega_j^{(k)}|,$$

so that

$$|N\gamma| = \prod_{k=1}^{n}|\gamma^{(k)}| \le r^n \prod_{k=1}^{n}\left\{\sum_{j=1}^{n}|\omega_j^{(k)}|\right\} = r^n M \le N\mathfrak{a} \cdot M,$$

and $\gamma \in \mathfrak{a}$ has $|N\gamma| \le M \cdot N\mathfrak{a}$, as claimed.

REMARK 12.   The above string of inequalities is clearly very wasteful. One would surmise that a much stronger result ought to hold. This is indeed the case and one may take a much smaller value for m, but then the statement becomes more difficult to prove and we shall not need the stronger result.

Lemma 3.   *Let $m = m(\mathbf{K})$ be defined as in Lemma 2; then in each class of ideals there exists an ideal $\mathfrak{a}$ such that $N\mathfrak{a} \le m$.*

PROOF.   Let $\mathfrak{U}$ be any class of ideals in **K**; select in $\mathfrak{U}$ an arbitrary ideal $\mathfrak{b}$. By Theorem 10, there exists an ideal $\mathfrak{c}$ in **K** such that $\mathfrak{c}\mathfrak{b}$ is principal. By Lemma 2 we may select in $\mathfrak{c}$ an integer γ such that $|N\gamma| \le m \cdot N\mathfrak{c}$. From $\gamma \in \mathfrak{c}$ follows $(\gamma) \subset \mathfrak{c}$, that is, $\mathfrak{c}|(\gamma)$; hence $\mathfrak{c}\mathfrak{a} = (\gamma)$ for some ideal $\mathfrak{a}$ in **K**. Two remarks are now in order: First by Lemma 1, from $\mathfrak{c}\mathfrak{a} \sim \mathfrak{c}\mathfrak{b}$ (indeed, both ideals are principal) follows that $\mathfrak{a} \sim \mathfrak{b}$; hence, $\mathfrak{a} \in \mathfrak{U}$. Second by Theorem 26, Theorem 24, and Lemma 2 it follows that $\mathfrak{c}\mathfrak{a} = (\gamma) \Rightarrow N\mathfrak{c} \cdot N\mathfrak{a} = N((\gamma)) = |N\gamma| \le m \cdot N\mathfrak{c}$, so that indeed $N\mathfrak{a} \le m$ and the lemma is proven.

PROOF OF THEOREM 30.   The proof of Theorem 30 is now rather trivial. By Theorem 28, there exist only finitely many ideals of a given norm; hence only finitely many whose norm does not exceed any given bound. Suppose that there are t ideals of norm $\le m$. By Lemma 3, in each class there exists at least one ideal of norm $\le m$; therefore there exist at most t classes of ideals.

DEFINITION 14.   The (finite) number of classes of ideals in **K** is called the *class number* of **K** and is denoted by $h = h(\mathbf{K})$.

Theorem 31.   *If h is the class number of the algebraic number field* **K**, *then for every ideal* $\mathfrak{a}$ *of* **K**, $\mathfrak{a}^h$ *is a principal ideal.*

PROOF.   Consider a set of h inequivalent ideals in **K**, $\mathfrak{a}_1, \mathfrak{a}_2, \ldots, \mathfrak{a}_h$. Their number being h, there is exactly one from each class in this set. If $\mathfrak{a}$ is any ideal of **K**, then $\mathfrak{a}\mathfrak{a}_1, \mathfrak{a}\mathfrak{a}_2, \ldots, \mathfrak{a}\mathfrak{a}_h$ is again a set of h inequivalent ideals (because by Theorem 10 and Lemma 1 it follows that $\mathfrak{a}\mathfrak{a}_j \sim \mathfrak{a}\mathfrak{a}_k \Rightarrow \mathfrak{b}\mathfrak{a}\mathfrak{a}_j \sim \mathfrak{b}\mathfrak{a}\mathfrak{a}_k \Leftrightarrow (c)\mathfrak{a}_j \sim (c)\mathfrak{a}_k \Leftrightarrow \mathfrak{a}_j \sim \mathfrak{a}_k$). Consequently, in the set $\{\mathfrak{a}\mathfrak{a}_j | 1 \le j \le h\}$ there is again exactly one ideal from each of the h classes. It now follows from Lemma 1 that $\mathfrak{a}_1\mathfrak{a}_2 \ldots \mathfrak{a}_h \sim \mathfrak{a}\mathfrak{a}_1 \cdot \mathfrak{a}\mathfrak{a}_2 \ldots \mathfrak{a}\mathfrak{a}_h = \mathfrak{a}^h \mathfrak{a}_1 \ldots \mathfrak{a}_h$, and once more using Theorem 10 and Lemma 1, $(1) \sim \mathfrak{a}^h$, as claimed.

The rational primes that do not divide the class number h will soon be of particular interest. In anticipation of that situation, we formulate:

Theorem 32.   *If* $p \nmid h$, *then* $\mathfrak{a}^p \sim \mathfrak{b}^p \Rightarrow \mathfrak{a} \sim \mathfrak{b}$.

PROOF.   By Lemma 1, for $k \in \mathbf{Z}^+$, $\mathfrak{a}^p \sim \mathfrak{b}^p \Rightarrow \mathfrak{a}^{pk} \sim \mathfrak{b}^{pk}$. Also, because of $(h, p) = 1$, there are positive integers m, k such that $kp - mh = 1$. Hence by Theorem 31, $\mathfrak{a}^h = (1)$, $\mathfrak{a}^{kp} = \mathfrak{a}^{mh+1} \sim \mathfrak{a}$; similarly, $\mathfrak{b}^{kp} \sim \mathfrak{b}$, and the result follows.

Corollary 32.1.   *If* $p \nmid h$ *and* $\mathfrak{a}^p$ *is principal in* **K**, *then so is* $\mathfrak{a}$.

PROOF.   Take $\mathfrak{b} = (1)$ in Theorem 32.

In developing the theory of ideals we have reached approximately the point corresponding to the first six theorems in Chapter 3 for the rational integers. It seems plain that the theory will not stop here. Indeed this chapter has only laid the foundations for the study of ideals in number fields; but we shall not pursue the matter further. On the one hand, it is likely that the reader will have gathered the general flavor of this theory. On the other hand, we now have at our disposal sufficient information to be able to handle the problems of Diophantine equations to be discussed in some of the chapters that follow. The interested reader, however, is advised not to stop here but to consult some of the excellent books either on algebraic numbers (such as [1], [4], or [5]) or on abstract ideal theory (see, e.g., [2] or [3]).

PROBLEMS

1. Prove Theorem 1.
2. Prove Theorem 3.
3. Justify Remark 5.

4. Prove Theorem 5.

5. Write out in detail the proof of Theorem 6.

6. Prove Corollary 6.1.

7. Let α be an integer of an algebraic number field **K** and let $\alpha^{(2)}, \alpha^{(3)}, \ldots, \alpha^{(n)}$ be its conjugates.

   Prove that $\alpha^{(2)} \cdot \alpha^{(3)} \cdot \ldots \cdot \alpha^{(n)}$ is also an integer in **K**.

8. Where does the proof of Theorem 12 break down if $\mathfrak{c} = 0$?

9. With reference to Remark 8, show that if an ideal of an algebraic number field **K** has the property of "primality," then it also has that of "irreducibility" or "maximality."

10. Prove Corollary 16.1

11. Give a detailed proof of Theorem 17.

12. Give a detailed proof of Corollary 17.1.

13. Write out the proof of Theorem 20 in detail.

14. Write out the proof of Theorem 21 in detail.

15. Write out the proof of Theorem 22 in detail.

16. Write out the proof of Corollary 22.1 in detail.

17. (a) Prove that in a field of degree n there are exactly $|a|^n$ residue classes mod a if $a \in \mathbf{Z}$.

    (b) If $\mathfrak{a} = (a)$ is a principal ideal, $a \in \mathbf{Z}$, $a > 0$, prove that there are a^n residue classes mod $\mathfrak{a}$.

18. Prove the following generalization of Fermat's theorem to ideals: For every $\alpha \in \mathbf{I}$ and $\mathfrak{p}$ a prime ideal of **K** $\ni \mathfrak{p} \nmid \alpha$, $\alpha^{N\mathfrak{p}-1} \equiv 1 \pmod{\mathfrak{p}}$.

BIBLIOGRAPHY

1. E. Landau, *Vorlesungen über Zahlentheorie*, Vol. 3. Leipzig: S. Hirzel, 1927.

2. N. H. McCoy, *Rings and Ideals* (Carus Monograph No. 8). La Salle, Illinois: Open Court, 1948.

3. D. G. Northcott, *Ideal Theory*. Cambridge: Cambridge University Press, 1953.

4. H. Pollard, *The Theory of Algebraic Numbers* (Carus Monograph No. 9). New York: Wiley, 1950.

5. E. Weiss, *Algebraic Number Theory*. New York: McGraw-Hill, 1963.

Chapter 12

Primes in Arithmetic Progressions

1 INTRODUCTION AND DIRICHLET'S THEOREM

In Chapter 3 (see Theorem 3.9) we learned that there are infinitely many primes. Almost the same proof (which goes back at least to Euclid) will show that there are infinitely many primes $p \equiv 3 \pmod 4$. Indeed suppose that there were only finitely many such primes and let $3, 7, 11, 19, \ldots, p_k$ be the complete list. Now consider the integer $N = 4 \cdot 3 \cdot 7 \cdot 11 \cdots p_k - 1$. Clearly $N \equiv 3 \pmod 4$. It cannot be a prime because $N > p_k$ and p_k is by assumption the largest prime $p \equiv 3 \pmod 4$. Hence N is a product of odd primes. Among them there is at least one prime $q \equiv 3 \pmod 4$, because the product of primes $p \equiv 1 \pmod 4$ is itself congruent to $1 \pmod 4$. However $q|N$ means that $q \nmid N + 1$; hence q is a prime not on the list, although $q \equiv 3 \pmod 4$. The list, however, was supposed to be complete, and so we have obtained a contradiction and our original assumption that the primes $q \equiv 3 \pmod 4$ are finite in number is not tenable.

A proof that there exist infinitely many primes $p \equiv 1 \pmod 4$ is somewhat more difficult. First we recall from Theorem 5.5 that if $p|N^2 + 1$ (so that $-1 \equiv N^2 \pmod p$ and -1 is a quadratic residue modulo p), then $p \equiv 1 \pmod 4$. Assume now that the number of primes $p \equiv 1 \pmod 4$ was finite and let $5, 13, 17, \ldots, p_k$ be the complete list of all these primes. Consider now the integer $N = 5 \cdot 13 \cdot 17 \cdots p_k$; then $N^2 + 1$ is divisible by some prime p ($p = N^2 + 1$ is not ruled out). This prime satisfies $p \equiv 1 \pmod 4$, as just observed. On the other hand, it is none of the primes on our list, because otherwise $p|N$ and $p|N^2 + 1$ imply $p|1$, which is false. Hence our list, no matter how long, did not contain this prime p and so was not complete. This proves that there are infinitely many primes $p \equiv 1 \pmod 4$.

In the case of the modulus 6, only the two residue classes 1 and 5 modulo 6 can possibly contain infinitely many primes, because otherwise for $r = 0, 2, 3, 4,$

$6m + r$ has an obvious nontrivial factor. So, e.g., if $r = 4$, $6m + 4 = 2(3m + 2)$ and cannot be a prime; for $r = 3$, $6m + 3 = 3(2m + 1)$ and can be a prime (namely 3) only for $m = 0$, etc.

We may construct proofs modeled on the preceding ones for $p \equiv 1 \pmod 4$ and $p \equiv 3 \pmod 4$ to show that both arithmetic progressions $p = 6m + 1$ and $p = 6m + 5$ contain infinitely many primes. Let us consider, however, the general situation.

Given an arithmetic progression $mk + r$, it is clear that for $(k, r) = d > 1$ this progression can contain at most one prime; indeed if $k = dk_1$, $r = dr_1$, then $mk + r = d(mk_1 + r_1)$ is not a prime, except perhaps if d is itself a prime and $r_1 = 1$, and even then only for the single value $m = 0$. If we look, however, at the $\phi(k)$ values of r that are coprime to k, it is legitimate to ask whether in each of these $\phi(k)$ arithmetic progressions $mk + r_j$ ($j = 1, 2, \ldots, \phi(k)$) there are or are not infinitely many primes. We proved the affirmative to be the case for $k = 4$ and very similar proofs work also for $k = 6$. Similar but more elaborate proofs can also be constructed for several other moduli, however not for all. The reason for this curious situation will become clearer in Section 6. Nevertheless, the result is true for all moduli, as has been shown by Dirichlet, who proved the following:

Theorem 1. *If $(k, r) = 1$, then the arithmetic progression $km + r$ contains infinitely many primes.*

Much more is in fact true. We remember (see Chapter 9) that $\pi(x)$, the number of primes up to x, is asymptotically equal to $x/\log x$. Now it has been shown that not only are there infinitely many primes in each of the $\phi(k)$ arithmetic progressions modulo k in which this is possible, but also that the primes are essentially evenly distributed over those progressions. Indeed, we have the following:

Theorem 2 (Siegel-Walfisz-Paige). *The number $\pi(x; k, r)$ of primes $p \leq x$, $p \equiv r \pmod k$ for each of the $\phi(k)$ values of r prime to k is asymptotically independent of r, and $\pi(x; k, r) \sim x/(\phi(k)\log x)$.*

The proof of this theorem is in principle similar to that of the prime number theorem in Chapter 9, but it requires far more care and the details are more complicated. We shall neither prove nor use it in this book, but proofs may be found in [2], [9], and [7]. We shall, however, give a proof of Theorem 1. The tools will be prepared in Section 2, on Characters, and in Section 3, on Dirichlet L-functions. Section 4 contains a sketch of the proof of Theorem 1. Section 5 treats some technical details and the proof is completed in Section 6. Section 7 discusses an alternative approach, while Section 8 states many interesting related results, without complete proofs.

2 CHARACTERS

In the proof of Theorem 1 we shall use a generalization of the Riemann zeta function $\zeta(s)$. This new function, denoted by $L(s, \chi)$ can be represented for σ ($= \mathrm{Re}\, s) > 1$ by the convergent series $\sum_{n=1}^{\infty} \chi(n) n^{-s}$; here $\chi(n)$ are certain functions that we shall immediately define precisely. Let us already mention here, however, that for $(n, k) = 1$ (k an arbitrary but fixed natural integer), $|\chi(n)| = 1$. It follows that if in particular $\chi(n) = 1$ for all integers (this corresponds to the choice $k = 1$), then $L(s, \chi)$ reduces to $\zeta(s)$; hence $L(s, \chi)$ is indeed a generalization of $\zeta(s)$.

DEFINITION 1. Let k be a fixed natural integer and let G be the group of prime residue classes modulo k. Then the functions that map G homomorphically into a group of roots of unity are called characters modulo k.

REMARK 1. This definition of the characters will be recognized as the usual definition of a one-dimensional group character.

REMARK 2. We may consider the characters as functions defined on all integers n coprime to k and constant on a given residue class modulo k. Indeed for $(a, k) = 1$, the set of all integers $n \equiv a \pmod k$ constitutes a prime residue class, say $\{a\}$, and χ maps it into a certain complex number of absolute value one. Instead of writing $\chi(\{a\})$, we shall write $\chi(a)$ or $\chi(n)$, because this is simpler and does not lead to any ambiguity.

REMARK 3. It simplifies matters a great deal if we extend the definition of $\chi(n)$ to all integers. This is done by setting $\chi(n) = 0$ for all n with $(k, n) > 1$.

Several properties of these functions χ follow immediately from Definition 1. So for instance, as this mapping has to be a homomorphism, it follows that if $n \cdot m = f$, then $\chi(n) \cdot \chi(m) = \chi(f) = \chi(n \cdot m)$ and has to be (even totally!) multiplicative. In particular, $\chi(n^a) = (\chi(n))^a$. Next, χ has to map the identity of G (i.e., the residue class $n \equiv 1 \pmod k$) into 1, so that we write (slightly abusively, see Remark 2) $\chi(1) = 1$ for all characters. The order of G is (see second proof of Theorem 4.10) $\phi(k)$; hence if $(n, k) = 1$, $n^{\phi(k)} \equiv 1 \pmod k$, as we already knew from Euler's Theorem 4.10. It follows that for $(n, k) = 1$, $\chi(n)^{\phi(k)} = \chi(n^{\phi(k)}) = \chi(1) = 1$. This shows that all nonvanishing values of all characters modulo k are $\phi(k)$th roots of unity. These considerations justify the following definition.

DEFINITION 2. For a given natural integer k, the arithmetical functions $\chi(n)$, totally multiplicative, with $\chi(n) = 0$ for $(n, k) > 1$ and $\chi(n)$ constant on all residue classes modulo k, and with values $\chi(n)$ in the set of $\phi(k)$th roots of unity for $(k, n) = 1$, are called *Dirichlet characters*.

It is particularly easy to actually construct all such characters in case G is cyclic, i.e., if k admits a primitive root. Indeed let g be a primitive root modulo k; then every integer n coprime to k satisfies $n \equiv g^j \pmod{k}$ for some j ($j = 0, 1, \ldots, \phi(k) - 1$). Next select any one among the $\phi(k)$th roots of unity, say $e^{2\pi i s/\phi(k)}$ ($s = 0, 1, \ldots, \phi(k) - 1$), and set $\chi(g) = e^{2\pi i s/\phi(k)}$ with $\phi(k)$ possible choices for the value of s. Once this choice is made, however, $\chi(n)$ is completely determined for all n with $(n, k) = 1$ by $\chi(n) = \chi(g^j) = \chi(g)^j = e^{2\pi i s j/\phi(k)}$. These s characters are distinct because they differ at least at some arguments (e.g., for $n = g$). It is also clear that there are no other characters besides the $\phi(k)$ already obtained, because $\chi(g)$ has to be one of the $\phi(k)$th roots of unity and there are only $\phi(k)$ of them, and once $\chi(g)$ is selected, $\chi(n)$ is uniquely determined. In the general case, $k = p_1^{a_1} p_2^{a_2} \cdots p_r^{a_r}$ and does not have a primitive root, so that G is no longer cyclic. It is, however, still abelian, and by use of the fundamental theorem on finite abelian groups (see [6] p. 109 or 204) we obtain the following general result, the proof of which some readers may wish to skip at a first reading.

Theorem 3.   *There exist $\phi(k)$ distinct characters modulo k, and no more.*

PROOF.   If k is odd, each factor p^a has a primitive root, say g, which generates a cyclic group of order $\phi(p^a) = p^a - p^{a-1}$. On it, we can define $\phi(p^a)$ distinct characters by choosing for $\chi(g)$ any one of the $\phi(p^a)$ different $\phi(p^a)$th roots of unity, say $\chi(g) = e^{2\pi i m/\phi(p^a)}$ ($m = 0, 1, \ldots, \phi(p^a) - 1$). By the fundamental theorem on finite abelian groups, G is the direct product of the mentioned cyclic groups, i.e., if $(n, k) = 1$, then $n \equiv g_1^{j_1} g_2^{j_2} \cdots g_r^{j_r} \pmod{k}$, where g_s is the primitive root corresponding to the factor $p_s^{a_s}$ of k. Then if $\chi(g_s) = e^{2\pi i m_s/\phi(p_s^{a_s})}$ ($s = 1, 2, \ldots, r$), $\chi(n) = \exp\{2\pi i \sum_{s=1}^{r} m_s j_s/\phi(p_s^{a_s})\}$ is uniquely determined and it is a simple matter to verify that it is a character on G. As the j_s are uniquely determined for a given n, and as there are $\phi(p_s^{a_s})$ choices for each m_s, it follows by the multiplicativity of Euler's ϕ-function that there are $\prod_{s=1}^{r} \phi(p_s^{a_s}) = \phi\{\prod_{s=1}^{r} p_s^{a_s}\} = \phi(k)$ different characters and no others.

In case k is even, there is one additional complication. Indeed while for $p = 2$, $a = 1$ or 2 everything goes through as before (because 2 and 4 have primitive roots), for $a \geq 3$ we remember (see Theorem 4.22) that $\phi(2^a) = 2^{a-1}$, while for any odd b, $b^{2^{a-2}} \equiv 1 \pmod{2^a}$, and 2^a has no primitive root. This shows that the abelian group of prime (i.e., odd) residue classes modulo 2^a is not cyclic but is itself a direct product of two cyclic groups, one of order 2^{a-2} generated, e.g., by $b = 5$ through its powers, and one of order 2. Indeed by repeating the proof of Theorem 4.22 with 5 instead of b, we conclude that $5^{2^{a-2}} = 1 + 2^a s$, with s odd. This shows that 5 actually belongs to the exponent 2^{a-2} modulo 2^a and yields, through its successive powers, exactly one half of the 2^{a-1} prime (i.e., odd) residue classes modulo 2^a. As $5 \equiv 1 \pmod{4}$, also $5^j \equiv 1 \pmod{4}$, and so all classes $n \equiv 1 \pmod{4}$ are represented modulo 2^a by

5^j ($j = 1, 2, \ldots, 2^{a-2}$). To obtain the other classes, all we have to do is consider the classes -5^j ($j = 1, 2, \ldots, 2^{a-2}$). It follows that all prime residue classes n modulo 2^a are obtained by $(-1)^e 5^j$ ($e = 0$ if $n \equiv 1 \pmod 4$), $e = 1$ if $n \equiv 3 \pmod 4$; $j = 1, 2, \ldots, 2^{a-2}$). One may now complete the definition of characters by choosing $\chi(-1) = +1$ or $= -1$, and by selecting for $\chi(5)$ any one of the roots of unity of order 2^{a-2}; this leads to $2 \cdot 2^{a-2} = 2^{a-1} = \phi(2^a)$ characters modulo 2^a and completes the proof of Theorem 3.

In our previous constructions, one of the possible choices for m_s was $m_s = 0$, which leads to $\chi(g_s) = 1$. If this choice is made for all s ($s = 1, 2, \ldots, r$), then $\chi(n) = 1$ for all $(n, k) = 1$. This function $\chi(n)$ satisfies all conditions for a character; it is called the *principal character* modulo k. We shall denote it by $\chi_0(n)$.

The characters themselves form a group under multiplication. Indeed if χ_1 and χ_2 are any two characters, then the function $f(n) = \chi_1(n)\chi_2(n)$, which we may denote by $\chi_1\chi_2(n)$, is also a character. Indeed, it is a $\phi(k)$th root of unity, it is totally multiplicative, it is constant on residue classes modulo k, and it maps these homomorphically into the roots of unity. This shows that the set is closed under multiplication. The principal character acts as identity element. Finally, if $\chi(n)$ is a character, also $f(n) = \overline{\chi(n)}$ (the complex conjugate value) is also a character. However, $\chi(n)\overline{\chi(n)} = |\chi(n)|^2 = 1$ for $(n, k) = 1$; hence $\chi(n)\overline{\chi(n)} = \chi_0(n)$, and if we denote $\overline{\chi(n)}$ by $\bar\chi(n)$, then $\bar\chi(n)$ is a character inverse to $\chi(n)$.

After these preliminaries, we can state and prove some of the more important theorems concerning characters.

Theorem 4. *Let $\{\chi(n)\}$ be the set of $\phi(k)$ characters modulo k; then*

$$\sum_{n \bmod k} \chi(n) = \begin{cases} \phi(k) & \text{if } \chi = \chi_0, \\ 0 & \text{otherwise;} \end{cases}$$

$$\sum_{\chi \bmod k} \chi(n) = \begin{cases} \phi(k) & \text{if } n \equiv 1 \pmod k. \\ 0 & \text{otherwise.} \end{cases}$$

PROOF. Let $(n, k) = 1$; then when m runs through a complete reduced residue system modulo k, so does mn (see first proof of Theorem 4.10). Consequently,

$$\sum_{m \bmod k} \chi(m) = \sum_{m \bmod k} \chi(nm) = \sum_{m \bmod k} \chi(n)\chi(m) = \chi(n) \sum_{m \bmod k} \chi(m),$$

so that $(1 - \chi(n))\sum_{m \bmod k}\chi(m) = 0$. In case $\chi(n) = \chi_0(n)$, this holds indeed, because $1 - \chi_0(n) = 0$. Also $\sum_{n \bmod k}\chi(n) = \sum_{n \bmod k}\chi_0(n) = \phi(k)$, and the first assertion of the theorem holds. If $\chi \neq \chi_0$, the first factor $1 - \chi_0(n) \neq 0$ for at least one n with $(n, k) = 1$, so that $\sum_{m \bmod k}\chi(m) = 0$ as claimed.

Similarly, if $n \equiv 1 \pmod{k}$, $\chi(n) = 1$ for all $\phi(k)$ characters, and $\sum_{\chi \bmod k} \chi(1) = \phi(k)$, as stated by the theorem. Otherwise, there is at least one character, say $\chi_1(n)$, for which $\chi_1(n) \neq 1$ (try to prove this formally); then with χ, $\chi\chi_1$ also runs through the complete set of characters modulo k. Hence

$$\sum_{\chi \bmod k} \chi(n) = \sum_{\chi \bmod k} \chi\chi_1(n) = \sum_{\chi \bmod k} \chi(n)\chi_1(n) = \chi_1(n) \sum_{\chi \bmod k} \chi(n),$$

so that $(1 - \chi_1(n))\sum_{\chi \bmod k} \chi(n) = 0$; however $1 - \chi_1(n) \neq 0$, and this finishes the proof of the theorem.

Theorem 5. *For* $(n, k) = 1$,

$$\sum_{\chi \bmod k} \chi(n)\bar{\chi}(r) = \begin{cases} \phi(k) & \text{if } n \equiv r \pmod{k}, \\ 0 & \text{if } n \not\equiv r \pmod{k}. \end{cases}$$

PROOF.   For $(r, k) = 1$, let r' be defined by $rr' \equiv 1 \pmod{k}$. Then $\chi(r)\chi(r') = \chi(1) = 1 = \chi_0(r)$, so that $\chi(r') = \chi^{-1}(r)$. By recalling also that $\bar{\chi} = \chi^{-1}$, we find that

$$\sum_{\chi \bmod k} \chi(n)\bar{\chi}(r) = \sum_{\chi \bmod k} \chi(n)\chi^{-1}(r) = \sum_{\chi \bmod k} \chi(n)\chi(r') = \sum_{\chi \bmod k} \chi(nr')$$

$$= \begin{cases} \phi(k) & \text{if } nr' \equiv 1 \pmod{k} \\ 0 & \text{if } nr' \not\equiv 1 \pmod{k} \end{cases}$$

by Theorem 4. However, $nr' \equiv 1 \pmod{k}$ means $n \equiv r \pmod{k}$, and the theorem is proved.

3   DIRICHLET'S L-FUNCTIONS

The functions introduced informally at the beginning of Section 2 and denoted by $L(s, \chi)$ were defined and studied by Dirichlet [3] for the express purpose of proving Theorem 1. In order to justify our preliminary definition, we need:

Theorem 6. *If* $\chi(n)$ *is a character modulo* k *and* $s = \sigma + it$ $(\sigma, t \in \mathbf{R})$, *then the series* $\sum_{n=1}^{\infty} \chi(n)n^{-s}$ *converges absolutely for* $\sigma > 1$ *and uniformly so for* $\sigma \geq 1 + \varepsilon$ $(\varepsilon > 0)$.

PROOF.   Let $\sigma = 1 + \varepsilon$, $\varepsilon > 0$; then

$$\left| \sum_{n=N+1}^{M} \chi(n)n^{-s} \right| \leq \sum_{n=N+1}^{M} |\chi(n)n^{-s}| \leq \sum_{n=N+1}^{M} n^{-\sigma} \leq (N+1)^{-(1+\varepsilon)}$$

$$+ \int_{N+1}^{M} x^{-(1+\varepsilon)} \, dx < (N+1)^{-(1+\varepsilon)} + \frac{(N+1)^{-\varepsilon}}{\varepsilon} - \frac{M^{-\varepsilon}}{\varepsilon}$$

$$< (N+1)^{-(1+\varepsilon)} + \varepsilon^{-1}(N+1)^{-\varepsilon}.$$

For fixed ε, this quantity can be made arbitrarily small, independently of M and s. Hence by Cauchy's criterion of convergence (see [1]), $\sum_{n=1}^{\infty}\chi(n)n^{-s}$ converges absolutely for $\sigma > 1$, and in fact, uniformly so for $\sigma \geq 1 + \varepsilon$ $(\varepsilon > 0)$, because the "error" is bounded by quantities independent of s.

DEFINITION 3. Let $\chi(n)$ be a Dirichlet character modulo k. The functions $L(s, \chi)$, defined for $\sigma > 1$ by the convergent series $\sum_{n=1}^{\infty}\chi(n)n^{-s}$, are called (*Dirichlet*) *L-functions*.

These functions, as well as Riemann's zeta function, are particular cases of functions defined by series of the general type $\sum_{n=1}^{\infty}a_n n^{-s}$, called *Dirichlet series*, that we already met with in Chapter 6. The regions of convergence of these series are half-planes $\sigma > \sigma_0$ and the real number σ_c, such that the series converges for $\sigma > \sigma_c$ and diverges for $\sigma < \sigma_c$ is called *abscissa of convergence*. For further details on Dirichlet series, see [4].

Theorem 7. *For $\sigma > 1$, the L-functions are represented by the Euler product*

$$L(s, \chi) = \prod_{p}(1 - \chi(p)p^{-s})^{-1}.$$

PROOF. The proof is identical with the corresponding one for $\zeta(s)$, by use of the total multiplicativity of $\chi(p)$. Indeed

$$\prod_{p \leq P}\left(1 - \frac{\chi(p)}{p^s}\right)^{-1} = \prod_{p \leq P}\left(1 + \frac{\chi(p)}{p^s} + \frac{\chi(p^2)}{p^{2s}} + \cdots\right) = \sum_{n \in S}\frac{\chi(n)}{n^s},$$

where **S** is the set of integers without prime factors larger than P. Each $n = p_1^{a_1}p_2^{a_2}\ldots p_j^{a_j}$ occurs at most once (in fact, exactly once, if max $p_j \leq P$), due to the theorem on the uniqueness of factorization in **Z** (see Theorems 3.3 and 8.2). Consequently,

$$L(s, \chi) - \prod_{p \leq P}(1 - \chi(p)p^{-s})^{-1} = \sum_{n \in T}\chi(n)n^{-s},$$

where **T** is the set of integers complementary to **S**. In particular, an integer in **T** contains certainly at least one prime factor larger than P, whence $n \in T \Rightarrow n > P$. Consequently,

$$\left|\sum_{n \in T}\chi(n)n^{-s}\right| \leq \sum_{n \in T}|\chi(n)n^{-s}| \leq \sum_{n > P}n^{-\sigma}.$$

From the convergence of $\sum n^{-\sigma}$ $(\sigma > 1)$, it follows that $\sum_{n > P}n^{-\sigma} \to 0$ as $P \to \infty$, and the proof is complete.

Theorem 8. *If $\chi = \chi_0$, the principal character modulo k, then for $\sigma > 1$,*

$$L(s, \chi_0) = \prod_{p|k}(1 - p^{-s}) \cdot \zeta(s). \tag{1}$$

PROOF.   For $\sigma > 1$, $L(s, \chi_0) = \Pi_p(1 - \chi_0(p)p^{-s})^{-1}$; however, $\chi_0(p) = 1$, except for $p|k$, when $\chi_0(p) = 0$. The result now follows from

$$\zeta(s) = \prod_p (1 - p^{-s})^{-1} = \prod_{p|k}(1 - p^{-s})^{-1} \prod_{p \nmid k}(1 - p^{-s})^{-1}$$

$$= \prod_{p|k}(1 - p^{-s})^{-1} \prod_{p \nmid k}(1 - \chi_0(p)p^{-s})^{-1} = \prod_{p|k}(1 - p^{-s})^{-1}L(s, \chi_0).$$

Corollary 8.1. $\lim_{s \to 1}(s - 1)L(s, \chi_0) = \Pi_{p|k}(1 - p^{-1}) = \phi(k)/k.$

PROOF. $\lim_{s \to 1}(s - 1)L(s, \chi_0) = \lim_{s \to 1}(s - 1)\zeta(s)\Pi_{p|k}(1 - p^{-s}) = \Pi_{p|k}(1 - p^{-1}) = \phi(k)/k$, by Corollary 8.6.2 and Theorem 6.6.

Corollary 8.2.   *Equation* (1) *and Corollary* 8.1 *yield the analytic continuation of* $L(s, \chi_0)$ *as a meromorphic function in the whole plane with a single singularity at* $s = 1$, *namely a pole of first order, with residue* $\phi(k)/k$.

PROOF.   The continuation follows from Theorems 8.6 and 8.7 and the discussion following Theorem 8.7. The value of the residue follows from Corollary 8.1.

Theorem 9.   *If* $\chi \neq \chi_0$, *the series* $\sum_{n=1}^{\infty}\chi(n)n^{-s}$ *converges for* $\sigma > 0$.

In the proof of Theorem 9 we shall need

Lemma 1.   *With previous notations, for* $\chi \neq \chi_0$, *the inequality* $|\sum_{n=a+1}^{b}\chi(n)| \leq \phi(k)$ *holds.*

PROOF.   By Theorem 4, for $\chi \neq \chi_0$, $\sum_{n=a+1}^{a+k}\chi(n) = \sum_{n \bmod k}\chi(n) = 0$; it follows that if $b = a + mk + r$, $0 \leq r < k$, then

$$\sum_{n=a+1}^{b}\chi(n) = \sum_{n=a+1}^{a+mk}\chi(n) + \sum_{n=a+mk+1}^{a+mk+r}\chi(n) = \sum_{n=a+1}^{a+r}\chi(n).$$

However, $|\sum_{n=a+1}^{a+r}\chi(n)| \leq \sum_{n=a+1}^{a+r}|\chi(n)| \leq \sum_{n \bmod k}|\chi(n)| = \phi(n)$.

PROOF OF THEOREM 9.   For any real w set $S(w) = \sum_{m \leq w}\chi(m)$; then $\chi(n) = S(n) - S(n - 1)$. Let $1 \leq u \leq v$, $u, v \in \mathbb{Z}$; then

$$\sum_{u \leq n \leq v}\frac{\chi(n)}{n^s} = \sum_{u \leq n \leq v}\frac{S(n) - S(n - 1)}{n^s}$$

$$= \sum_{n=u}^{v-1}S(n)\left(\frac{1}{n^s} - \frac{1}{(n + 1)^s}\right) + \frac{S(v)}{v^s} - \frac{S(u - 1)}{u^s}.$$

By Lemma 1, $|S(u)| \leq \phi(k)$ and $|S(v)| \leq \phi(k)$; hence the last two terms are each at most $\phi(k)/u^{\sigma}$ in absolute value, and if $\sigma > 0$, vanish in the limit as $u \to \infty$. It follows that each of them is less than any preassigned value $\eta/4$, with $\eta > 0$, if $u \geq u_1(\eta)$. The first term can be written as

$$s \sum_{n=u}^{v-1} S(n) \int_n^{n+1} \frac{dx}{x^{s+1}} = s \sum_{n=u}^{v-1} \int_n^{n+1} \frac{S(x)}{x^{s+1}} dx = s \int_u^v \frac{S(x)}{x^{s+1}} dx.$$

The absolute value of this term does not exceed $|s|\phi(k)\int_u^v dx/x^{\sigma+1} = (|s|/\sigma)\phi(k)(u^{-\sigma} - v^{-\sigma}) \leq (|s|/\sigma)\phi(k)u^{-\sigma}$. Also, this term vanishes in the limit if $u \to \infty$ (recall: $\sigma > 0$, fixed). Hence for u larger than some $u_2(\eta)$, this value is less than $\eta/2$. If $u_0(\eta) = \max(u_1(\eta), u_2(\eta))$, then for $u \geq u_0(\eta)$ and any $v \geq u$, $|\sum_{n=u}^v \chi(n)/n^s| < \eta$, and the series converges by Cauchy's criterion. The proof of Theorem 9 is complete.

Corollary 8.2 and Theorem 9 extend the regions in which the L-functions are defined by the series $\sum_{n=1}^\infty \chi(n)n^{-s}$. Beyond that, Theorem 8 permits us to write down a functional equation satisfied by $L(s, \chi_0)$. Indeed by Theorem 8.7,

$$\zeta(s) = 2^s \pi^{s-1} \zeta(1-s)\Gamma(1-s)\sin\frac{\pi s}{2}.$$

If we replace here $\zeta(s)$ by $L(s, \chi_0)\prod_{p|k}(1 - p^{-s})^{-1}$, we obtain

$$L(s, \chi_0)\prod_{p|k}(1 - p^{-s})^{-1} = 2^s \pi^{s-1} L(1-s, \chi_0)\prod_{p|k}(1 - p^{s-1})^{-1}$$

$$\cdot \Gamma(1-s)\sin\frac{\pi s}{2}.$$

This leads by routine manipulations to:

Theorem 10. *If χ_0 is the principal character modulo k, then $L(s, \chi_0)$ satisfies the functional equation*

$$L(s, \chi_0) = 2^s \pi^{s-1} \prod_{p|k} \frac{1 - p^{-s}}{1 - p^{s-1}} \cdot \Gamma(1-s)\sin\frac{\pi s}{2} \cdot L(1-s, \chi_0),$$

or equivalently,

$$L(1-s, \chi_0) = 2^{1-s} \pi^{-s} \prod_{p|k} \frac{1 - p^{s-1}}{1 - p^{-s}} \Gamma(s)\cos\frac{\pi s}{2} L(s, \chi_0). \tag{2}$$

As in the case of the zeta function, this functional equation permits us to study $L(s, \chi_0)$ in the half-plane $\sigma < 0$ by using the values of $L(s, \chi_0)$ with $\sigma > 1$, where the defining series converges absolutely. For the question of the location of the zeros of $L(s, \chi_0)$, however, we don't need anything beyond (1). Indeed it is clear that $L(s, \chi_0)$ vanishes precisely at the union of the set of zeros of $\zeta(s)$ and that of $\prod_{p|k}(1 - p^{-s})$. The last product is finite for every integer k and $s \in \mathbf{C}$; hence there is no pole of the last product that could cancel some zero of $\zeta(s)$. As for the zeros of the product itself, these are easily found. The product vanishes precisely at the union of the zeros of its (finitely many) factors, and the typical factor $1 - p^{-s}$, $p|k$ vanishes precisely at the complex values of s for which $p^{-s} = e^{-s\log p} = 1$, i.e., for $s = 2\pi im/\log p (m \in \mathbf{Z})$. All these zeros are on the imaginary axis, i.e., $s = it$ with $t = 2\pi m/\log p (m \in \mathbf{Z}, p|k)$. In particular, $s = 0$ is always a zero of $L(s, \chi_0)$ for $k > 1$, in contrast to the case of $\zeta(s)$, for which we recall that $\zeta(0) = -1/2$.

In view of the great advantage of a functional equation of type (2), for the study of these functions in $\sigma < 1/2$ it is natural to ask about similar equations for $L(s, \chi)$, $\chi \neq \chi_0$. Such functional equations have indeed been established. The corresponding proofs, however, are more involved; as we shall not need to know anything about $L(s, \chi)$ for $\chi \neq \chi_0$ and $\sigma \leq 0$, and as their series representation remains valid for $\sigma > 0$, we shall here only state the corresponding theorem without proof (a proof may be found, e.g., in [9]). In Theorem 11 the technical term *primitive character* modulo k occurs; as we shall not need it in any other context, it will suffice if we say here that it means roughly that the character cannot be defined modulo a proper divisor of k (see [5] or [9]).

Theorem 11.   *For every character $\chi(\mathrm{mod}\ k)$, let $\bar{\chi}$ be the complex conjugate character. Then*

$$L(1 - s, \bar{\chi}) = \varepsilon(\chi)\eta_\chi(k, s)2^{1-s}\pi^{-s}k^{s-1/2}\cos\frac{\pi(s - a)}{2}\Gamma(s)L(s, \chi); (3)$$

here $a = 0$ if $\chi(-1) = 1$, $a = 1$ if $\chi(-1) = -1$; $\varepsilon(\chi)$ is a constant that depends only on the character χ with $|\varepsilon(\chi)| = 1$; also, if χ is a primitive character $(\mathrm{mod}\ k)$, then $\eta_\chi(k, s) = 1$, identically.

REMARK 4.   For $k > 1$, the principal character χ_0 is *not* primitive but it is real, so that $\bar{\chi}_0 = \chi_0$. For simplicity, let us consider a squarefree modulus k. The product $\prod_{p|k}(1 - p^{s-1})/(1 - p^{-s})$ that occurs in (2) may be written as

$$\prod_{p|k} p^{s-1/2} \cdot \eta_{\chi_0}(k, s) = k^{s-1/2}\eta_{\chi_0}(k, s), \text{ with } \eta_{\chi_0}(k, s) = \prod_{p|k} \frac{p^{1/2} - p^{s-1/2}}{p^s - 1}.$$

We see that $\eta_{\chi_0}(k, \frac{1}{2}) = 1$ identically in k, but $\eta_{\chi_0}(k, s)$ is not identically equal to one for all k and s. We also saw that $L(s, \chi_0) = \prod_{p|k}(1 - p^{-s}) \cdot \zeta(s)$, so that the zeros of the denominator of $\eta_{\chi_0}(s, k)$ are cancelled by those of $L(s, \chi_0)$ on $s = it (t \in \mathbf{R})$, and the only pole of $L(1 - s, \chi_0)$ occurs at $s = 0$, which is a pole of $\Gamma(s)$. This case, while particular, is rather typical of the general one.

From Theorem 11 immediately follows:

Corollary 11.1.   *For $\chi \neq \chi_0$, the functions $L(s, \chi)$ are entire functions.*

PROOF. If $\sigma > 0$, and in particular for $\sigma \geq 1/2$, it follows from Theorem 9 that $L(s, \chi)$ is holomorphic. For $\sigma \leq 1/2$, let $s = \sigma + it$ in (3). Then $\mathrm{Re}(1 - s) \leq 1/2$, and (3) defines $L(s', \bar{\chi})$ $(s' = 1 - s)$ in the half-plane $\sigma \leq 1/2$. All factors on the right of (3) are holomorphic in $\sigma \geq 1/2$, so that $L(s', \bar{\chi})$ is holomorphic in $\sigma' = 1 - \sigma \leq 1/2$. Here $\bar{\chi}$ may be taken as any nonprincipal character, and this finishes the proof of the corollary.

4 PROOF OF THEOREM 1

Before going into details, let us first sketch briefly the idea of the proof of Theorem 1.

For $\sigma > 1$, it follows from $L(s, \chi) = \prod_p (1 - \chi(p)p^{-s})^{-1}$ that $\log L(s, \chi) = \sum_p \sum_{m=1}^{\infty} \chi(p^m)/mp^{ms}$ or, by the absolute convergence of the double series, that $\log L(s, \chi) = \sum_p \chi(p)/p^s + \sum_{m=2}^{\infty} \sum_p \chi(p^m)/mp^{ms}$. Here the second term will be shown to be bounded as $s \to 1$. It will then follow, if for $(r, k) = 1$ we multiply by $\bar{\chi}(r)$ and sum over all characters, that

$$\sum_\chi \bar{\chi}(r)\log L(s, \chi) = \sum_\chi \bar{\chi}(r) \sum_p \chi(p)p^{-s} + f(s, k) = \sum_p p^{-s} \sum_\chi \bar{\chi}(r)\chi(p)$$

$$+ f(s, k) = \phi(k) \sum_{p \equiv r(\mathrm{mod}\ k)} p^{-s} + f(s, k);$$

here use has been made of Theorem 5 and of the absolute convergence of $\sum_p \chi(p)p^{-s}$ for $\sigma > 1$. The function $f(s, k)$ is bounded, and in fact $|f(s, k)| \leq \phi(k)$. On the other hand, $\sum_\chi \bar{\chi}(r)\log L(s, \chi) = \log L(s, \chi_0) + \sum_{\chi \neq \chi_0} \bar{\chi}(r)\log L(s, \chi)$. We shall show that for $s \to 1^+$ the second term stays bounded, and it will then follow from (1) that the first term behaves essentially like $\log 1/(s - 1)$. This shows that for $s \to 1^+, \sum_{p \equiv r\ (\mathrm{mod}\ k)} p^{-s}$ differs from $\log(s - 1)^{-1}$ by a bounded term. If the number of primes $p \equiv r\ (\mathrm{mod}\ k)$ were finite, then for $s \to 1^+, \sum_{p \equiv r\ (\mathrm{mod}\ k)} p^{-s} \to \sum_{p \equiv r\ (\mathrm{mod}\ k)} p^{-1}$, a finite (even rational) number, while $\log(s - 1)^{-1} \to \infty$, an obvious contradiction.

5 SOME AUXILIARY RESULTS

We now take up the proof in detail. We recall (see Theorems 8 and 9) that all $\phi(k)$ functions $L(s, \chi)$ are holomorphic in $\sigma > 0$, except for $\chi = \chi_0$ at $s = 1$; indeed (1) and Corollary 8.6.2 show that

$$L(s, \chi_0) = \frac{\phi(k)/k}{s - 1} + c_0 + c_1(s - 1) + c_2(s - 1)^2 + \cdots \qquad (4)$$

with constants $c_0, c_1, \ldots$ and $\sum_{\nu=0}^{\infty} c_\nu(s - 1)^\nu$ convergent for all $s \in \mathbf{C}$. Next, we know that $L(s, \chi_0)$ does not vanish except at the zeros of $\zeta(s)$ (discussed in Chapter 8), or on the imaginary axis. In fact, the following theorem, entirely analogous to Theorem 8.9, holds for all characters.

Theorem 12.   *For all natural integers k, all characters χ (mod k), and all real t, $L(1 + it, \chi) \neq 0$.*

REMARK 5.   For $\chi = \chi_0$, $L(s, \chi_0)$ has a pole of first order at $s = 1$, so that one cannot speak properly of the value of $L(s, \chi_0)$ at $s = 1$. It is, however, quite clear that, $s = 1$ being a pole of the function, $L(1 + it, \chi_0)$ does not have a zero at $t = 0$, and that is the meaning of Theorem 12 for $\chi = \chi_0$.

In what follows, we shall have to use the following two statements, which we formulate as Lemmas.

Lemma 2.   *For all characters modulo k, $\chi(-1) = \pm 1$.*

PROOF. $\chi(-1)^2 = \chi((-1)^2) = \chi(1) = 1$, and the Lemma follows.

Lemma 3.   *If $\chi(n)$ is real for all n, then $\chi(n) = \pm 1$ for $(n, k) = 1$ and $\chi^2 = \chi_0$.*

PROOF.   The only real roots of unity are ± 1, and that proves the first statement. For $(n, k) > 1$, $\chi^2(n) = \chi_0(n) = 0$; for $(n, k) = 1$, $\chi^2(n) = (\pm 1)^2 = 1 = \chi_0(n)$, and the Lemma is proved.

For $\sigma > 1$, $L(s \cdot \chi) \neq 0$; hence we can define a single-valued branch of $\log L(s, \chi)$. By Theorem 7, the branch that vanishes for $s \to +\infty$ is represented by the series

$$\log L(s, \chi) = -\sum_p \log\left(1 - \frac{\chi(p)}{p^s}\right) = \sum_p \sum_{m=1}^{\infty} \frac{\chi(p^m)}{mp^{ms}}.$$

For $\sigma \geq 1 + \varepsilon$ the double series is absolutely and uniformly convergent; therefore, we may differentiate termwise and obtain

$$\frac{L'}{L}(s) = \frac{L'}{L}(\sigma + it) = -\sum_p \sum_{m=1}^{\infty} \frac{\chi(p^m)\log p}{p^{ms}}$$

$$= -\sum_{n=1}^{\infty} \frac{\Lambda(n)\chi(n)}{n^s} = -\sum_{n=1}^{\infty} \frac{\Lambda(n)\chi(n)}{n^{\sigma}} n^{-it}, \qquad (5)$$

again uniformly and absolutely convergent for $\sigma \geq 1 + \varepsilon$. In analogy with the proof of Theorem 8.9, we now consider for $\sigma > 1$ the function $S(s) = |L^3(\sigma, \chi_0)L^4(\sigma + it, \chi)L(\sigma + 2it, \chi^2)|$. Then

$$\log S(s) = \operatorname{Re}\{3\log L(\sigma, \chi_0) + 4\log L(\sigma + it, \chi) + \log L(\sigma + 2it, \chi^2)\}$$

or, differentiating with respect to σ and using (5),

$$\frac{\zeta'(s)}{\zeta(s)} = -\sum_{n=1}^{\infty} \frac{\Lambda(n)}{n^{\sigma}} \operatorname{Re}\{3\chi_0(n) + 4\chi(n)n^{-it} + \chi^2(n)n^{-2it}\} = G(\sigma),$$

say. Let $\phi = \arg \chi(n)n^{-it}$; then for $(n, k) = 1$, the factor of $\Lambda(n)n^{-\sigma}$ reduces to $3 + 4\cos\phi + \cos 2\phi = 2(1 + \cos\phi)^2 \geq 0$, so that $G(\sigma) \leq 0$ for $\sigma > 1$ and also for $\sigma \to 1^+$. On the other hand, by (1) and Corollary 8.6.1, $(L'/L)(\sigma, \chi_0) = -(\sigma - 1)^{-1} + 0(1)$. If we now assume that $L(s, \chi)$ has an a-fold zero at $s = 1 + it(0 < a \in \mathbf{Z})$, then $(L'/L)(\sigma + it, \chi) = a/(\sigma - 1) + 0(1)$. Finally, $L(\sigma + 2it, \chi^2)$ is holomorphic at $\sigma = 1$, except if $t = 0$ and $\chi^2 = \chi_0$. We postpone for a moment the consideration of this exceptional case. Hence, for $\sigma \to 1^+$, $L(\sigma + 2it, \chi^2)$ approaches a finite constant, perhaps zero, so that $(L'/L)(\sigma + 2it, \chi^2) = b/(\sigma - 1) + 0(1)$, where the integer b is the order of the zero (if any) of $L(\sigma + 2it, \chi^2)$ at $\sigma = 1$; if $L(1 + 2it, \chi^2) \neq 0$, then $b = 0$. It follows that for $\sigma \to 1^+$,

$$G(\sigma) = -\frac{3}{\sigma - 1} + \frac{4a}{\sigma - 1} + \frac{b}{\sigma - 1} + 0(1) = \frac{1}{\sigma - 1}(4a + b - 3) + 0(1).$$

Here, by assumption, $a \geq 1, b \geq 0$ are integers. Consequently, for $\sigma \to 1^+, G(\sigma) \geq 1/(\sigma - 1) + 0(1)$ is positive (in fact, $G(\sigma) \to +\infty$); this contradicts the previous result that $G(\sigma) \leq 0$, and shows that $L(s, \chi)$ cannot have a zero $s = 1 + it$ unless, perhaps, we are in the omitted exceptional case $t = 0$, and χ is real.

In order to handle this most difficult case, we still need two more results.

Lemma 4. *If χ is real, then $\zeta(s) \cdot L(s, \chi) = \sum_{n=1}^{\infty} a_n n^{-s}$ has only non-negative coefficients a_n and for all n, $a_{n^2} \geq 1$.*

Lemma 5 (Landau; see [8]). *If the series $\sum_{n=1}^{\infty} a_n n^{-s}$ has non-negative coefficients and its abscissa of convergence σ_c is finite, then the function represented for $\sigma > \sigma_c$ by the series has a singularity at $s = \sigma_c$.*

PROOF OF LEMMA 4. By Theorems 7 and 8.2, $\zeta(s)L(s, \chi) = \prod_p(1 - p^{-s})^{-1}\prod_p(1 - \chi(p)p^{-s})^{-1}$. As χ is real, $\chi(p) = \pm 1$ by Lemma 3. If $\chi(p) = +1$, then $(1 - p^{-s})^{-1}(1 - \chi(p)p^{-s})^{-1} = (1 - p^{-s})^{-2} = \sum_{m=0}^{\infty}(m + 1)p^{-ms}$; if $\chi(p) = -1$, then $(1 - p^{-s})^{-1}(1 - \chi(p)p^{-s})^{-1} = (1 - p^{-2s})^{-1} = \sum_{m=0}^{\infty}p^{-2ms}$. In both cases the coefficients are non-negative and a product of series with non-negative coefficients has non-negative coefficients. Furthermore, the coefficient of p^{-2ms} equals $2m + 1$ in the first case and 1 in the second case; hence the coefficient of n^{-2s} is a positive integer as product of factors of the form either $2m + 1$ or 1.

Lemma 5 is a theorem due to Landau and complete proofs may be found in [8] and [9]. Here is a sketch of a proof.

Let $g(s) = \sum_{n=1}^{\infty} a_n n^{-s}$ converge for $\sigma > \sigma_c$ and diverge for $\sigma < \sigma_c$; then $g(s)$ is holomorphic in the half-plane $\sigma > \sigma_c$, and its derivatives are $g^{(m)}(s) = \sum_{n=1}^{\infty}(-1)^m a_n n^{-s} \log^m n$. For any $\sigma_0 > \sigma_c$, one has the Taylor expansion $g(s) = \sum_{m=0}^{\infty}((-1)^m/m!)(s-\sigma_0)^m \sum_{n=1}^{\infty} a_n n^{-\sigma_0}\log^m n$. If $g(s)$ is also holomorphic at σ_c, then the circle of convergence with center σ_c has positive radius, so that the last series also converges for some $s = \sigma < \sigma_c$, i.e.,

$$g(\sigma) = \sum_{m=0}^{\infty} \frac{(-1)^m}{m!}(\sigma - \sigma_0)^m \sum_{n=1}^{\infty} a_n n^{-\sigma_0}\log^m n$$

$$= \sum_{m=0}^{\infty} \frac{(\sigma_0 - \sigma)^m}{m!} \sum_{n=1}^{\infty} a_n n^{-\sigma_0}\log^m n.$$

Now, however, all terms of the double series are positive; hence the series is actually absolutely convergent, and by Weierstrass' theorem, we may change the order of summation without affecting either the convergence of the series or the value of its sum. It follows that $g(s) = \sum_{n=1}^{\infty} a_n n^{-\sigma_0}\sum_{m=0}^{\infty}((\sigma_0 - \sigma)^m/m!)\log^m n$. In the inner sum we recognize now the series expansion of $e^{(\sigma_0 - \sigma)\log n} = n^{\sigma_0 - \sigma}$, so that $g(s) = \sum_{n=1}^{\infty} a_n n^{-\sigma_0}n^{\sigma_0 - \sigma}\sum_{n=1}^{\infty} a_n n^{-\sigma}$, with the series converging at σ. However, $\sigma < \sigma_c$, so that we have obtained a contradiction to the assumption that the series diverges for $\sigma < \sigma_c$; Lemma 5 is proved.

We now return to the proof of Theorem 12 in the previously excluded case of a real character $\chi \neq \chi_0$ and at $s = 1$.

Let $f(s) = \zeta(s)L(s,\chi) = \sum_{n=1}^{\infty} a_n n^{-s}$, say, with χ real but $\chi \neq \chi_0$. For real $s > 1$, it follows from $a_n \geq 0$ and $a_m \geq 1$ for $m = n^2$ that $f(s) \geq \sum_{n=1}^{\infty} a_{n^2} n^{-2s} \geq \sum_{n=1}^{\infty} n^{-2s}$ and diverges for $\sigma < 1/2$. The series for $\zeta(s)$ and $L(s,\chi)$ converge absolutely for $\sigma > 1$, so that $\sum_{n=1}^{\infty} a_n n^{-s}$ has an abscissa of convergence σ_c, with $\frac{1}{2} \leq \sigma_c \leq 1$. By Lemma 5, $s = \sigma_c$ is a singularity of $f(s)$. However, $f(s)$ can have only one singularity in $\sigma > 0$, namely the pole of $\zeta(s)$ of first order at $s = 1$, because $L(s,\chi)$ is holomorphic for $\sigma > 0$. Consequently, $s = 1$ is a singular point of $f(s) = \zeta(s)L(s,\chi)$. If now $L(1,\chi) = 0$, then this zero of $L(s,\chi)$ at $s = 1$ would cancel the pole of first order of $\zeta(s)$ and $f(s)$ would have no singularity, at least for $\sigma > 0$. Then, however, the Dirichlet series $\sum_{n=1}^{\infty} a_n n^{-s}$ of $f(s)$ with non-negative coefficients a_n could not have an abscissa of convergence σ_c in the interval $\frac{1}{2} \leq \sigma_c \leq 1$. This contradiction shows that $L(1,\chi) \neq 0$ for real χ, $\chi \neq \chi_0$, and this finishes the proof of Theorem 12.

Let us remark also that, for real $s > 1$, $\zeta(s)L(s,\chi)$ and $\zeta(s)$ are both positive; hence so is $L(s,\chi)$, and also $\lim_{s \to 1^+} L(s,\chi) = L(1,\chi)$.

6 END OF THE PROOF OF THEOREM 1

The proof of Theorem 1 is now immediate. Indeed,

$$\log L(s, \chi) = -\sum_p \log(1 - \chi(p)p^{-s}) = \sum_p \sum_{m=1}^{\infty} \frac{\chi(p^m)}{mp^{ms}}.$$

For $\sigma > 1$ the double series is absolutely convergent; therefore we may invert the order of summation and obtain

$$\log L(s, \chi) = \sum_{m=1}^{\infty} \sum_p \frac{\chi(p^m)}{mp^{ms}} = \sum_p \frac{\chi(p)}{p^s} + R(s, \chi).$$

Here

$$|R(s, \chi)| = \left| \sum_{m=2}^{\infty} \sum_p \frac{\chi(p^m)}{mp^{ms}} \right| \leq \sum_p \sum_{m=2}^{\infty} \frac{1}{mp^{m\sigma}} \leq \frac{1}{2} \sum_p \sum_{m=2}^{\infty} \frac{1}{p^{m\sigma}}$$

$$= \frac{1}{2} \sum_p \frac{1}{p^{2\sigma}(1 - p^{-\sigma})} \leq \sum_p \frac{1}{p^{2\sigma}}.$$

But for $\sigma > 1$,

$$\sum_p \frac{1}{p^{2\sigma}} < \sum_p \frac{1}{p^2} < \sum_{n \geq 2} \frac{1}{n^2} = \frac{\pi^2}{6} - 1 < 1$$

by (10) of Chapter 8. It follows that, on the one hand,

$$\sum_{\chi \bmod k} \bar\chi(r)\log L(s, \chi) = \log L(s, \chi_0) + \sum_{\chi \neq \chi_0} \bar\chi(r)\log L(s, \chi);$$

and on the other hand,

$$\sum_{\chi \bmod k} \bar\chi(r)\log L(s, \chi) = \sum_{\chi \bmod k} \bar\chi(r)\left\{ \sum_p \frac{\chi(p)}{p^s} + R(s, \chi) \right\}$$

$$= \sum_p \frac{1}{p^s} \sum_{\chi \bmod k} \bar\chi(r)\chi(p) + f(s, k),$$

where $|f(s, k)| = |\sum_{\chi \bmod k} \bar\chi(r)R(s, \chi)| \leq \phi(k)$. Also, by Theorem 5,

$$\sum_{\chi \bmod k} \bar\chi(r)\chi(p) = \begin{cases} \phi(k) & \text{if } p \equiv r \pmod k \\ 0 & \text{if } p \not\equiv r \pmod k, \end{cases}$$

and equating the two expressions we obtain

$$\phi(k) \sum_{p \equiv r \pmod k} \frac{1}{p^s} + f(s, k) = \log L(s, \chi_0) + \sum_{\chi \neq \chi_0} \bar\chi(r)\log L(s, \chi). \quad (6)$$

We now let $s \to 1^+$; then all $L(s, \chi)(\chi \neq \chi_0)$ approach finite nonzero (in fact, positive) values $L(1, \chi)$. Consequently, $\sum_{\chi \bmod k} \bar{\chi}(r) \log L(s, \chi)$ approaches a finite value as $s \to 1^+$. Finally, by Corollary 8.1, $\log L(s, \chi_0) = \log\{(\phi(k)/k)/(s - 1) + c_0 + c_1(s - 1) + \cdots\} = \log(s - 1)^{-1} + g(s)$, with $g(s)$ bounded as $s \to 1^+$ (in fact, $g(1) = \log(\phi(k)/k)$). By also using $|f(s, k)| \leq \phi(k)$, it follows from (6) that for $s \to 1^+$, $\sum_{p \equiv r(\bmod k)} p^{-s} = \log(s - 1)^{-1} + F(s)$, with $|F(s)|$ bounded. If the number of primes $p \equiv r \pmod{k}$ were finite, then $\lim_{s \to 1^+} \sum_{p \equiv r \pmod{k}} p^{-s} = \sum_{p \equiv r \pmod{k}} p^{-1}$ would be finite (in fact, even rational), while on the right, $\log(s - 1)^{-1} + F(s)$ would increase indefinitely as $s \to 1^+$. This is of course not possible because the two sides are equal. This shows that the number of primes $p \equiv r \pmod{k}$ cannot be finite, and finishes the proof of Theorem 1.

It may be worthwhile to remark at this point that for the divisors of 24 and only for them, all characters are real. That explains why it is possible to give elementary proofs of Theorem 1 for them and for no other moduli. By this we mean proofs of the type seen for $k = 4$, e.g.; other types of technically elementary proofs are known for all moduli.

7   DIRICHLET'S APPROACH

The most difficult and complicated part of the proof of Theorem 1 was the proof of Theorem 12 in the special case of real $\chi \neq \chi_0$ and $s = 1$, i.e., the proof that $\lim_{s \to 1^+} L(s, \chi) = L(1, \chi) \neq 0$. Dirichlet, in his original proof of this theorem, proceeded differently. First let us observe that if $\chi(n)$ is a real character modulo k, there are only two possibilities. The simplest is $\chi(n) = \chi_0(1) = 1$ for $(n, k) = 1$, $= 0$ otherwise. As χ has to be real, any other possible choice for χ can assign to $\chi(n)$ only one of the values ± 1 if $(n, k) = 1$, $\chi(n) = 0$ if $(n, k) > 1$; also, $\chi(n)$ has to have period k, i.e., $\chi(n + k) = \chi(n)$. This strongly suggests that perhaps this choice coincides with (k/n). This first guess turns out to be almost—but not entirely—correct. In the first place, n takes on all integer values, including even ones. Hence the symbol has to be interpreted as a Kronecker symbol. Once this is established, if k is odd the symbol is defined only if $k \equiv 1 \pmod 4$. This forces us to consider for $k \equiv 3 \pmod 4$ $(-k/n)$ rather than (k/n). Finally if, e.g., $k = \pm 4d$, with $d \equiv 1$ or $d \equiv 0 \pmod 4$, then $(\pm k/n) = (d/n)$. If the reader refers back to Problems 17 and 18 of Chapter 5, he will convince himself that as a function of n the Kronecker symbol (d/n) has the requisite period k, and indeed a nonprincipal real character modulo k can be defined by $\chi(n) = (d/n)$ with $|d| \| k$, $d > 0$ or $d < 0$.

Let us assume first that $d > 0$ and consider the real quadratic field $\mathbf{K} = \mathbf{Q}(\sqrt{d})$. It has a fundamental unit $\varepsilon > 1$, and the discriminant d (recall: $d \equiv 0$ or $\equiv 1$ modulo 4). If $d < 0$, the imaginary quadratic field $\mathbf{K} = \mathbf{Q}(\sqrt{d}) = \mathbf{Q}(\sqrt{-|d|})$ of discriminant d has, in general, only the $w = 2$ roots of unity ± 1, but for $d = -4$ it has $w = 4$, and for $d = -3$ it has $w = 6$ roots of unity. Let us define a new parameter κ, in the case $d > 0$ by $\kappa = (2 \log \varepsilon)/d$ and in the case $d < 0$ by $\kappa = 2\pi/(w\sqrt{|d|})$. With these notations, Dirichlet proved the following amazing fact:

If the quadratic field $\mathbf{K}$ has class number h (i.e., if the number of classes of ideals as defined in Chapter 11 is equal to h), then

$$L(1, \chi) = h \cdot \kappa. \tag{7}$$

As, quite obviously, $h > 0$ and $\kappa > 0$, the desired result $L(1, \chi) > 0$ immediately follows, and this takes care of the most difficult part of the proof of Theorem 1.

The complete proof of this result of Dirichlet would lead us too far, but the interested reader will find a lucid proof in, e.g., [5] and a (very incomplete) sketch of it in the next section.

8 CONCLUDING REMARKS

There are many other topics that now come up naturally. It is tantalizing that for many of them we already possess most of the tools needed to treat them. Unfortunately, in general, small additional ingredients (different for each item) are needed in order to give satisfactory proofs. Often, each new lemma needed would require rather long excursions, far afield. The results, however, are important as well as interesting, and (as is so often the fortunate case in number theory) can be easily understood without even the knowledge (let alone the proof) of those lemmas. For that reason, we list here some of those questions, results, etc. without proofs.

Let $\mathfrak{A}$ be a class of ideals in the ring $\mathbf{I}$ of integers of the algebraic number field $\mathbf{K}$ of degree n over the rational field $\mathbf{Q}$, where $n = r_1 + 2r_2$, with the usual notations. Then we may ask for the number $Z(\mathfrak{A}, t)$ of ideals $\mathfrak{m}$ in $\mathfrak{A}$, with norm $N\mathfrak{m} \leq t$. It appears reasonable to assume that $Z(\mathfrak{A}, t)$ will increase with t and will depend (in addition to its dependence on the parameters that characterize $\mathbf{K}$) also on $\mathfrak{A}$. We find that $Z(\mathfrak{A}, t)$ is indeed roughly proportional to t, but also the perhaps rather surprising result that, for large t at least, $Z(\mathfrak{A}, t)/t$ is essentially the same for all classes of ideals. If we set $\lim_{t \to \infty}(Z(\mathfrak{A}, t))/t = \kappa$, then $\kappa = (2^{r_1 + r_2}\pi^{r_2}R)/w\sqrt{|d|}$ depends only on the field $\mathbf{K}$. This leads naturally to the perhaps even more natural question of the size of $Z(t)$, the total number of ideals in $\mathbf{I}$ of norm not in excess of t. By a previous result, the answer to this question is

$$\lim_{t \to \infty} Z(t)/t = \kappa h, \tag{8}$$

where h is the number of classes of ideals in **K**, as defined in Definition 11.13 and which we know to be finite by Theorem 11.30.

We also know (Theorem 11.28) that the number of ideals of **I** of norm m is finite, say $F(m)$. Hence $Z(t) = \sum_{m=1}^{t} F(m)$. The reader has already had the opportunity to observe (see, e.g., Chapter 7 on Partitions) that when we want to study a sequence of constants, say a_n, it is often convenient to form either a Taylor series $\sum_{n=0}^{\infty} a_n x^n$ or a Dirichlet series $\sum_{n=1}^{\infty} a_n n^{-s}$, and to study the properties of the function represented by the series. In the present case, let us consider the Dirichlet series $\sum_{m=1}^{\infty} F(m) m^{-s}$. By the definition of $F(m)$, the series may also be written as $\sum_{\mathfrak{a} \subset \mathbf{K}} 1/N\mathfrak{a}^s$. This way of writing immediately suggests a similarity with Riemann's zeta function $\zeta(s) = \sum_{n=1}^{\infty} 1/n^s$, for $\sigma > 1$, to which it reduces if $\mathbf{K} = \mathbf{Q}$. In fact, like the series for $\zeta(s)$, the new series also converges for $\sigma > 1$ and defines there a function of the complex variable s, denoted by $\zeta_{\mathbf{K}}(s)$, the *Dedekind zeta function* of the algebraic number field **K**,

$$\zeta_{\mathbf{K}}(s) = \sum_{\mathfrak{a} \subset \mathbf{K}} \frac{1}{N\mathfrak{a}^s}.$$

This function can be continued analytically as a meromorphic function in the whole complex plane, holomorphic everywhere except for $s = 1$, where it has a pole of first order. And (miraculous coincidence!) the residue of $\zeta_{\mathbf{K}}(s)$ at $s = 1$ turns out to be precisely the product $\kappa \cdot h$ that occurs in (8). It follows that $\lim_{s \to 1^+} (s - 1)\zeta_{\mathbf{K}}(s) = \kappa h$. This suggests a method for the computation of the class number h of **K**.

Exactly as in the case of Riemann's function $\zeta(s)$, it is shown that for $\sigma > 1$,

$$\zeta_{\mathbf{K}}(s) = \prod_{\mathfrak{p} \subset \mathbf{K}} \left(1 - \frac{1}{N\mathfrak{p}^s}\right)^{-1}, \tag{9}$$

where $\mathfrak{p}$ ranges over all prime ideals of **K**.

We now recall that in Section 5 we met the function $f(s) = \zeta(s)L(s, \chi)$, for real nonprincipal χ (i.e., for $\chi(n) = (d/n)$). From $L(s, \chi) = \prod_p (1 - \chi(p)p^{-s})^{-1}$ it follows that

$$f(s) = \prod_p (1 - p^{-s})^{-1}(1 - \chi(p)p^{-s})^{-1} = \prod_{\chi(p) = +1} (1 - p^{-s})^{-2}$$

$$\times \prod_{\chi(p) = -1} (1 - p^{-2s})^{-1} \times \prod_{\chi(p) = 0} (1 - p^{-s})^{-1}.$$

In the quadratic field $\mathbf{K} = \mathbf{Q}(\sqrt{d})(d \lessgtr 0)$ of discriminant d, if p is the norm of an ideal, say $N\mathfrak{p} = p$, then $\mathfrak{p}$ is a prime ideal. Indeed if $N\mathfrak{a} = N(\mathfrak{b} \cdot \mathfrak{c}) = N\mathfrak{b} \cdot N\mathfrak{c} = p$, then one of the factors of p equals one, so that one of the factors of the ideal $\mathfrak{a}$ is the unit ideal and $\mathfrak{a}$ is a prime ideal. By Theorem 11.27, $\mathfrak{p}|(p)$, so that if we write (slightly abusively, but simpler) p for both the integer and the prime ideal (p), then $p = \mathfrak{p}\mathfrak{p}'$. Indeed it is a fact (which we did not prove) that in a quadratic field a principal ideal generated by a prime p can split into the product of at most two ideals. Clearly, $Np = p^2 = N\mathfrak{p} \cdot N\mathfrak{p}'$, so that as $N\mathfrak{p} \neq 1 \neq N\mathfrak{p}'$, it follows that $N\mathfrak{p} = N\mathfrak{p}' = p$. It may also be shown that when this happens, $\chi(p) = (d/p) = 1$. On the other hand, the converse does not hold, and the norm of a prime (either number or ideal) need not be a rational prime. Indeed if $\chi(p) = (d/p) = -1$, the ideal (p) is itself a prime ideal in $\mathbf{K}$, so that $Np = p^2$. Finally, in the case of the finitely many prime divisors of d, $\chi(p) = (d/p) = 0$. In this case one may show that $\mathfrak{p} = \mathfrak{p}'$, so that $p = \varepsilon\mathfrak{p}^2 = N\mathfrak{p}$ holds[†] for a single prime ideal $\mathfrak{p}$. We now see that the three factors of $f(s)$ can be written as follows:

(i) $\displaystyle\prod_{\chi(p) = +1} (1 - p^{-s})^{-2} = \prod_{\substack{\mathfrak{p}|p \\ \chi(p)=1}} (1 - (N\mathfrak{p})^{-s})^{-1}$, because $\mathfrak{p}$ and $\mathfrak{p}'$ each
furnish one factor $(1 - N\mathfrak{p}^{-s})^{-1}$.

(ii) $\displaystyle\prod_{\chi(p) = -1} (1 - p^{-2s})^{-1} = \prod_{\substack{\mathfrak{p}=p \\ \chi(p)=-1}} (1 - (N\mathfrak{p})^{-s})^{-1}$.

(iii) $\displaystyle\prod_{\substack{\chi(p)=0 \\ (p|d)}} (1 - p^{-s})^{-1} = \prod_{\mathfrak{p}|d} (1 - (N\mathfrak{p})^{-s})^{-1}$.

If we put this together, we obtain $f(s) = \prod_{\mathfrak{p}} (1 - (N\mathfrak{p})^{-s})^{-1}$, and on account of (9),

$$\zeta_{\mathbf{K}}(s) = f(s) = \zeta(s)L(s, \chi). \qquad (10)$$

If we now multiply (10) by $s - 1$ and let $s \to 1^+$, by taking into account Corollary 8.6.1, we obtain

$$\lim_{s \to 1^+} (s - 1)\zeta_{\mathbf{K}}(s) = L(1, \chi) = \sum_{n=1}^{\infty} \frac{\chi(n)}{n}.$$

We recall that $\chi(n)$ stands here for (d/n), with d the positive or negative discriminant of the quadratic field $\mathbf{K}$. On the other hand, the first member stands for the residue of $\zeta_K(s)$ at the pole $s = 1$ of first order, and as seen, equals κh. As κh is clearly different from zero, this justifies (modulo much "handwaving") Dirichlet's result (proved here by a very different method) used in Section 6, that for real nonprincipal character χ, $L(1, \chi) \neq 0$.

[†] Here ε is a unit. Care is needed when equalities of ideals are written as equalities among integers. Consider the case $\mathbf{K} = \mathbf{Q}(i)$, $d = -4$, $\mathfrak{p}|d \Rightarrow p = 2$, $\mathfrak{p} = (1 + i) \Rightarrow \mathfrak{p}|d$, but $(1 + i)^2 = 2i \neq 2$.

In order to obtain this result in the form quoted in Section 6, we consider separately the cases $d > 0$ and $d < 0$. If $\mathbf{K}$ is a real quadratic field, then $r_1 = 2, r_2 = 0, r = r_1 + r_2 - 1 = 1$, so that there exists exactly one fundamental unit, say ε, and the regulator R reduces to $\log \varepsilon$. Also $w = 2$, as there are in $\mathbf{K}$ only the two real units ± 1. Consequently, $\kappa = (2^2 \pi^0 \log \varepsilon)/2\sqrt{d} = (2 \log \varepsilon)/\sqrt{d}$. If $d < 0$, on the other hand, $r_1 = 0$ and $r_2 = 1$, so that $r = r_1 + r_2 - 1 = 0$, and there is no fundamental unit, i.e., $R = 1$. Also in general $w = 2$, but $w = 4$ if $d = -4$, and $w = 6$ if $d = -3$. Consequently $\kappa = (2^1 \pi^1 \cdot 1)/w\sqrt{|d|} = 2\pi/w\sqrt{|d|}$, and we have recovered the results of Dirichlet quoted in Section 7.

The present sketch is of course very incomplete and has many gaps. The most puzzling for a reader may well be the apparent miraculous coincidence between the identical values of the apparently unrelated limits

$$\lim_{t \to \infty} \frac{1}{t} \sum_{m=1}^{t} F(m) \quad \text{and} \quad \lim_{s \to 1^+} (s - 1)\zeta_{\mathbf{K}}(s).$$

In fact, by a general theorem of Perron, the sum $\sum_{m=1}^{t} F(m)$ of the coefficients of the Dirichlet series of $\zeta_{\mathbf{K}}(s)$ is given by a certain line integral over $\zeta_{\mathbf{K}}(s)$, and this clarifies the mystery completely (but see also [5], where this result is proved by a very different method). While the respective proofs would require more space than can be spared here, the careful reader may have already discovered the analogy with the pair of equations (1), (5) of Chapter 9 and with Problems 2 and 3 of Chapter 8. One observes, indeed, that $\sum_{m=1}^{t} F(m)$, just like $M(t) = \sum_{m=1}^{t}\mu(m)$ and $\psi(t) = \sum_{m=1}^{t}\Lambda(m)$, is the sum of the first t coefficients of the Dirichlet series that represents the function in s, which occurs in the respective equalities.

To add to the miracles, let us mention that the series $L(1, \chi) = \sum_{n=1}^{\infty}\chi(n)/n$ can be summed in finite terms. If that is done and use is made of (7) and the indicated values of κ, we obtain the following formulae for the class number of quadratic fields:

$$h = -\frac{1}{|d|} \sum_{n=1}^{|d|-1} n\left(\frac{d}{n}\right) \quad \text{for} \quad d < -4;$$

$$h = \frac{1}{2 \log \varepsilon}\log \prod_{(n, d)=1} \left\{\sin\frac{\pi n}{d}\right\}^{-\chi(n)} \quad \text{for } d > 0.$$

Before we finish this chapter, one more remark may be in order. For that, we return to (10), which may also be written as

$$\frac{\zeta_{\mathbf{K}}(s)}{\zeta(s)} = L(s, \chi).$$

We recall that $L(s, \chi)$ is an entire function, hence holomorphic for all $s \in \mathbf{C}$. This means in particular that every zero of $\zeta(s)$ is cancelled by a zero of at least the same multiplicity of $\zeta_{\mathbf{K}}(s)$ at the same spot. Something similar seems to be true in a much more general framework. If we start from an arbitrary algebraic number field $\mathbf{L}$ and extend it to a larger algebraic field $\mathbf{K}, \mathbf{K} \supset \mathbf{L}$, is it still true that $\zeta_{\mathbf{K}}(s)/\zeta_{\mathbf{L}}(s)$ is an entire function? The affirmative answer constitutes the so-called Artin Conjecture. This has been proved in many cases (e.g., if the group of automorphisms of $\mathbf{K}$ that leave the elements of $\mathbf{L}$ invariant is an abelian group), but the general problem is still open.

PROBLEMS

1. Show that if $k|24$, all characters modulo k are real. (Hint: you may want to start by showing that for the divisors k of 24, if $(n, k) = 1$, then $n^2 \equiv 1 \pmod{k}$.)

2. Prove that there are infinitely many primes $p \equiv 5 \pmod 6$.

3*. Prove that there are infinitely many primes $p \equiv 1 \pmod 6$.

4*. Show that if $k \nmid 24$, then there exists a character with complex values.

5. Prove Dirichlet's Theorem by elementary methods for as many moduli k ($k|24$) as you can. If the going gets rough, you may want to look into the Dissertation (Copenhagen, 1937) of A. S. Bang (fluency in Danish is recommended) and the paper by P. T. Bateman and M. E. Low, Prime numbers in arithmetic progression with difference 24, *Amer. Mathem. Monthly* 72 (1965) 139–143.

6. Prove that if $k \geq 3$, $(n, k) = 1$ and $n \not\equiv 1 \pmod k$, then there exists a character $\chi \pmod k$, such that $\chi(n) \neq 1$.

7. In Lemma 1, we proved and used the fact that for χ a character modulo k, $\chi \neq \chi_0$, we have $|\sum_{n=a+1}^{b}\chi(n)| \leq \phi(k)$. Prove that, in fact, even $|\sum_{n=a+1}^{b}\chi(n)| \leq \phi(k)/2$ holds.

8. Determine all characters modulo 5 and verify that with any $\chi(n)$, $\bar{\chi}(n)$ also occurs as a character.

9. Write down the L-series corresponding to the character $\chi(n)$ modulo 5 for which $\chi(2) = i$. Use this series to find the approximate value of the sum of the conditionally convergent series $L(1, \chi) = \sum_{n=1}^{\infty}\chi(n)/n$.

10. Complete the details in the proof of (2).

11. Consider the nonprincipal character modulo 4 and write down the corresponding L-series. Find its sum in finite terms. (Hint: Consider the function $f(x) = x - x^3/3 + x^5/5 - \cdots$ and observe that $f'(x) = 1/(1 + x^2)$.)

12. Verify Lemma 1 on the characters modulo 5 of Problem 9 and on the character modulo 4 of Problem 11.

13. Verify Lemma 3 on the characters modulo 4.

14. Find the Dedekind zeta function for the *Gaussian field* $\mathbf{K} = \mathbf{Q}(\sqrt{-1})$ $(= \mathbf{Q}\sqrt{-4}$ with $d = -4$, because $-1 \equiv 3 \pmod 4$) and verify that it satisfies (10). (Hint: If $p \equiv 3 \pmod 4$, then p is a prime in $\mathbf{K}$ and $Np = p^2$. If $p \equiv 1 \pmod 4$, then $p = \mathfrak{p}\mathfrak{p}'$, $N\mathfrak{p} = N\mathfrak{p}' = p$; if $p = 2$, then $ip = \mathfrak{p}^2$, where $\mathfrak{p} = (1 + i)$, $N\mathfrak{p} = 2$.)

BIBLIOGRAPHY

1. T. Apostol, *Mathematical Analysis*. Reading, Mass.: Addison-Wesley, 1957.

2. R. Ayoub, *An Introduction to the Analytic Theory of Numbers* (Mathematical Surveys). Providence, R.I.: American Mathem. Soc., 1963.

3. L. Dirichlet, Recherches sur diverses applications de l'Analyse infinitésimale à la Théorie des Nombres, *Journal für die reine und angewandte Mathematik* (Crelle) 19 (1839) 324–369; 21 (1840) 1–12, 134–155; *Werke* (1889), vol. 1, pp. 411–496.

4. G. H. Hardy and M. Riesz, *The General Theory of Dirichlet Series* (Cambridge Tracts in Mathematics and Mathematical Physics No. 18). Cambridge: Cambridge University Press, 1915.

5. E. Hecke, *Vorlesungen über die Theorie der algebraischen Zahlen*. New York: Chelsea Publishing Company, 1948.

6. I. N. Herstein, *Topics in Algebra*, 2nd ed. Lexington, Mass.: Xerox College Publ. (Ginn & Co.), 1975.

7. E. Landau, *Handbuch der Lehre von der Verteilung der Primzahlen*, vol. 1, 2nd ed. New York: Chelsea Publ. Co., 1953.

8. E. Landau, *Mathematische Annalen* 61 (1905) 527–550.

9. K. Prachar, *Primzahlverteilung* (Die Grundlehren der Math. Wissenschaften in Einzeldarstellung, vol. 91). Berlin: Springer Verlag, 1957.

Chapter 13

Diophantine Equations

1 INTRODUCTION

The following is the introductory section of the chapter on Diophantine Equations of the first edition of the present book. It was written about 20 years ago and is reproduced here in full:

"In Chapter 4 we considered linear congruences, or equivalently, linear Diophantine equations, and found that the questions one may be interested to ask generally have simple, straightforward answers. Therefore, it may come as something of a surprise to the reader that for nonlinear Diophantine equations hardly any general results were known even 40–50 years ago. Actually, even today we are reduced in most cases to studying individual equations rather than classes of equations, and one can hardly consider the results very satisfactory.

As a matter of fact, we do not have a general method by which to decide whether a given Diophantine equation has any solutions, or how many solutions. Partial answers to some of these questions are contained in theorems first proven by mathematicians who are still very active. So for instance, there exists a set of results due to A. Thue, A. Ostrowski, E. Landau, and C. L. Siegel (see [16] pp. 52–60), of which a typical theorem reads as follows: Let $a, b, c, d, n \in \mathbf{Z}$, $ad \neq 0$, $b^2 - 4ac \neq 0$, $n \geq 3$; then $ay^2 + by + c = dx^n$ has only finitely many solutions.

The determination of the number of solutions of the equation $\sum_{j=0}^{r} a_j x_j^{n_j} = b$ over a finite field (this contains as a particular case that of a congruence of the same form mod p—why?) is the object of a beautiful paper by A. Weil (see [33]; weaker results, found earlier, are also quoted in [33]). L. J. Mordell proved many important results concerning the number of integral and rational solutions of Diophantine equations, but a famous conjecture of his (which will not even be quoted, because it involves the concept of genus of an algebraic curve, not defined in this book) has not yet been proven in its full generality. H. Hasse has shown that a certain type of Diophantine equation will have solutions, unless this is "obviously" impossible, either on account of the fact that the two sides of the equation cannot even be congruent modulo some prime, or else on account of some sign condition (e.g., a sum of squares set equal to a negative number). Many other mathematicians have contributed to the increase of our knowledge of Diophantine equations. An excellent account, at least of the work done before 1938, may be found in T. Skolem's monograph [29]. However, just to illustrate the general situation as it is today, let it be said that such simple-looking Diophantine equations as $y^2 - k = x^3$ or $2^y - k = x^2 (k \in \mathbf{Z})$ are the object of current investigations (see Ramanujan [25] p. 327 for a conjecture proven by Skolem, Chowla, and Lewis [18] pp. 241–244, and Ljunggren [19])."

Few arguments can better justify the publication of a new edition of the present book than the reading of this one page. Indeed while the statements, taken individually, may perhaps still be defensible from the perspective of the year 1960, to the contemporary (i.e., 1980s) reader the emphasis will appear wrong on about every item, and the general impression left by the reading of this old introduction is hopelessly false.

So much has been learned during these last 20 years (perhaps 25 years is more accurate; Roth's breakthrough came in 1955, but had, unfortunately, been ignored in the first edition), that the feeling of relative helplessness ("...even today we are reduced in most cases to studying individual equations rather than classes of equations...") has been replaced by one of optimism and diligent, even feverish activity.

Not everything that was achieved during these short 20 or 25 years confirmed the previously prevailing conjectures, which shows the highly nontrivial nature of the respective achievements.

It is not possible, in these few pages, even to start giving a full account of what has been done recently in the field of Diophantine equations. Also, as it turns out, these problems are connected with deep results in logic, in algebraic geometry, and in the theory of Diophantine approximations, among others. A knowledge of these fields is a prerequisite for the full understanding of these often very subtle and difficult proofs, and the reader of the present book is not expected to be an expert in those fields. Fortunately, however, as so often in number theory, the statements of the results are rather easy to understand. Sometimes a few definitions are needed, but not too many. These results are so beautiful and important that it appears worthwhile to state them here, even if the treatment they can receive will sometimes be less than satisfactory.

2   HILBERT'S TENTH PROBLEM

Perhaps the most important single discovery, and one that throws an entirely new light upon the statement "...we do not have a general method by which to decide whether a given Diophantine equation has any solutions...", found in the old introduction, was the result of work by several mathematicians and logicians.

In this case, it may be worthwhile to go back to the year 1900, when Hilbert (1862–1943), perhaps the most influential mathematician of his time, gave a speech at the International Congress in Paris, in which he outlined what seemed to him the most important and challenging mathematical problems of the coming 20th century. Most people agree that Hilbert's evaluation was surprisingly, almost prophetically, accurate.

The tenth among the problems he stated was the search for an algorithm that would permit us to decide, in a finite number of steps, whether an arbitrary Diophantine equation has solutions, and if in the affirmative, how many. In the ensuing years, mathematicians tried very hard, but always without success, to find such an algorithm. At a given moment, some mathematicians started wondering whether perhaps such an algorithm simply cannot exist. In fact, in 1961, Martin Davis, Hilary Putnam, and Julia Robinson wrote a paper [11] in which they proved, by a most ingenious combination of formal logic and number theory, and on the basis of a rather plausible (but quite technical) assumption, that no such algorithm could exist. In 1970 the then very young Russian mathematician Y. Matijasievič proved that said conjecture was in fact true [20]. A key role in his proof was played by the *Fibonacci numbers*, defined recursively by $a_0 = a_1 = 1$, $a_n = a_{n-1} + a_{n-2}$ ($n \geq 2$), which constitute a very interesting sequence in their own right. By his work, Matijasievič placed the last brick (or, perhaps better, keystone) on the edifice of this wonderful proof.

What the statement claims (and now this is a proved theorem) is not that we *don't have* an algorithm to decide whether a given Diophantine equation has or does not have solutions, but the amazing fact that *there cannot exist* any such algorithm. This statement does not prevent us from deciding the existence of solutions for *some* such equations, but rather means that there can exist no such algorithm that will work, in a finite number of steps, for *every* Diophantine equation. While it is not possible to say more here about this fascinating topic, there are now available two superb presentations of this problem. One of them, by Martin Davis [9], requires relatively little background, while the other one, by Martin Davis, Y. Matijasievič, and J. Robinson [10] is more technical; both can be highly recommended to the motivated reader.

The joint work of this group of mathematicians and logicians has given a definitive, albeit negative, solution to Hilbert's tenth problem.

3   EFFECTIVE METHODS

To introduce the next recent discovery of fundamental importance, let us quote again, but now completely, the same sentence of the old introduction by which we started the discussion of Hilbert's Tenth Problem:

> As a matter of fact, we do not have a general method by which to decide whether a given Diophantine equation has any solutions, or how many solutions.

To determine the number of solutions, it is of course sufficient to find all of them. The task is reduced to a purely numerical verification if we know upper bounds for the absolute values of the variables involved.

In the first edition, the text continues to mention the names of A. Thue, A. Ostrowski, E. Landau, and C. L. Siegel, who had all made important contributions to this problem. Their methods were based on ideas of Diophantine approximations, initiated already by Liouville (1809–1882). Indeed, Liouville showed that if ξ is an algebraic number of degree not higher than n, then the inequality $|\xi - p/q| < c/q^n$, with a constant c independent of n, can have only a finite number of solutions with $p, q, \in \mathbf{Z}$. This result has been improved successively by Thue, who reduced the exponent from n to $(n/2) + 1$; by Siegel, who reduced it further to $\min_\sigma((n/\sigma + 1) + \sigma) + \varepsilon$; and by Dyson, who obtained $\sqrt{2n}$. The final and best possible result was obtained by K. F. Roth, who showed in 1955 [26] that the result holds for any exponent larger than 2. In other words, if the real number ξ can be approximated so well by a sequence of rationals, say p_m/q_m, that the inequality $|\xi - (p_m/q_m)| < c/q^{2+\varepsilon}$ ($\varepsilon > 0$, constant) has infinitely many solutions in integers p_m, q_m, then ξ cannot be algebraic, but must be *transcendental*. We recall that a number is algebraic (over $\mathbf{Q}$) if it satisfies a polynomial equation with integer coefficients; any number that is not algebraic is called *transcendental*.

By use of these results of *Diophantine approximation*, it was possible to show that a large class of Diophantine equations can have only finitely many solutions in integers. All these methods, however, had one shortcoming in common, namely they were not *effective*. By this we mean that they led to formulae that contained constants of unknown (and not computable) values, so that although one could prove that the number of solutions is finite, one could not find bounds for their values. No matter how far we go when we use such a non-effective method in checking for solutions, it is always possible that we miss a still (absolutely) larger solution.

In the 1960s, A. Baker started a sequence of papers in which he remedied the situation by obtaining effective theorems. This means that he obtained effectively computable upper bounds for the size of any solution variable—bounds that depend only on the numerical parameters of the given Diophantine equation. If these bounds are, say $|x| \leq X, |y| \leq Y$, then we can try out

$x = 0, x = \pm 1, \ldots, x = \pm X$, and for each value of x we take successively $y = 0, y = \pm 1, \ldots, y = \pm Y$. Then if we make a list of those couples (x, y) for which the Diophantine equation is satisfied, we are certain to have obtained all the solutions, or if no such couples (x, y) satisfied the equation, to have proved that it has no solution. This completely disposes of the statement quoted at the beginning of this section.

Many classical problems, on which mathematicians had been working for many years with only limited success, could not be attacked successfully, and in many cases solved completely. The number of such instances is too large for enumeration, but among them one may quote the following two, which were outstanding successes.

1. The complete solution of the problem of determining all imaginary quadratic fields with class numbers $h = 1$ and $h = 2$ (see A. Baker [3a], [3b] and H. Stark [30a], [30b]; the case $h = 1$ had essentially been solved earlier by a different approach, by K. Heegner [15]; see also M. Deuring [12] and C. L. Siegel [28]).

2. The work of R. Tijdeman (alone or with coauthors, among them A. Schinzel, T. N. Shorey, A. van der Poorten, C. A. Grimm, and others) on special types of Diophantine equations, such as $y^m = P(x)$ ($P \in \mathbf{Q}[x]$), or $ax^p - by^q = k$. The simplest instance of the last equation, $x^p - y^q = 1$, is known as *Catalan's equation*. It has trivial solutions, like $x = 1, y = 0$; $x = -1$ (for even p), $y = 0$; $x = 0, y = -1$ (for q odd), etc. It also has the nontrivial solutions $x = \pm 3, y = 2, p = 2, q = 3$, but does it have any other solutions with p, q, x, y integers all greater than one? The problem appears to have been posed first by Catalan in 1844, and had resisted all efforts at its solution for over 130 years. Cassels conjectured that it had only finitely many solutions and this was proved to be true by G. V. Čudnowski and by R. Tijdeman (working independently). Then Tijdeman (see [31]) proved[†] that $p = y = 2, q = x = 3$ is in fact the only nontrivial solution of Catalan's equation. This work greatly impressed the mathematical world by its ingenuity and by the illustration it gave of the power of Baker's method, but perhaps also by the fact that it had overcome a difficulty that had for so long resisted all efforts toward a solution.

4   THE GENUS OF A CURVE

In the preceding section mention was made of a certain class of Diophantine equations to which results of Siegel, Baker, and others apply. This class is of considerable importance, so that it is worthwhile to define it clearly. In the

---

[†]As of this writing, some finite numerical work seems to be needed for the completion of the proof.

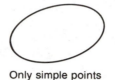

Only simple points

Figure 1

simplest case, we are interested in Diophantine equations of the form

$$f(x, y) = 0 \tag{1}$$

where $f(x, y)$ is a polynomial in the two variables x and y with rational integral coefficients. These polynomials have a degree n, which is the highest total degree of any monomial of $f(x, y)$. By total degree we understand the sum of the exponents of x and y, so that the total degree of $x^a y^b$ is $a + b$. We may think of equation (1) as representing a plane curve. If $g(x, y) = 0$ is another similar equation, we say that the two curves are *birationally equivalent* if the following is true: There exist two pairs of functions, $\phi(\xi, \eta)$, $\psi(\xi, \eta)$ and $\mu(x, y)$, $\nu(x, y)$, all rational over **Q**, such that if $x = \phi(\xi, \eta)$, $y = \psi(\xi, \eta)$, then $f(x, y) = 0$ becomes $g(\xi, \eta) = 0$; and if $\xi = \mu(x, y)$, $\eta = \nu(x, y)$, then $g(\xi, \eta) = 0$ becomes $f(x, y) = 0$.

Let us now consider the curve represented by (1) somewhat more closely. It may consist only of simple points, or it may have double, or even multiple points and also cusps (see Figures 1 and 2). Any multiple point can be resolved into a certain number of double points by slight local deformations. A glance at Figure 3 will clarify what is meant by this sentence better than a thousand words. Let δ be the number of double points so obtained and let κ be the number of cusps. Then if n is the degree of $f(x, y)$, we shall define the *genus* of

double points

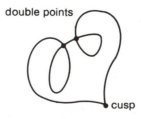

cusp

Figure 2

Triple point resolved into three double points

Figure 3

the curve represented by (1) as follows:

DEFINITION 1. The non-negative number $g = \frac{1}{2}(n-1)(n-2) - \delta - \kappa$ is called the *genus* of the curve (1).

That $g \geq 0$ is of course a theorem; we shall accept it here. The reader may also be mystified by the relevance of such a definition. The importance of this concept of a genus is due to a large extent to the fact that it remains invariant under birational transformations. This permits one to reduce a given equation to some simple standard form in which the problem under consideration can be solved and the result can then be retransposed to the original equation.

The conic sections have $n = 2$, hence $g = (1/2)(2-1)(2-2) = 0$ and they can therefore have no multiple points or cusps, as $g \geq 0$. If a conic section represented by (1) has even one point P with rational coordinates, then it has infinitely many, and all can be obtained by considering the points of intersection with the conic of the chords through P with rational slope (see Problem 4).

If (1) represents a cubic curve (i.e., $n = 3$) with one double point P, then $g = (1/2)(3-1)(3-2) - 1 = 0$, and previous reasoning still holds if P has rational coordinates (see Problem 5).

Cubic curves without a double point have $g = 1$. For all curves of positive genus we have the following noneffective theorem of C. L. Siegel (see [27]):

Theorem 1. *If* (1) *represents a curve of genus $g \geq 1$, then it can have at most finitely many integral points* (*i.e., points with integral coordinates*).

In the particular case of $g = 1$, L. J. Mordell [21] proved and A. Weil [32a, b] generalized the following theorem, whose exact meaning will be discussed in Section 14.

Theorem 2. *Under the operation of "chords and tangents" the rational points on a curve of genus 1 form a finitely generated abelian group.*

Mordell further stated the following conjecture, which has still not been settled as of this writing.

CONJECTURE (Mordell [21]).   The set of rational points on a curve of genus $g \geq 2$ with rational coefficients is finite.[†]

5   THE THEOREMS AND CONJECTURES OF A. WEIL

The work of A. Weil quoted in the old introduction to this chapter inspired much of the most exciting research of the 30 years since its publication [33]. To understand it better, let us recall (see Section 4.2) that there exist finite fields. Indeed, the congruence classes modulo any prime p form a finite field, say $\mathbf{F}_p$, of p elements under addition and multiplication modulo p. It may be shown that every field of p elements is isomorphic to $\mathbf{F}_p$. Next we recall that the field $\mathbf{Q}$ can be enlarged by adjoining to it a zero, say ξ, of a polynomial $f \in \mathbf{Q}[x]$ irreducible over $\mathbf{Q}$. If the degree of f is m, say $\partial^0 f = m$, then the elements α of $\mathbf{Q}(\xi)$ can be represented by

$$\alpha = A_0 + A_1\xi + \cdots + A_{m-1}\xi^{m-1}, \quad A_j \in \mathbf{Q} \quad (j = 0, 1, \ldots, m-1).$$

In exactly the same way one may define polynomials $f \in \mathbf{F}_p[x]$. If $\partial^0 f = m$ and f is irreducible over $\mathbf{F}_p$, then the elements $a_0 + a_1\xi + \cdots + a_{m-1}\xi^{m-1}$ $(a_j \in \mathbf{F}_p, j = 0, 1, \ldots, m-1; f(\xi) = 0)$ form a field $\mathbf{K}$, with $\mathbf{F}_p \subset \mathbf{K}$. As each of the m coefficients can take exactly p values, the number of elements of $\mathbf{K}$ is $q = p^m$, and it may be shown that all finite fields are isomorphic to one of this type; hence for all finite fields, the number q of their elements is a prime power.

We now consider equations of the form

$$a_0 x^{n_0} + a_1 x^{n_1} + \cdots + a_r x^{n_r} + b = 0, \quad a_j \in \mathbf{K}, \quad j = 0, 1, \ldots, r \quad (2)$$

where, without loss of generality, $b = 0$, or $b = 1$. In [33], Weil determines the asymptotic value N of the number of solutions of (2). While his result is more general, we shall quote here only the important particular case with $n_0 = n_1 = \cdots = n_r = n$, when we have

Theorem 3.   *With the above notations, the estimate*

$$|N - q^r| < Cq^{r/2} \tag{3}$$

holds, with C depending only on n.

If we consider, furthermore, extensions of degrees s of K and denote by N_s the number of solutions of (2) (with all exponents equal to n) in the extension $\mathbf{K}_s$ of $\mathbf{K}$, $[\mathbf{K}_s : \mathbf{K}] = s$, then the function $Z(U) = \exp\{\sum_{s=1}^{\infty}(N_s/s)U^s\}$ is of

[†] Note added in proof (July 1983). At this moment news arrives that Mordell's Conjecture had just been proved by the young German mathematician Gerd Faltings. His proof has not yet been published, but it has been seen by several mathematicians and appears to be correct.

particular interest. It has several properties that remind us of Riemann's $\zeta(s)$. So, e.g., $Z(U)$ satisfies the functional equation

$$Z\left(\frac{1}{q^n U}\right) = \pm(q^{n/2}U)^{\chi}Z(U).$$

(Here χ is an integer of geometric significance for the geometric object S represented by (2).) Next, $Z(U)$ is rational in U (like a finite Euler product):

$$Z(U) = \frac{P_1(U)P_3(U)\ldots P_{2n-1}(U)}{P_0(U)P_2(U)\ldots P_{2n}(U)},$$

with $P_0(U) = 1 - U$, $P_{2n}(U) = 1 - q^n U$, $P_h(U) = \prod_{i=1}^{B_h}(1 - \alpha_{ih}U)$ ($h = 1, 2, \ldots, 2n - 1$), where the degrees B_h of the $P_h(U)$ generalize certain topologically relevant parameters of S. For the number theorist, however, it is of particular interest to learn that the α_{ih} are algebraic integers of absolute value $q^{h/2} = \exp((h \log q)/2)$, so that $\log|\alpha_{ih}|/\log q^h = 1/2$. This is the analog of the statement Re{complex zeros of $\zeta(s)$} = 1/2, i.e., the Riemann hypothesis. In [33], A. Weil proves (3) in full generality and the other statements in certain particular cases. The general validity of each of these "Weil Conjectures", which became almost instantaneously famous, has been proved one by one, by the concerted efforts of several eminent mathematicians, among whom one may quote Dwork, Artin, Grothendieck, and Lubkin. The last remaining conjecture was proved only quite recently, by the young mathematician P. Deligne. It is the proof of $|\alpha_{ih}| = q^{h/2}$, i.e., of the analog of the Riemann hypothesis.

6   THE HASSE PRINCIPLE

In its roughest form, this principle states that a Diophantine equation (at least a polynomial one) can be solved unless it is obvious that it cannot be.

In a more precise formulation, it states that in certain cases the obviously *necessary* conditions that $f(x_1, x_2, \ldots, x_r) = 0$ (f a polynomial over **Z** in $x_1, x_2, \ldots, x_r$) be solvable in real numbers $x_1, x_2, \ldots, x_r$ and also as a congruence modulo all prime powers (i.e., in an equivalent formulation, as equations over the so called *p-adic fields*, which were not discussed in this book), are also *sufficient* conditions for solvability in rational integers $x_1, x_2, \ldots, x_r$.

The exact extent of the validity of this "principle" is still not known. That it is not universally valid follows from the fact that instances are known when such an equation is solvable both in the real field and in all *p*-adic fields (i.e., as a congruence modulo all prime powers), but has no solutions in rational integers. So, e.g., Cassels and Guy have shown (see [6]) that the equation

$$5x^3 + 9y^3 + 10z^3 + 12w^3 = 0 \tag{4}$$

has solutions with $(x, y, z, w) \neq (0,0,0,0)$, both in the reals and also as a congruence modulo p^r for all primes p and exponents $0 < r \in \mathbf{Z}$, but has no solution in rational integers not all zero. This is somewhat surprising, in view of the fact that the principle holds for equations $ax^3 + by^3 + cz^3 + dw^3 = 0$ $(a, b, c, d \in \mathbf{Z})$ if $ab = cd$, a condition violated by (4). On the other hand, an important classical case where the Hasse principle holds will be presented in Sections 10 and 11.

7   LINEAR AND QUADRATIC DIOPHANTINE EQUATIONS

The purpose of the preceding sections has been to give a broad panorama of some of the accomplishments of the last 25 or so years. Only isolated results could be mentioned as illustrations, and no proofs were attempted.

While, as already mentioned, an exhaustive treatment of the subject of Diophantine equations would require a multivolume treatise, an attempt will be made in the following sections to proceed somewhat more systematically, prove most results stated, and give at least some precise statements of theorems whose proofs exceed the framework of the present book.

We may start by thinking of linear Diophantine equations and systems. These were already discussed in Chapter 4, and the generalization to any number of variables is rather immediate. Some of the procedures to be used are at least suggested in Section 1.2 (see also [23] pp. 133–137).

Next we consider Diophantine equations of the second degree. Let

$$Q(x_1, x_2, \ldots, x_r) = \sum_{\substack{1 \le i \le r \\ 1 \le j \le r}} a_{ij} x_i x_j$$

be a homogeneous polynomial of the second degree with rational integral coefficients in the r variables $x_1, x_2, \ldots, x_r$. Such homogeneous polynomials are also often called *forms* of the respective degree. We say that Q is a *quadratic form* in r variables. Similarly, $L(x_1, x_2, \ldots, x_r) = \sum_{1 \le j \le r} b_j x_j$ is a *linear form* in the same variables. Also let $c (\in \mathbf{Z})$ be a constant. Then

$$Q(x_1, x_2, \ldots, x_r) + L(x_1, x_2, \ldots, x_r) + c = 0 \tag{5}$$

is the most general Diophantine equation of the second degree. If we set $x_j = y_j + z_j$ and substitute in (5), one can in general determine the z_j in such a way that the first-degree terms all vanish (see Problem 7), and (5) reduces to

$$Q(x_1, x_2, \ldots, x_r) = n. \tag{6}$$

The theory of quadratic forms is highly developed. It had already been studied by Gauss (see [13]), and for a modern presentation the reader may want to consult O'Meara [24]. Here we shall consider only a few particular cases, all of the *diagonal type*, i.e., where the mixed terms such as xy, yz, $x_i x_j$, etc. are missing.

We are in general faced with two problems:

(a) Given $Q(x_1, x_2, \ldots, x_r)$, we want to *characterize those integers n* for which (6) has solutions in integers $x_1, x_2, \ldots, x_r$; such integers n are said to be *representable* by $Q(x_1, x_2, \ldots, x_r)$.

(b) Given $Q(x_1, x_2, \ldots, x_r)$ and a representable integer n, we ask for the *number* of solutions of (6) in integers.

8   REPRESENTATIONS BY SUMS OF TWO SQUARES

In (6) let us take $Q(x, y) = x^2 + y^2$, so that (6) becomes

$$x^2 + y^2 = n. \tag{7}$$

With reference to the first question, simple examples like $n = 3, 6, 7, \ldots$, show that (7) is not solvable for every n. On the other hand, (7) is solvable for $n = 2, 5, 8, \ldots$, and in particular, every square $n = m^2$ admits the solutions $x = \pm m, y = 0$ and also $x = 0, y = \pm m$. We do consider these four solutions as distinct.

Solutions of (7) with $(x, y) = 1$ are called *primitive*. If $(x, y) = d > 1$ is a solution of (7), then $x = dx_0$, $y = dy_0$, $n = d^2 n_0$, and x_0, y_0 is a primitive solution of $x_0^2 + y_0^2 = n_0$. We obtain a (nonprimitive) solution of (7) by setting $x = dx_0, y = dy_0$, and it follows that all solutions of (7) can be obtained from primitive solutions of (7), with n replaced by n/d^2, for any d with $d^2 | n$. If $xy \neq 0$, then we consider the four couples of solutions $\pm x, \pm y$ as four distinct solutions of (7); moreover if $x \neq y$, we shall consider the four solutions obtained by interchanging x with y as another set of four distinct solutions. So, e.g., $x^2 + y^2 = 5$ has the eight solutions $(1, 2)$, $(-1, 2)$, $(1, -2)$, $(-1, -2)$, $(2, 1)$, $(2, -1)$, $(-2, 1)$, $(-2, -1)$.

We observe that (7) has no solutions if $n \equiv 3 \pmod 4$, because x^2 and y^2 are each congruent to either 0 or 1 modulo 4. On the other hand, if $n = n_1 n_2^2$ and $x_0^2 + y_0^2 = n_1$, then $(n_2 x)^2 + (n_2 y)^2 = n_1 n_2^2 = n$, and n_2 may well be congruent to 3 (mod 4) or contain divisors of that kind. It is clear, however, that a prime $q \equiv 3 \pmod 4$ that divides n must in fact divide n_2, and so occur in n with an even exponent. This result will be obtained again by an entirely different method.

In order now to approach the second problem, that of determining the number $r_2(n)$ of representations of n by the sum of two squares, we shall use our knowledge of the Gaussian field. This is one of several possible approaches, and follows in general the presentation in [14].

Let $n = 2^f \prod_{p \equiv 1(4)} p^r \prod_{q \equiv 3(4)} q^s$, with $f, r, s \in \mathbf{Z}$; then as we know from Chapter 10, $n = \{(1 + i)(1 - i)\}^f \prod \{(a + bi)(a - bi)\}^r \prod q^s$ is the unique factorization of n into Gaussian primes. Without loss of generality we may take $a, b > 0$, and we also have $a \neq b$; indeed, if $a = b$, then $p = a^2 + b^2 = (a + bi)(a - bi) = a^2(1 + i)(1 - i) = 2a^2$, with $p \equiv 1 \pmod 4$, a contradiction. If $n = u^2 + v^2 = (u + iv)(u - iv)$, then

$$u + iv = i^t(1 + i)^{f_1}(1 - i)^{f_2}\prod\{(a + ib)^{r_1}(a - ib)^{r_2}\}\prod q^{s_1}$$

$$u - iv = i^{-t}(1 + i)^{f_2}(1 - i)^{f_1}\prod\{(a + ib)^{r_2}(a - ib)^{r_1}\}\prod q^{s_2}$$

with $t = 0, 1, 2,$ or 3; $f_1 + f_2 = f$; $r_1 + r_2 = r$; $s_1 + s_2 = s$. The change $i \to -i$ does not affect the purely real factors q, so that $s_1 = s_2$ and $s = 2s_1$ is even, as we expected it to be by a previous reasoning. There are four choices for t, $f + 1$ choices for f_1 (and these also determine $f_2 = f - f_1$), and $r + 1$ choices for r_1 (with $r_2 = r - r_1$). It follows that the number of ways of splitting up the Gaussian primes between the two factors of $u^2 + v^2$ is $4(1 + f)\prod(r + 1)$, with the last product taken over the primes $p|n$, $p \equiv 1 \pmod 4$. However, not to all of these different allocations of Gaussian primes correspond different values of those two factors. Indeed, when we replace $1 + i$ by $1 - i$, we simply multiply the respective factor by $(1 - i)/(1 + i) = i^3$ and this is already taken into account by the choice of t. Hence we actually obtain only $4\prod(r + 1)$ different pairs of factors $u + iv$, $u - iv$, and each such pair determines u and v uniquely. Consequently, $r_2(n)$ is four times the number of divisors of $n_1 = \prod_{p^r \| n} p^r$, say $4d(n_1)$ (see Corollary 6.15.2; here the symbol $p^r\|n$ means that p^r is the exact power of p that divides n). From a factor $u + iv$, that corresponds to a representation $u^2 + v^2$, we obtain by multiplication by $i^t (t = 1, 2, 3)$ the three other factors $iu - v$, $-u - iv$, and $-iu + v$; these correspond to the representations $n = (-v)^2 + u^2 = (-u)^2 + (-v)^2 = v^2 + (-u)^2$, respectively. The further (trivial) modification, which corresponds to an exchange of u and v in the above representations, is obtained by replacing a given factor $u + iv$ by its complex conjugate $u - iv$, and so we also have $n = v^2 + u^2 = u^2 + (-v)^2 = (-v)^2 + (-u)^2 = (-u)^2 + v^2$. These, however, were already counted by $4d(n_1)$, because they correspond to an exchange of r_1 and r_2.

In the particular case $n = p \equiv 1 \pmod 4$, we have $r = 1$ and find that there are $4(1 + 1) = 8$ representations (all trivial variations of what is essentially a single one), as we already saw in the case $p = 5$.

The expression $d(n_1)$ can be given an interesting interpretation in terms of n itself. Let us write $n = 2^f n_1 n_2$, $n_1 = \prod_{p \equiv 1 \,(\mathrm{mod}\,4)} p^r$, $n_2 = \prod_{q \equiv 3 \,(\mathrm{mod}\,4)} q^s$. Denote by $\chi(d)$ the nontrivial character modulo 4; i.e., set $\chi(d) = 1$ if $d \equiv 1$ (mod 4), $\chi(d) = -1$ if $d \equiv 3$ (mod 4), with $\chi(d) = 0$ for even d. Then $\sum_{d|n} \chi(d) = d_1(n) - d_3(n)$, where $d_1(n)$ is the number of divisors d of n, with $d \equiv 1$ (mod 4) and $d_3(n)$ the number of divisors $d|n$, $d \equiv 3$ (mod 4).

From $n_1 n_2 = \prod(1 + p + \cdots + p^r)\prod(1 + q + \cdots + q^s)$ ($p^r \| n$, $p \equiv 1$ (mod 4), $q^s \| n$, $q \equiv 3$ (mod 4)), it is clear that a divisor is counted by $d_1(n)$ if it contains an even number of factors q; otherwise it is counted by $d_3(n)$. Hence we obtain $d_1(n) - d_3(n)$ if we set in the last product $p = 1$ and $q = -1$. Then the factor $1 + p + \cdots + p^r$ becomes $1 + r$ and $1 + q + \cdots + q^s = 0$ or 1, according to whether s is odd or even. Hence the product becomes $\prod_p (1 + r)\prod_q (1 + (-1)^s)/2$, and this equals, as seen, $d_1(n) - d_3(n)$. On the other hand, the product vanishes if even a single one of the components s is odd; otherwise the product over the qs equals one, and that over the ps equals $\prod(r + 1)$. We conclude that $\prod(r + 1) = d_1(n) - d_3(n)$, and may formulate the results obtained as:

Theorem 4. *The number $r_2(n)$ of representations of n as a sum of two squares is zero if the product of the prime divisors q of n with $q \equiv 3$ (mod 4) is not a perfect square. If that product is a square, then $r_2(n) = 4(d_1(n) - d_3(n))$.*

9   REPRESENTATIONS BY SUMS OF THREE SQUARES

It is a curious fact that, while the problem of the representation of natural integers as sums of two or four squares can be treated by rather elementary methods, and in any case the results can be expressed by using nothing more sophisticated than divisor functions such as $d(n)$, $d_1(n)$, $d_3(n)$, etc., the problem of the representation by sums of three squares involves much more subtle considerations, and even to only state the final result it is convenient to use the concept of class number.

The first representation problem (see Section 7), namely the characterization of those integers n which can be represented as sums of three squares, is still rather easy.

Clearly from $a^2 \equiv 0$ or 1 (mod 8) it follows that $a^2 + b^2 + c^2 = 0, 1, 2, 3, 4, 5, 6$, (mod 8) are all possible by choosing appropriately the parities of a, b, and c. However, no such choice leads to $a^2 + b^2 + c^2 \equiv 7$ (mod 8). Furthermore, if all summands are even, one can factor out of them some power of 4. If 2^a is the largest power of 2 that divides each of a, b, and c, then $a^2 + b^2 + c^2 = 4^a(a_1^2 + b_1^2 + c_1^2)$, with at least one of a_1, b_1, c_1 odd. Hence

$a_1^2 + b_1^2 + c_1^2 \equiv 1, 2, 3, 5,$ or $6 \pmod 8$ are the only possibilities. With this we have proved the first part of the following theorem.

Theorem 5. *If the rational integer n is of the form $n = 4^a(8k + 7)$, then the equation*

$$x^2 + y^2 + z^2 = n \tag{8}$$

has no solutions in rational integers x, y, z. For every integer n not of that form, (8) has rational integral solutions.

The proof of the last statement will not be given here. As Mordell [22] rightly observes, no really elementary treatment is known. One of the most accessible ones may be found in [22] itself.

Concerning now the second question to be considered, namely that of the number $r_3(n)$ of solutions of (8) for representable $n(\neq 4^a(8k + 7))$, this is contained in the following:

Theorem 6. *The number of representations of the natural integer n as a sum of three squares is given by the formula*

$$r_3(n) = \frac{16}{\pi}\sqrt{n}\,K(-4n)\psi(n)P(n),$$

where for $n = 4^a k, 4 \nmid k$ one has

$$\psi(n) = \begin{cases} 0 \text{ if } k \equiv 7 \pmod 8 \\ 2^{-a} \text{ if } k \equiv 3 \pmod 8 \\ 3 \cdot 2^{-a-1} \text{ if } k \equiv 1, 2, 5, 6 \pmod 8; \end{cases}$$

$K(-4n) = \sum_{m=1}^{\infty}(-4n/m)m^{-1} = L(1, \chi)$ with $\chi(n) = (-4n/m)$, the Jacobi symbol (by Definition 5.4' this is equal to zero for m even) and

$$P(n) = \prod_{p^{2b}\|n}\left(1 + \frac{1}{p} + \frac{1}{p^2} + \cdots + \frac{1}{p^{b-1}} + \frac{1}{p^b\left(1 - \left(\frac{-n}{p}\right)\frac{1}{p}\right)}\right).$$

Proofs of Theorem 6 may be found, e.g., in [5], [2], and [16].

The formula simplifies considerably for squarefree numbers, because then $P(n) = 1$. The result simplifies further if $n \equiv 3 \pmod 8$ and $n \geq 11$, when one computes that $K(-4n) = (3/2)\pi n^{-1/2}h(n)$, where $h(n)$ is the class number of the quadratic field $Q(\sqrt{-n})$. In that case, Theorem 6 becomes

Theorem 7. *For squarefree $n \equiv 3 \pmod 8$, $n \geq 11$, one has*

$$r_3(n) = 24\,h(n),$$

where $h(n)$ is the number of ideal classes of the imaginary quadratic field $Q(\sqrt{-n})$.

As an example, let us consider $n = 19 \equiv 3 \pmod{18}$ and squarefree. We recall (see Chapter 10) that the fields $Q(\sqrt{-n})$ have $h(n) = 1$ for $n = 3, 11, 19, 43, 67, 163$, all congruent to 3 (mod 8) (and also for $n = 1, 2$, and 7, which are not congruent to 3 (mod 8)); hence, Theorem 7 yields $r_3(19) = 24$. Equation (8) with $n = 19$ has essentially only one solution, namely, $1^2 + 3^2 + 3^2 = 19$. This leads to 24 different solutions by permutations of the summands and by changes of their signs, namely

$$(\pm 1)^2 + (\pm 3)^2 + (\pm 3)^2 = (\pm 3)^2 + (\pm 1)^2 + (\pm 3)^2$$
$$= (\pm 3)^2 + (\pm 3)^2 + (\pm 1)^2.$$

The relevance of the restrictions (e.g., $n \geq 11$) in Theorem 7 becomes clear if we observe that, e.g., $n = 3$ leads to the solutions $(\pm 1)^2 + (\pm 1)^2 + (\pm 1)^2 = 3$ and $r_3(3) = 8$ and not 24, as would follow from the formula of Theorem 7. On the other hand, Theorem 6 holds for every natural integer n.

10 A THEOREM OF LEGENDRE

We obtain a more general quadratic form than the simple sum of three squares if we allow each square to also have a numerical coefficient. Here, however, instead of considering an arbitrary natural integer n, we shall restrict ourselves to the possibility of representing $n = 0$ by a diagonal form in three variables. In other words, we shall study the solvability of the equation

$$ax^2 + by^2 + cz^2 = 0 \tag{9}$$

in integers x, y, z, not all zero. This problem has been solved by Legendre [17], who proved the following theorem.

Theorem 8 (Legendre). *Let a, b, c be squarefree rational integers, $a \cdot b \cdot c \neq 0$, $(a, b) = (b, c) = (c, a) = 1$. Then the Diophantine equation (9) has nontrivial solutions (i.e., solutions different from $x = y = z = 0$) if and only if the following two conditions are satisfied:*

(i) *a, b, and c do not all have the same sign; and*

(ii) *$-ab, -bc$, and $-ca$ are quadratic residues modulo c, a, and b, respectively.*

The requirement that a, b, c be squarefree is not a real restriction. If $a = a_1 a_2^2$ (a_1 squarefree), then the term $ax^2 = a_1 x_1^2$ with $x_1 = a_2 x$ and a_1 squarefree, and similarly for b and c. Also the condition $a \cdot b \cdot c \neq 0$ follows from $(a, b) = (b, c) = (c, a) = 1$, except for some trivial cases when the result is obvious.

REMARK 1.   With x, y, z, also mx, my, mz is a solution of (9); therefore it is sufficient to find only *primitive* solutions, i.e., solutions with $(x, y, z) = 1$. In this case, we may observe that x, y, z are actually coprime in pairs. Indeed, if $p|x$, $p|y$, then $p^2|ax^2 + by^2$ or $p^2|cz^2$; but c is squarefree, so that $p|z$, contrary to our assumption that $(x, y, z) = 1$.

REMARK 2.   The necessity of the first condition of Theorem 8 is already obvious for the existence of solutions in the field **R** of reals because the sum of three quantities all of the same sign cannot vanish.

REMARK 3.   The second condition of the theorem also has a simple interpretation. On account of $c \neq 0$, equation (9) has solutions if and only if $acx^2 + bcy^2 + c^2z^2 = 0$ has solutions. Now let $p|a$; then the last equation shows that

$$-bcy^2 \equiv (cz)^2 \,(\text{mod } p). \tag{10}$$

If x, y, z is a primitive solution of (9), from $(a, c) = 1$ and $p|a$ it follows that $p \nmid c$. Hence if $p|y$, then $p|z$, which is false for a primitive solution; consequently, $p \nmid y$ and the congruence $yy_1 \equiv 1 \,(\text{mod } p)$ has a solution y_1. We now multiply (10) by y_1^2, and setting $cyz = u$, obtain $-bc \equiv u^2 \,(\text{mod } p)$. This result holds for each prime divisor of the squarefree number a; hence $-bc$ is a quadratic residue modulo a. Permutation of the letters yields the other two conditions. We may now observe that these three conditions simply mean that (9) can be solved as a congruence modulo all prime divisors of $a \cdot b \cdot c$, and this is obviously a necessary condition for the solvability of (9) in integers. On the other hand, these conditions are automatically satisfied if we are told that (9) can be solved as a congruence modulo *all* prime powers—which again is clearly necessary if (9) is to have solutions in **Z**. The conditions of the theorem will be satisfied if we require that (9) be solvable both in the real field **R** and as a congruence modulo all prime powers. By using the language of Section 6, we shall say that we require the solvability of (9) both in the field **R** of the reals and in all p-adic fields. The conclusion of the theorem then states that these (obviously) necessary conditions are also sufficient. Thus Legendre's theorem illustrates an instance of the validity of Hasse's Principle.

    A last remark: The conditions of the theorem appear weaker than the requirements listed here; in fact, however, one may show that they already imply solvability as a congruence modulo all prime powers.

11   PROOF OF LEGENDRE'S THEOREM

    The necessity of the conditions has been established. We now proceed to show that they are also sufficient. The proof is somewhat complicated and is based on two different ideas; therefore it appears advisable to start with a general sketch of it.

It would be easy to prove the theorem if we could represent the quadratic form (9) as a product of two linear polynomials, because equations of the form $a_1 x + b_1 y + c_1 z = 0$ are easy to solve. This is, however, not possible in general. The next best thing is to obtain such a factorization as a congruence, and this is the first idea of the proof. Indeed, we shall show that there exist integers a_i, b_i, c_i ($i = 1, 2$), such that

$$ax^2 + by^2 + cz^2 \equiv (a_1 x + b_1 y + c_1 z)(a_2 x + b_2 y + c_2 z) \pmod{a \cdot b \cdot c}.$$
$$(11)$$

It is easy to find nontrivial solutions to a congruence like $a_1 x + b_1 y + c_1 z \equiv 0$ (mod $a \cdot b \cdot c$), and this has an immediate consequence that also $ax^2 + by^2 + cz^2 \equiv 0$ (mod $a \cdot b \cdot c$). This is of course not sufficient to allow us to conclude that (9) itself holds. It is here that the second idea of the proof comes in. We shall show that the last congruence has solutions with *small* values of the variables x, y, z such that by omitting a few special values of the coefficients we obtain

$$-2abc < ax^2 + by^2 + cz^2 < abc. \qquad (12)$$

By also using the congruence satisfied by the trinomial, it follows that $ax^2 + by^2 + cz^2 = 0$, or $= -abc$. In the first case, we have obtained a solution of (9); in the second case we set $x' = -by + xz$, $y' = ax + yz$, $z' = z^2 + ab$, and verify that

$$ax'^2 + by'^2 + cz'^2 = a(b^2 y^2 - 2bxyz + x^2 z^2)$$
$$+ b(a^2 x^2 + 2axyz + y^2 z^2) + c(z^4 + 2abz^2 + a^2 b^2)$$
$$= ab(ax^2 + by^2 + cz^2) + xyz(-2ab + 2ab)$$
$$+ z^2(ax^2 + by^2 + cz^2) + cab(z^2 + ab)$$
$$= -ab \cdot abc - z^2 \cdot abc + z^2 \cdot abc + cab \cdot ab = 0,$$

so that in this case x', y', z' is a solution of (9). In the few omitted cases, solutions are fairly obvious, and this finishes the proof of the theorem.

To execute this program, we first have to prove (11); we do this in two steps. We show that if $(u, v) = 1$, then from a factorization modulo u and a factorization modulo v we obtain a factorization modulo uv. Next we show that such a factorization actually holds, say modulo a. A similar proof will show that a similar, but in general different, factorization holds modulo b and still another one modulo c. As $(a, b) = 1$, we obtain a factorization modulo ab, and as $(ab, c) = 1$, we obtain a factorization modulo $a \cdot b \cdot c$, which is the desired result (11).

So let us assume that we know u_i, v_i, w_i and r_i, s_i, t_i $(i = 1, 2)$ such that

$$ax^2 + by^2 + cz^2 = (u_1 x + v_1 y + w_1 z)(u_2 x + v_2 y + w_2 z) \pmod{u},$$

$$ax^2 + by^2 + cz^2 = (r_1 x + s_1 y + t_1 z)(r_2 x + s_2 y + t_2 z) \pmod{v}.$$

By use of the Chinese Remainder Theorem 4.16 we may select a_1 such that $a_1 \equiv u_1 \pmod{u}$ and $a_1 \equiv r_1 \pmod{v}$ hold simultaneously, and similarly for the other coefficients b_1, c_1 and a_2, b_2, c_2. With these choices, $ax^2 + by^2 + cz^2 \equiv (a_1 x + b_1 y + c_1 z)(a_2 x + b_2 y + c_2 z) \pmod{uv}$ obviously holds. In order to obtain (11), it is therefore sufficient that such factorization holds, e.g., modulo a. By $(a, b) = 1$ it follows that there exists b' such that $bb' \equiv 1 \pmod{a}$. Next $-bc$ is a quadratic residue modulo a, so that there exists an integer g that satisfies the congruence $g^2 \equiv -bc \pmod{a}$. It now follows that, modulo a,

$$ax^2 + by^2 + cz^2 \equiv by^2 + cz^2 \equiv bb'(by^2 + cz^2)$$
$$\equiv b'(b^2 y^2 + bcz^2) \equiv b'(b^2 y^2 - g^2 z^2)$$
$$\equiv b'(by + gz)(by - gz) \equiv (y + b'gz)(by - gz)$$
$$\equiv (0 \cdot x + y + (b'g)z)(0 \cdot x + by - gz),$$

the desired factorization. This finishes the proof of (11).

We now proceed to show that linear congruences in 3 variables can be solved with relatively small values of the variables. More precisely, we shall prove the following:

Lemma 1. *Let α, β, γ be positive reals such that $\alpha\beta\gamma = m \in \mathbf{Z}$; then the congruence $a_1 x + b_1 y + c_1 z \equiv 0 \pmod{m}$ has a nontrivial solution with $|x| \leq \alpha, |y| \leq \beta, |z| \leq \gamma$.*

The proof uses *Dirichlet's Box Principle*. If you have k objects and have to place them into $n < k$ boxes, then there must be at least one box with at least 2 objects. In the present case, let x take on the values $0, 1, \ldots, [\alpha]$, y the values $0, 1, \ldots, [\beta]$, and z the values $0, 1, \ldots, [\gamma]$. This leads to $(1 + [\alpha])(1 + [\beta])(1 + [\gamma]) > \alpha \cdot \beta \cdot \gamma = m$ triplets (x, y, z). There are, however, only m distinct residue classes modulo m, so that there are at least two triplets, say (x_1, y_1, z_1) and (x_2, y_2, z_2), that lead to the same residue class for $a_1 x + b_1 y + c_1 z$, i.e., $a_1 x_1 + b_1 y_1 + c_1 z_1 \equiv a_1 x_2 + b_1 y_2 + c_1 z_2 \pmod{m}$. This means if we set $x = x_1 - x_2, y = y_1 - y_2, z = z_1 - z_2$, that $a_1 x + b_1 y + c_1 z \equiv 0 \pmod{m}$, and by the ranges of the $x_i, y_i, z_i (i = 1, 2)$, $|x| \leq \alpha, |y| \leq \beta, |z| \leq \gamma$, as claimed.

We now return to (11) and want to have, e.g.,

$$a_1 x + b_1 y + c_1 z \equiv 0 \pmod{a \cdot b \cdot c} \tag{13}$$

with "small" x, y, z. We choose $\alpha = \sqrt{|bc|}, \beta = \sqrt{|ca|}, \gamma = \sqrt{|ab|}$. Then (13) holds with $x^2 \le |bc|$, $y^2 \le |ca|$, $z^2 \le |ab|$. In fact by the coprimality in pairs of a, b, c and the fact that they are squarefree, it follows that we have strict inequalities everywhere unless two of the coefficients have absolute value one. We may of course ignore the trivial case $|a| = |b| = |c| = 1$, as we know well that $x^2 = y^2 + z^2$ has infinitely many solutions. Also, as the coefficients are not all of the same sign, we may assume without loss of generality that $a > 0$, $b < 0$, $c < 0$. If we also temporarily ignore the possibility that $b = c = -1$, the inequalities become explicitly $x^2 < bc$, $y^2 \le -ac$, $z^2 \le -ab$. It now follows, recalling the signs of a, b, c that $ax^2 + by^2 + cz^2 < a \cdot bc +$ (nonpositive) $< abc$; and also $ax^2 + by^2 + cz^2 \ge$ (positive) $+ b \cdot (-ac) + c(-ab) > -2abc$, and this finishes the proof of (12), under the assumption that $b = c = -1$ is not the case.

We already know how to obtain the theorem from (12), except for one possibility. It is conceivable that $ax^2 + by^2 + cz^2 = -abc$, and that when we compute x', y', z' all these vanish and we obtain only the trivial solution. This, however, can happen only for $a = -b = 1$, when $x = 1 = -y$, $z = 0$ is a solution. Indeed if $z' = z^2 + ab = 0$, then $z^2 = -ab$, and as $(a, b) = 1$, a and b squarefree, $-ab$ can be a perfect square with $a > 0$ only if $a = 1 = -b$.

It only remains to consider the excluded case $b = c = -1$. We recall that $-bc$ is a quadratic residue modulo a. Consequently if $p|a$, then $(-1/p) = +1$, $p \equiv 1 \pmod 4$, and (see Section 8) a can be represented by a sum of two squares, say $a = y_0^2 + z_0^2$. It follows that if we set $x = 1$, $y = y_0$, $z = z_0$, we obtain that $ax^2 + by^2 + cz^2 = ax^2 - y^2 - z^2 = a - a = 0$, so that $1, y_0, z_0$ is a solution of (9). This finishes the proof of Legendre's Theorem.

12 LAGRANGE'S THEOREM AND THE REPRESENTATION OF INTEGERS AS SUMS OF FOUR SQUARES

As we have seen, not all integers are sums of either two or three squares. In other words, equations (7) and (8) do not always have solutions. It is therefore of considerable interest to know that the equation

$$x_1^2 + x_2^2 + x_3^2 + x_4^2 = n \tag{14}$$

does have solutions for any integer $n \ge 0$. This is the content of the following theorem due to Lagrange (1736–1813).

Theorem 9. *Every non-negative integer is the sum of four squares of integers.*

FIRST PROOF.   It is sufficient to prove the statement for $n \not\equiv 0 \pmod 4$. Indeed if $n_0 = 4^a n$ and $n = \sum_{i=1}^4 x_i^2 \not\equiv 0 \pmod 4$, then $n_0 = \sum_{i=1}^4 (2^a x_i)^2$. Hence we may assume without loss of generality that $n \equiv 1, 2, 3, 5, 6,$ or $7 \pmod 8$. Except for $n \equiv 7 \pmod 8$, we know from Theorem 5 that (14) has solutions, even with $x_4 = 0$. If, however, $n \equiv 7 \pmod 8$, then $n - 1 \equiv 6 \pmod 8$, $n - 1 = x_1^2 + x_2^2 + x_3^2$ has solutions by Theorem 5, and hence $n = x_1^2 + x_2^2 + x_3^2 + 1^2$.

It is also possible to prove Theorem 9 without reference to Theorem 5, and this is desirable because Theorem 5 has not been proved here.

In order to avoid trivial special cases, we observe immediately that for $x_1 = x_2 = x_3 = x_4 = 0$ we obtain by (14) a representation of zero; similar representations of $n = 1$ and $n = 2$ are also evident. For $n > 2$, the proof consists of two parts:

(a) One shows that every odd prime can be represented by (14).

(b) One proves that if n and m have representations as sums of four squares, then so does their product nm.

The proof of (b) is essentially trivial and consists merely in the verification of the identity

$$\left(x_1^2 + x_2^2 + x_3^2 + x_4^2 \right)\left(y_1^2 + y_2^2 + y_3^2 + y_4^2 \right)$$

$$= \left(x_1 y_1 + x_2 y_2 + x_3 y_3 + x_4 y_4 \right)^2 + \left(x_1 y_2 - x_2 y_1 + x_3 y_4 - x_4 y_3 \right)^2$$

$$+ \left(x_1 y_3 - x_3 y_1 + x_4 y_2 - x_2 y_4 \right)^2 + \left(x_1 y_4 - x_4 y_1 + x_2 y_3 - x_3 y_2 \right)^2$$

$$\tag{15}$$

Indeed both sides are equal to $\sum_{i=1}^{4} \sum_{j=1}^{4} x_i^2 y_j^2$; as Littlewood used to say, every identity is trivial, once somebody shows it to you!

To complete the proof, it only remains to show that (14) always has solutions if $n = p$ an odd prime. The proof of the last statement is facilitated by the use of the following Lemma.

Lemma 2. *For all primes, the equation*

$$x^2 + y^2 + 1 = mp$$

can be solved with an integer $0 < m < p$.

PROOF. For $p = 2$, take $x = 1, y = 0, m = 1$. For $p \equiv 1 \pmod 4$, take $y = 0$. Indeed we know (see Corollary 5.5.1) that $(-1/p) = +1$, so that $x^2 + 1 \equiv 0 \pmod p$ for some integer x. This may be selected as absolutely least residue modulo p, so that $|x| \le (p - 1)/2$ and $0 < x^2 + 1 \le ((p - 1)/2)^2 + 1 = p^2 - (3p^2 + 2p - 5)/4 < p^2$ (the last inequality holds for all $p > 1$). If $p \equiv 3 \pmod 4$, we can no longer take $y = 0$, because $(-1/p) = -1$ and $x^2 + 1 \equiv 0$

(mod p) has no solution. However if we consider as before the $(p + 1)/2$ distinct values of x^2 for $x = 0, 1, 2, \ldots, (p - 1)/2$, and also the $(p + 1)/2$ distinct values of $-(y^2 + 1)$ for $y = 0, 1, \ldots, (p - 1)/2$, then the union of these two sets contains $p + 1$ integers. Hence by Dirichlet's "Box Principle" (see Section 11), two of them must belong to the same among the only p different residue classes modulo p. For those particular values of x and y, we have, therefore, that $x^2 \equiv -1 - y^2$ (mod p), or that $x^2 + y^2 + 1 = mp$. On the other hand, $0 \leq x^2 \leq ((p - 1)/2)^2, 1 \leq 1 + y^2 \leq 1 + ((p - 1)/2)^2, x^2 + y^2 + 1 \leq 2((p - 1)/2)^2 + 1 < p^2$, whence $m < p$, as claimed.

SECOND PROOF OF THEOREM 9. This is modeled after the presentation in [14]. We already know from Lemma 2 that

$$mp = x_1^2 + x_2^2 + x_3^2 + x_4^2 \tag{16}$$

always has a solution, even under the restriction $x_3 = 1$, $x_4 = 0$, and with $0 < m < p$. Here we want to show that if we drop the conditions on x_3 and x_4, then we may take $m = 1$. Let us assume instead that min $m = m_0 > 1$. If m_0 is even, then the number of odd summands in (16) must be even; hence we can pair them off into pairs of same parity, say x_1 with x_2 and x_3 and x_4. Then, however, $((x_1 + x_2)/2)^2 + ((x_1 - x_2)/2)^2 + ((x_3 + x_4)/2)^2 + ((x_3 - x_4)/2)^2 = \frac{1}{2}m_0 p$, which contradicts the definition of m_0. It follows that m_0 is odd, and if $m_0 > 1$, then $m_0 \geq 3$. Also, m_0 cannot divide all x_is, as otherwise $m_0^2 | m_0 p$, which is impossible for $3 \leq m_0 < p$. Now select $y_i \equiv x_i$ (mod m_0) as absolutely least residues modulo m_0, i.e., $y_i = x_i - a_i m_0, |y_i| < m_0/2$, so that $0 < \sum_{i=1}^{4} y_i^2 < 4(m_0/2)^2 = m_0^2$ and $\sum_{i=1}^{4} y_i^2 \equiv 0$ (mod m_0). Hence $\sum_{i=1}^{4} y_i^2 = m_0 m_1$ with $m_1 < m_0$, and if we multiply this by (16) we obtain, also using (15), that

$$m_0^2 m_1 p = \sum_{i=1}^{4} z_i^2, \tag{17}$$

with the z_is given by (15). We easily check, on hand of (15), that all z_is are divisible by m_0. So, e.g., $z_1 = \sum_{i=1}^{4} x_i y_i = \sum_{i=1}^{4} x_i(x_i - a_i m_0) \equiv \sum_{i=1}^{4} x_i^2 = m_0 p \equiv 0$ (mod m_0), and similarly for $z_2, z_3,$ and z_4. If we now set $z_i = m_0 x_i'$, (17) becomes $m_1 p = \sum_{i=1}^{4}(x_i')^2$, $m_1 < m_0$, contrary to the definition of m_0. This shows that $m_0 = 1$, and finishes the proof of the theorem. Much more is known about equation (14). In fact the following theorem holds:

Theorem 10. *The number $r_4(n)$ of solutions of the Diophantine equation (14), (i.e., the number of representations of n as a sum of four squares) equals 8 times the sum of the divisors of n that are not multiples of 4.*

REMARK 4. As until now, one considers representations as distinct even if they differ only by the order or by the sign of the x_is. ($i = 1, 2, 3, 4$).

EXAMPLES

1. Let $n = 101$; then 1 and 101 are the only divisors of n and neither of them is divisible by 4; hence $r_4(n) = 8(101 + 1) = 816$. Indeed we find for 101 the five essentially distinct solutions of (14):

$$101 = 1^2 + 10^2 + 0^2 + 0^2 = 1^2 + 6^2 + 8^2 + 0^2 = 2^2 + 4^2 + 9^2 + 0^2$$
$$= 4^2 + 6^2 + 7^2 + 0^2 = 2^2 + 5^2 + 6^2 + 6^2.$$

By changing the order of the summands, the first representation leads to twelve different sums, and each of them, by changes of signs, to four distinct solutions counted by $r_4(101)$; hence the first sum counts for 48 distinct representations. Similarly, in the second sum the zero may occupy any one of the four positions; the three nonvanishing summands have 6 distinct permutations, and each of these summands is the square of a positive or of a negative integer, so that the second representation of 101 counts for $4 \cdot 6 \cdot 8 = 192$ solutions of (14); the same number of solutions is also yielded by the other representations, for a total of $48 + 4 \cdot 192 = 816$ solutions, as expected.

2. Let $n = 92$; it has the divisors 1, 2, 4, 23, 46, and 92. The sum of the divisors not divisible by 4 is $1 + 2 + 23 + 46 = 72$, so that the number of solutions of (14) with $n = 92$ is $r_4(92) = 8 \cdot 72 = 576$. Indeed, $92 = 1^2 + 1^2 + 3^2 + 9^2 = 2^2 + 4^2 + 6^2 + 6^2 = 3^2 + 3^2 + 5^2 + 7^2$, and there are no other essentially different representations. Each of these three representations has only nonvanishing summands, with one summand repeated; hence each leads to 12 different permutations, each of which (by changes of signs) corresponds to $2^4 = 16$ different solutions of (14). Altogether we obtain $3 \cdot 12 \cdot 16 = 576$ solutions, as expected.

There exist several proofs of Theorem 10, some of which are technically elementary, but not really simple. One of them, of a rather computational nature, may be found in [14]. However a real insight into the problem is more likely to come from the theory of elliptic functions.

Let us observe that if we write $f(x) = \sum_{m=-\infty}^{\infty} x^{m^2}$, then

$$\{f(x)\}^4 = \sum_{\substack{m=-\infty \\ i=1,2,3,4}}^{\infty} x^{m_1^2 + m_2^2 + m_3^2 + m_4^2} = \sum_{n=0}^{\infty} a(n)x^n,$$

so that $a(n) = r_4(n)$. Now, $f(x) = 1 + 2x + 2x^4 + 2x^9 + \cdots + 2x^{m^2 + \cdots}$ is an instance of a *theta function*, which occurs in the theory of elliptic functions —we already met them in Chapter 7. In fact, $f(x)$ is the same function that occurs in Theorem 7.8 for $z = 1$. By use of the theory of theta functions, it is possible to obtain expressions for the coefficients $a(n) = r_4(n)$ of $\{f(x)\}^4$ and this yields the statement of Theorem 10, in an elegant and transparent way.

13   REPRESENTATIONS OF INTEGERS AS SUMS OF r SQUARES

It is clear that if (6) always has solutions for $Q = \sum_{i=1}^{r} x_i^r$, with $r = 4$, it has, *a fortiori*, solutions for $r > 4$, and this renders the first question of Section 7 trivial. On the other hand, it is of considerable interest to consider in these cases the second problem stated in Section 7, namely that of the *number* of solutions of the Diophantine equation (6), with $Q(x_1, \ldots, x_r) = \sum_{i=1}^{r} x_i^2$. For the small even values of $r = 2s$, such as $r = 6, 8, \ldots, 24$, methods based on the theory of elliptic (and more generally, modular) functions lead to formulae for $r_{2s}(n)$ somewhat similar to those of Theorem 10 (see [14] for more bibliographic and historic information).

A different and very powerful method has been introduced by Hardy, Ramanujan, and Littlewood, with later contributions by Rademacher, Vinogradov, Davenport, and others. It works equally well for odd numbers of squares. It is based on the evaluation of the coefficients $a_k(n)$ of $\{f(x)\}^k = \{\sum_{m=-\infty}^{\infty} x^m\}^k = \sum_{n=0}^{\infty} a_k(n)x^n$ by Cauchy's integral $a_k(n) = (2\pi i)^{-1}\int_{|z|=R}(\{f(z)\}^k/z)\,dz$ and the remark that $a_k(n) = r_k(n)$, the number of representations of n as a sum of k squares. While this idea is very appealing and seems to be simplicity itself, the effective computation of the integral, taken along a circle of radius R, is anything but simple. The interested reader will find an excellent exposition of this "circle method" in Davenport's book [8].

14   THE MORDELL-WEIL THEOREM

The case of Diophantine equations of first degree offered no serious difficulties. The case of second-degree Diophantine equations leads to the study of quadratic forms, and while some of them could be handled completely, the treatment of these equations in Sections 7–13 was far from complete, especially if more than two unknowns were to be determined.

It will not surprise the reader that for higher degrees the difficulties will be still greater. Nevertheless, at least the case of the Diophantine equations in two variables of genus one deserves a more thorough discussion here. Although Theorem 2 will not be proved, the reader will at least obtain a thorough understanding of its meaning, and one hopes, of its relevance.

Let us recall that in a (not necessarily abelian) group $\mathbf{G}$, we say that a (finite or infinite) set of elements, say $x_1, x_2, \ldots, x_n, \ldots$ forms a set of generators if every element $g \in \mathbf{G}$ can be written as a finite "word" in those generators, i.e., $g \in \mathbf{G} \Rightarrow g = x_{i_1} x_{i_2} \ldots x_{i_r}$. Here the subscripts are selected from among

$1, 2, \ldots, n, \ldots$, but are not necessarily distinct. Even if we collect identical adjacent generators and write them as powers, we obtain, say $g = x_{j_1}^{a_1} x_{j_2}^{a_2} \ldots x_{j_t}^{a_t}$ and $j_1, j_2, \ldots$ need not be all distinct, unless the given generators of $\mathbf{G}$ commute. It is obvious that if $\mathbf{G}$ is a finite group, then it also has a finite set of generators. In fact as we have not required that the generators be independent, we may select all elements of $\mathbf{G}$ as generators. However, it may well happen that even if $\mathbf{G}$ is infinite it still possesses a finite set of generators. This is possible for a nonabelian group $\mathbf{G}$ even if all generators are of finite order, as well as in the case of an infinite abelian group, provided that at least one of its generators is of infinite order.

Let us now assume that a Diophantine equation of the form $y^2 = Ax^3 + Bx^2 + Cx + D$ is given with A, B, C, D rationals, and we are interested in *rational* (not necessarily integral) solutions. It is clear that we may assume without loss of generality that the coefficients are actually integers. Indeed if m is a common multiple of the denominators of A, B, C, and D, we obtain, upon multiplication by m^2, that $(my)^2 = ax^3 + bx^2 + cx + d$, with $a, b, c, d \in \mathbf{Z}$. Finally, if we set $y_1 = my$, the equation has the original form with integral coefficients, and clearly y and y_1 are simultaneously rationals. Hence we shall consider from here on the Diophantine equation

$$y^2 = ax^3 + bx^2 + cx + d, \qquad a, b, c, d \in \mathbf{Z}, \tag{18}$$

and try to find rational solutions (all, if possible) of it.

Let us think now of the graph of (18). This is a cubic curve, which we denote by $\mathfrak{C}$. As we have already discussed linear and quadratic Diophantine equations, we shall assume that the graph of (18) does not split into that of a conic and a straight line, i.e., that the cubic is *irreducible*. As already observed in Section 4, in case $\mathfrak{C}$ has no cusps or multiple points, its genus is $g = \frac{1}{2}(3 - 2)(3 - 1) = 1$. As $g \geq 0$, it is clear that we can have at most one double point (or cusp), in which case $g = 0$.

Let us also suppose that we already happen to know two points, say P_1, P_2, of $\mathfrak{C}$ with rational coordinates. Then the chord $P_1 P_2$ cuts the cubic in exactly one more point, say P_3' (see Fig. 4). Let the equation of the line $P_1 P_2$ be

$$y = Ax + B, \qquad A, B \in \mathbf{Q}. \tag{19}$$

Indeed, as P_1 and P_2 have rational coordinates, the slope A is rational and so is $B = y_1 - Ax_1$. We obtain the coordinates of P_3' by replacing y in (18) by its value from (19). This leads to a third-degree equation in x with rational coefficients whose three roots are the x-coordinates of the three points of intersection of (19) with $\mathfrak{C}$. Of these we already know two roots, namely the x-coordinates of P_1 and P_2, say x_1 and x_2, both rational. If we divide out of the cubic equation in x the factors $x - x_1$ and $x - x_2$, we remain with a linear factor with rational coefficients; hence also P_3', the third point of intersection, has a rational abscissa. The y-coordinate of P_3', say y_3', is now obtained as a

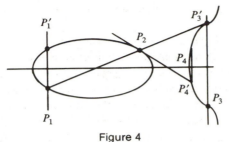

Figure 4

rational number from (19). However, with P_3', also P_3 of coordinates x_3, $y_3 = -y_3'$ belongs to $\mathfrak{C}$ as is obvious from (18), and it turns out that it is more convenient to consider P_3, rather than P_3', as the end result of our operation. In short, we may say that from the knowledge of two points P_1, P_2 with rational coordinates, we have obtained a new point P_3 on $\mathfrak{C}$, also with rational coordinates.

As a matter of fact, we don't even need two *distinct* points P_1 and P_2 for this construction. Indeed if P_2 coincides with P_1, then the secant P_1P_2 becomes the tangent to $\mathfrak{C}$ at P_2. If the coordinates x_2, y_2 of P_2 are rationals, then the equation of the tangent at P_2 has rational coefficients. As this tangent contains two (coincident) points of $\mathfrak{C}$, it cuts the curve in exactly one more point. We show as before that the coordinates of this new point are again rational. If these are x_4, y_4', then as before we observe that x_4, $y_4 = -y_4'$ are also the rational coordinates of a point P_4 on $\mathfrak{C}$, symmetric to P_4' with respect to the x axis.

As an example, let us consider the Diophantine equation

$$y^2 = x^3 + x + 6. \qquad (20)$$

We verify that $x = -1$, $y = 2$ is a solution. The slope of the tangent at the point $P_1(-1, 2)$ is computed easily: $2yy' = 3x^2 + 1$, $y' = (3x^2 + 1)/2y = (3(-1)^2 + 1)/2 \cdot 2 = 1$, so that the equation of the tangent is $y - 2 = x + 1$, or $y = x + 3$. If we substitute this in (20), we obtain the equation

$$x^3 - x^2 - 5x - 3 = 0.$$

We expect $x = -1$ to be a double root, and indeed $x^3 - x^2 - 5x - 3 = (x + 1)^2(x - 3)$. It follows that $x = 3$ is the abscissa of the new point, say P_2', of intersection of the curve $\mathfrak{C}$ given by (20), with the tangent $y = x + 3$ of $\mathfrak{C}$ at $P_1(-1, 2)$. The corresponding ordinate of P_2' is $y = x + 3 = 6$, and as indicated, we choose for P_2 the point of coordinates $(3, -6)$, also a solution of (20). We may now use P_1 and P_2 to obtain a third point on $\mathfrak{C}$, also with rational coordinates. The chord P_1P_2 has the equation $y = -2x$. Substituting this in (20), we obtain the equation $x^3 - 4x^2 + x + 6 = 0$, of which we know the two roots $x = -1$ and $x = 3$. The third root is $x = 2$ and $y_3' = -2x = -4$, so that

P_3 has the coordinates $(2, 4)$. We may now continue this process by using either chords or tangents.

By starting with a single point with rational coordinates of $\mathfrak{C}$, say P, we may continue this construction indefinitely. The outcome may be only one of the following two: After having found a certain number of points with rational coordinates, each chord through any two of them and each tangent to $\mathfrak{C}$ at each of them crosses $\mathfrak{C}$ again in one of the previously obtained points; or we can continue to obtain more and more new points with rational coordinates.

Before going any further, let us make one more preliminary remark. In Figure 4, if $P_1(x_1, y_1)$ is a point with rational coordinates on $\mathfrak{C}$, then the symmetric point $P_1'(x_1, -y_1)$ also has rational coordinates. If we perform the chord construction on (P_1, P_1'), we obtain as a new point the "point at infinity," say P_∞. For this reason, it is convenient to attribute to P_∞, obtained from a rational x, also a rational y-coordinate, although strictly speaking this statement has no concrete meaning. Here P_∞ is to be considered as the intersection of the cubic $\mathfrak{C}$ given by (18) or (20), with a line $x = x_0 \in \mathbf{Q}$. Having now adjoined P_∞ to the set of rational points on $\mathfrak{C}$, let us apply the chord process to P_∞ and P_1, say. Then the third point of intersection of (P_∞, P_1) is P_1', and we obtain as the result of the construction the "new" point P_1 again! Presumably by now the reader starts guessing why we define as the result of an operation (P_1, P_2) not P_3' but P_3, with $y_3 = -y_3'$.

The "chord and tangent process" can be formalized. If we start from P_1 and P_2 and obtain P_3, let us write $P_1 \oplus P_2 = P_3$. Clearly $P_2 \oplus P_1 = P_1 \oplus P_2 = P_3$ also. Furthermore, if we have a fourth point P_4 and consider $P_3 \oplus P_4 = P_5$, then it may be verified that also $P_1 \oplus (P_2 \oplus P_4) = P_5$. The operation $\oplus$ is therefore both associative and commutative. Also, $P_1 \oplus P_\infty = P_\infty \oplus P_1 = P_1$, so that the operation with P_∞ leaves the elements unchanged. Finally, $P_1 \oplus P_1' = P_\infty$, so that P_1 and P_1' may be considered as elements "inverse" to each other with respect to the "identity element" P_∞.

In the case of a tangent process, we may replace $P \oplus P$ by $2P$. By use of associativity, it is clear that if we start from a point P with rational coordinates, first with a tangent, then all points so obtained may be denoted by nP. We conclude that the rational points (i.e., points with rational coordinates) that we obtain when we start from any given point P with rational coordinates form a cyclic group, with P_∞ as the identity element and points P and P', with the same abscissa, as elements inverse to each other.

Even if this process goes on indefinitely and produces more and more new rational points on $\mathfrak{C}$, this does not mean that in this way we necessarily obtain *all* rational points on $\mathfrak{C}$. Let us assume that there exists on $\mathfrak{C}$ a rational point, say P_2, that we never reach if we start out with P_1. Then we may start the whole process again with P_2 and obtain another (in general infinite) cyclic group nP_2. If we select a point, say mP_1 of the first group, and a point nP_2 of the second one, we may perform the operation $mP_1 \oplus nP_2$ and obtain again a point with rational coordinates not obtained before. The identity element for both cyclic groups is of course the same P_∞. Starting from both P_1 and P_2, we observe that the points obtained form an abelian group that contains previously described cyclic groups as subgroups. In fact, the group so obtained is the direct product (or if you prefer, sum, as we write the groups additively) of those cyclic subgroups.

In general, there is no reason to expect that those cyclic groups are finite, and in general, they are not. Similarly, one may expect that in general the number of generators $P_1, P_2, \ldots$ need not be finite. Here we have, however, the rather unexpected theorem, already stated in Section 4 as:

Theorem 2 (Mordell [21], Weil [32]). *Under the operation of "chords and tangents" the rational points on a curve of genus 1 form a finitely generated abelian group.*

The neatest proof of this theorem may presumably be found in [32]—it uses elliptic functions. Under certain restrictions, there exists an entirely elementary proof by Chowla [7], but it takes 15 pages of computations. Mordell's original proof is also essentially elementary, but not very simple. As no really satisfactory proof seems to exist that covers the situation in sufficient generality and that is both elementary and transparent, one may suggest to the reader anxious to see a proof that he learn the elements of the theory of elliptic functions and then read [32].

15 BACHET'S EQUATION

One of the simplest particular cases of Diophantine equations that correspond to curves of genus 1 is the equation

$$y^2 = x^3 + k, \tag{21}$$

often called *Bachet's equation.*

Everything said about it in Section 14 applies, but here we shall be interested not in rational but in integral solutions. According to Siegel's Theorem 1, we expect only a finite number of integral solutions, i.e., of points with integral coordinates on the curve represented by (21). Evidently, for $k = 0$ the curve has a cusp at the origin, so that it is of genus zero, and indeed we see that for every $n \in \mathbf{Z}$ the formulae $x = n^2$, $y = \pm n^3$ yield an infinite set of integral points. For $k \neq 0$, however, the genus is $g = 1$ and the set of integral solutions is finite. In fact, for many values of k there are no integral solutions at all. In many cases, this can be shown by simple considerations of congruences. In other cases, one has to appeal to all the resources of ideal theory.

Let us consider, e.g., equation (21) with $k = 7$. If x is even, then (21) shows that $y^2 \equiv 7 \pmod 8$, which is impossible. Similarly, $x \equiv 3 \pmod 4$ yields

$y^2 \equiv 3 + 7 \equiv 2 \pmod 8$, equally impossible. Hence any solution of (21) with $k = 7$ must have $x \equiv 1 \pmod 4$. Let us assume then that this is the case, and add 1 on both sides of (21); we obtain $y^2 + 1 = x^3 + 8 = (x + 2)(x^2 - 2x + 4)$. However, for $x \equiv 1 \pmod 4$, $x^2 - 2x + 4 \equiv 3 \pmod 4$. Consequently, there exists a prime $p \equiv 3 \pmod 4$ which divides $x^2 - 2x + 4$ (a product of primes $p \equiv 1 \pmod 4$ is itself congruent to 1 $\pmod 4$), and hence p also divides $y^2 + 1$. That means $y^2 \equiv -1 \pmod{: p}$ or -1 is a quadratic residue modulo p. This is not possible for $p \equiv 3 \pmod 4$, as we know from Theorem 5.2, and hence there are no solutions to (21) with $k = 7$.

The same reasoning holds for $k = 11$ (to test $x \equiv 1 \pmod 4$, add 16 on both sides of the equation), for $k = 23$ (here you add 4), etc.

It is not difficult to verify that (21) can have no solutions with $k \equiv 3 \pmod 4$ and either even x (because this would lead to $y^2 \equiv 3 \pmod 4$) or $x \equiv -1 \pmod 4$ (which would lead to $y^2 \equiv 2 \pmod 4$). To show that (21) cannot hold for $x \equiv 1 \pmod 4$ either, we try to find an integer y_1 such that $k + y_1^2 = x_1^3$ with $x_1 \equiv 2$ or $x_1 \equiv 3 \pmod 4$; i.e., we attempt to solve

$$y_1^2 = x_1^3 - k \tag{21'}$$

with $x_1 \equiv 2$ or $x_1 \equiv 3 \pmod 4$. Clearly (21') is entirely analogous to (21) and is in principle no easier to solve. On the other hand, all we need is a *single* solution, and for many values of k not too large, we can find such x_1, y_1 by inspection. Once we have found them, we add (21') to (21) and obtain

$$y^2 + y_1^2 = x^3 + x_1^3 = (x + x_1)(x^2 - xx_1 + x_1^2).$$

For $x \equiv 1 \pmod 4$ and $x_1 \equiv 2$ or 3 $\pmod 4$, $x^2 - xx_1 + x_1^2 \equiv 1 - x_1 + x_1^2 \equiv 3 \pmod 4$, and there exists a prime $p \equiv 3 \pmod 4$, $p | x^2 - xx_1 + x_1^2$, $p | y^2 + y_1^2$, so that $y^2 \equiv -y_1^2 \pmod p$, or if $p \nmid y_1$, $(-y_1^2/p) = (-1/p) = +1$, which contradicts $p \equiv 3 \pmod 4$. It only remains to check, for the finitely many primes $p | y_1$, that $y = py_1$ also does not lead to any solutions.

As an example of the last possibility, let us consider the equation

$$y^2 = x^3 + 47.$$

If we add on both sides $y_1^2 = 13^2 = 169$, we obtain

$$y^2 + 13^2 = x^3 + 216 = (x + 6)(x^2 - 6x + 36).$$

As before, the possibilities $x \equiv 0, 2, 3 \pmod 4$ are eliminated by $y^2 \equiv 3$ or 2 $\pmod 4$, and for $x \equiv 1 \pmod 4$, $x^2 - 6x + 36 \equiv 3 \pmod 4$. Hence $p | x^2 - 6x + 36$ for some $p \equiv 3 \pmod 4$ and $y^2 \equiv -13^2 \pmod p$. If $13 \nmid y$, then $y^2 \equiv -1 \pmod 4$, $(-1/p) = +1$, which is false for $p \equiv 3 \pmod 4$. If $13 | y$, so that $y = 13 y'$, we obtain $(13 y')^2 \equiv -13^2 \pmod p$, which leads to either $p = 13$, false by $3 \equiv p = 13 \equiv 1 \pmod 4$, or $(-1/p) = +1$, false by $p \equiv 3 \pmod 4$.

Often such simple approaches do not work. Sometimes we do not find an appropriate y_1^2 to add on both sides of (21), but we find that by *subtracting* a square we obtain $k - y_1^2 = x_1^3$, a perfect cube, and variations of the previous approach succeed.

A more general method, however, that permits us not only to rule out solutions for certain values of k but also to solve (21) for some values of k for which solutions exist, will now be sketched.

Equation (21) may be written as $y^2 - k = x^3$. If $k = a^2$, a perfect square, then there always exist at least the two solutions $x = 0$, $y = \pm a$. Others may also exist, as, e.g., for $k = 3^2$ the solutions $x = 3$, $y = \pm 6$. (Indeed $6^2 = 36 = 3^3 + 3^2$. What else can you say about this equation and its solutions?) More interesting, however, is the case $k \neq a^2$, and for simplicity of the presentation in this sketch, we shall in fact consider only the case in which $|k|$ is even assumed squarefree and $k \not\equiv 1 \pmod 4$. From these assumptions it follows in particular that $k \equiv 2$ or $\equiv 3 \pmod 4$. This in turn implies for any solution (x, y) of (21) that x is odd; otherwise $y^2 \equiv 2$ or $3 \pmod 4$, which is impossible. We also now have $(x, k) = 1$; otherwise, $p|x$, $p|k \Rightarrow p|y \Rightarrow p^2|y^2$, $p^3|x^3$, and hence $p^2|k$, contradicting the assumption that k is squarefree.

Let us now factor $y^2 - k$ in $\mathbf{K} = \mathbf{Q}(\sqrt{k})$. We obtain $(y - \sqrt{k})(y + \sqrt{k}) = x^3$. The simplest situation is of course that in which the ring $\mathbf{I}$ of integers of $\mathbf{K}$ is a domain with uniqueness of factorization. We also verify that the two factors $y - \sqrt{k}$ and $y + \sqrt{k}$ are coprime. Indeed, if δ is their greatest common divisor in $\mathbf{K}$, then $\delta|2\sqrt{k}$, $\delta|x$; as x is odd, $\delta|\sqrt{k}$, so that $\delta = \sqrt{c}$ with $c|k$. For $1 < c < k$, however, $\sqrt{c} \notin \mathbf{I}$, so that $\delta = 1$ or $\delta = \sqrt{k}$. If $\delta = \sqrt{k}$, then $\delta^2 = k$ is a divisor of x^3, contrary to $(x, k) = 1$, and so $\delta = 1$. It now follows from the uniqueness of factorization that $y + \sqrt{k} = \alpha^3$, $\alpha \in \mathbf{I}$. As $k \equiv 2$ or $3 \pmod 4$, $\alpha = a + b\sqrt{k}$, $a, b \in \mathbf{Z}$, and $y - \sqrt{k} = \bar{\alpha}^3 = (a - b\sqrt{k})^3$; hence $y^2 - k = x^3 = (a^2 - b^2k)^3$ and $x = \varepsilon(a^2 - b^2k)$, ε a unit of $\mathbf{I}$. Indeed now we have to start taking into account the units of $\mathbf{I}$. If $k < 0$ this is easy, at least if we may ignore the case $k = -1$. This we may do, because for $k = -1$, (21) reduces to a particular case of Catalan's equation, and we know its solutions from Section 3.

If $-1 \neq k < 0$, $k \equiv 2$ or $3 \pmod 4$, then $\mathbf{I}$ contains only the units ± 1. Consequently, we obtain from $y + \sqrt{k} = a^3 + 3ab^2k + \sqrt{k}(3a^2b + kb^3)$ that $y = a(a^2 + 3b^2k)$ and $1 = (3a^2 + b^2k)b$. The last equation shows that $b = \pm 1$. According to the sign of b, $3a^2 = -k \pm 1$ or $a = \pm\sqrt{(-k \pm 1)/3}$. At most one of the two signs under the radical can lead to an integral value for a; if none does, then there are no solutions. If one is an integer, then $y = \pm|a|(a^2 + 3k)$ is also integral. Finally, $x = a^2 - k = -(4k \pm 1)/3$ is also (for the proper choice of sign) an integer, and we have found the two solutions of (21)

$$x_0 = -(4k \pm 1)/3, \qquad y_0 = \pm\sqrt{(-k \pm 1)/3}\,\{(8k \pm 1)/3\} \qquad (22)$$

(for only one choice of the sign of ± 1). There are no other solutions.

We now remember (see Chapter 10) that the only imaginary quadratic fields with uniqueness of factorization are $\mathbf{Q}(\sqrt{k})$ with $k = -1, -2, -3, -7, -11, -19, -43, -67, -163$, and all except $k = -2$ fail to satisfy the conditions on k. For $k = -2$, we find $a = \pm 1$, $x_0 = 3$, $y_0 = \pm 5$, and indeed $y^2 = x^3 - 2$ is satisfied by these values, as $(\pm 5)^2 = 3^3 - 2$.

For our rather complicated procedure, we have found, so far, only one single application. Therefore it is of interest to observe that the method succeeds, in fact, for all imaginary quadratic fields $\mathbf{K} = \mathbf{Q}(\sqrt{k})$, $k < 0$, $k \equiv 2, 3 \pmod 4$, k squarefree, unless the class number $h \equiv 0 \pmod 3$. Indeed we proceed exactly as before, except that now we have to factor into ideal factors:

$$(y - \sqrt{k}) \cdot (y + \sqrt{k}) = (x^3) = (x)^3.$$

We show, essentially as before, that the two ideal factors on the left are coprime, and conclude that the principal ideal $(y + \sqrt{k}) = \mathfrak{a}^3$, $\mathfrak{a} \subset \mathbf{I}$. We now apply Corollary 11.32.1 (it is here and only here that we use the assumption $3 \nmid h$) and conclude that $\mathfrak{a} = (a + b\sqrt{k})$, a principal ideal. As before, the only units in $\mathbf{K}$ are ± 1, so that now $y = \pm a(a^2 + 3b^2 k)$, $1 = \pm(3a^2 + b^2 k)b$. From here on, the reasoning continues essentially as before, with some care for signs and with the same conclusion.

EXAMPLES.

1. Let $k = -57$; then $h(\mathbf{Q}(\sqrt{-57})) = 4$, $-57 \equiv 3 \pmod 4$, so that the results obtained apply. Also, $(\pm 1 - k)/3 \notin \mathbf{Z}$; hence (21) has no solutions for $k = -57$.

2. Let $k = -109 (\equiv 3 \pmod 4)$, $3 \nmid h(\mathbf{Q}(\sqrt{-109})) = 8$, and we obtain by (22) $y = \pm 6(6^2 - 3 \cdot 109) = \pm 1746$ and $x = 6^2 - (-109) = 145$. We verify that these values are indeed solutions of the equation $y^2 = x^3 - 109$, because $1746^2 = 3,048,516 = 145^3 - 109$.

Essentially the same procedures also work for $k > 0$. There is, however, one essential additional difficulty, in that now $\mathbf{I}$ has a fundamental unit ε and contains all its positive and negative powers. The way to treat this case will not be discussed here, but the interested reader will find a very clear and thorough presentation in [1].

Even the present approach, which made use of the theory of ideals, yielded results at best only if $k \equiv 2$ or $3 \pmod 4$ and was squarefree (some of the cases with $h(K) \equiv 0 \pmod 3$ or with k not squarefree can still be handled by this method, or variations of it), i.e., in less than half the cases.

The complete solution (at least theoretically) of equation (21) came only in 1968 through the work of A. Baker, already mentioned in Section 3. Indeed in [4], Baker proved that for all $0 \neq k \in \mathbf{Z}$, if (x, y) is a solution of (21), then $\max(|x|, |y|) < \exp(10^{10}|k|)^{10^4}$. Hence at least in principle, all solutions of (21) for all values of k can be found with a finite amount of computation. While the bound appears so large as to discourage any attempt to effectively compute all solutions of (21) (and this indeed may well be impossible at present for large $|k|$), it turns out that an even larger bound, valid for a more general situation, permitted Tijdeman to solve among others, Catalan's equation, where fortunately $|k| = 1$ (see Section 3).

16 FINAL REMARKS

As already observed, on the one hand the Diophantine equations that can be studied at present are very numerous, and on the other hand, there is no hope of ever finding a uniform method that should permit us to handle all of them. For that reason, instead of listing more and more equations with indications of what is known about them we shall return to the one most famous of them all, to the equation already mentioned in Chapter 1 and alluded to more than once in these pages, i.e., to Fermat's equation. The next (and last) chapter of this book is dedicated to its study.

PROBLEMS

1. The *Fibonacci numbers* were defined by the recurrence relation $F_n = F_{n-1} + F_{n-2}$ with $F_1 = F_2 = 1$.
   (a) Show that $F_n = \dfrac{1}{\sqrt{5}}\left\{\left(\dfrac{1 + \sqrt{5}}{2}\right)^n - \left(\dfrac{1 - \sqrt{5}}{2}\right)^n\right\}$.
   (b) Show that the formula given for F_n in (a) represents an integer for all n. (Hint: Check for $n = 0$ and $n = 1$ and use induction on n.)

2. Prove that consecutive Fibonacci numbers are coprime.

3. Show that if $n|m$, then $F_n|F_m$.

4. Find all rational points (i.e., points with rational coordinates) on the conic section

   $$y^2 = x(x + 1).$$

5. Find all rational points on the cubic $y^2 = x^2(x + 1)$.

6. Find all rational points on the cubic $(y - 1)^2 = (x + 2)^3$.

7. Consider the second degree surface

   $$x^2 + 2y^2 + 3z^2 + xy - yz - 2x - 6y - 6z + 1 = 0.$$

   Make a change of variables and reduce the equation to the form $Q(x_1, y_1, z_1) = c$, where $Q(x_1, y_1, z_1)$ is a (not necessarily diagonal) form in x_1, y_1, z_1 and c is a constant.

8. Find the number of solutions of $x^2 + y^2 = n$ for all integers n, with $1 \leq n \leq 8$ and also for $n = 18$. In each case determine the number of essentially distinct solutions as well as the total number (solutions obtained by permutations of the summands or by changes of signs are counted as distinct). Compare with Theorem 4.

9. Find the number of representations of n as a sum of two squares for $n = 100$, 101, and 102; how many essentially distinct representations are there in each case and how many different representations correspond to each essentially distinct one?

10. Determine directly the number of representations as a sum of three squares of $n = 8$, 19, 20, 29, 31, and 33; in each case indicate the number of representations that correspond to each essentially distinct one. Check, whenever appropriate, against Theorem 7.

11. Consider the following equations:

$$\text{(i)} \quad 3x^2 - 5y^2 + 7z^2 = 0;$$

$$\text{(ii)} \quad x^2 + 2y^2 + 3z^2 = 0;$$

$$\text{(iii)} \quad -x^2 + y^2 - 3z^2 = 0.$$

(a) By use of Theorem 8, indicate which ones do and which ones don't have solutions.

(b) Solve those equations that have nontrivial solutions.

12. Show that if the conditions of Theorem 8 are satisfied, then (9) is solvable as a congruence modulo all prime powers.

13. Find a solution of $2x + 3y + 4z = 0 \pmod{24}$, with $|x| \leq 3, |y| \leq 4$, $|z| \leq 2$. (Observe: the product of the bounds equals 24.)

14. Verify Theorem 10 by computing directly $r_4(n)$ for $n = 7$, 10, 20, and 25.

15. Find $r_4(10)$ by use of the function $f(x) = 1 + 2\sum_{n=1}^{\infty} x^{n^2} = \sum_{-\infty}^{\infty} x^{n^2}$.

16. Consider the cubic $y^2 = x^3 + x - 1$. Find by inspection a point with rational coordinates, start with a "tangent process" to find another rational point, and continue with either tangents or chords to find at least 5 rational points.

17. Study the following Diophantine equations and either solve them (in integers, naturally!) or show that they have no solutions.

$$\text{(i)} \quad y^2 = x^3 + 11;$$

$$\text{(ii)} \quad x^3 - 3x^2 - y^2 + 3x - 2y + 21 = 0.$$

BIBLIOGRAPHY

1. W. W. Adams and L. J. Goldstein, *Introduction to Number Theory*. Englewood Cliffs, N.J.: Prentice-Hall, 1976.

2. P. Bachmann, *Die Arithmetik von quadratischen Formen*. Leipzig, 1898.

3. A. Baker, (a) *Mathematika* 13(1966) 204–216; *ibid.* 14(1967) 102–107, 220–228; 15(1968) 204–216. (b) *Annals of Mathematics* (2) 94(1971) 139–152.

4. A. Baker, *Philosphical Transactions of the Royal Soc. of London*, Ser. A 263(1968) 173–208.

5. P. T. Bateman, *Transactions of the American Mathem. Soc.* 71(1951) 70–101.

6. J. W. S. Cassels and M. J. T. Guy, *Mathematika* 13(1966) 111–120.

7. S. Chowla, *The Riemann Hypothesis and Hilbert's Tenth Problem*. London: Blackie, 1965.

8. H. Davenport, *Analytic Methods for Diophantine Equations and Diophantine Inequalities*. University of Michigan, 1962; Ann Arbor, Mich.: Ann Arbor Publishers, 1963.

9. M. Davis, *Amer. Mathem. Monthly* 80(1973) 233–269.

10. M. Davis, Y. Matijasevič, and J. Robinson, *Proc. Symposia in Pure Mathem.*, Amer. Mathem. Soc. 28(1976) 323–378.

11. M. Davis, H. Putnam, and J. Robinson, *Annals of Math.* 74(1961) 425–436.

12. M. Deuring, *Inventiones Mathem.* 5(1968) 169–179.

13. C. F. Gauss, *Disquisitiones Arithmeticae*, Translated from the second Latin edition of 1870 by A. A. Clarke. New Haven Conn.: Yale Univ. Press, 1966.

14. G. H. Hardy and E. M. Wright, *An Introduction to the Theory of Numbers*, 3rd ed. Oxford: Clarendon Press, 1954.

15. K. Heegner, *Mathematische Zeitschrift* 56(1952) 227–253.

16. E. Landau, *Vorlesungen über Zahlentheorie*. Leipzig: S. Hirzel, 1927.

17. A. M. Legendre, *Essai sur la Théorie des Nombres*. Paris: Duprat, 1808. (Legendre gives credit for many results to Sophie Germaine.)

18. D. Lewis, *Report on the Institute in the Theory of Numbers*, Univ. of Colorado, Boulder, Colorado, 1959.

19. W. Ljunggren, *Acta Arithmetica* 8(1962/63) 451–463.

20. Y. Matijasevič, *Doklady Akad. Nauk SSSR* 19(1970) 279–282.

21. L. J. Mordell, *Proc. Cambridge Philos. Soc.*, 21(1922) 179–192.

22. L. J. Mordell, *Diophantine Equations*. London: Academic Press, 1969.

23. I. Niven and H. S. Zuckerman, *The Theory of Numbers*, 4th ed. New York: Wiley, 1980.

24. O. T. O'Meara, *Introduction to Quadratic Forms* (Grundlehren der Math. Wiss, vol. 117). Berlin: Springer, 1963.

25. S. Ramanujan, *Collected Papers*, edited by G. H. Hardy, P. V. S. Aiyar, and B. M. Wilson. Cambridge: Cambridge University Press, 1927.

26. K. F. Roth, *Mathematika* 2(1955) 1–20; corrigendum p. 168.

27. C. L. Siegel, *Abhandlungen Preuss. Akad. Wiss.*—Phys.-Mathem. Klasse (1929) No. 1, *Ges. Abh.* vol. 1, pp. 209–266, Springer 1966.

28. C. L. Siegel, *Inventiones Mathematicae* 5(1968) 180–191; *Ges. Abh.* vol. 4, pp. 41–52, Springer 1979.

29. T. Skolem, *Diophantische Gleichungen* (Ergebnisse der Mathem. u.i. Grenzgebiete, vol. 5). Berlin: Springer, 1938.

30. H. Stark, (a) *Michigan Mathem. Journal* 14(1967) 1–27. (b) *Mathem. of Computation* 29(1975) 289–302.

31. R. Tijdeman, *Acta Arithmetica* 29(1976) 197–208.

32. A. Weil, (a) *Acta Mathematica* 52(1928) 281–315; also *Collected Works*, vol. 1, pp. 11–45. (b) *Bull. Sc. Mathem.* (II) 54(1929) 182–191; also *Collected Works*, vol. 1, pp. 47–56.

33. A. Weil, *Bull. Amer. Math. Soc.* 55(1949) 497–508; *Coll. Works*, vol. 1, pp. 399–410.

Chapter 14

Fermat's Equation

1 INTRODUCTION

Of all Diophantine equations, by far the most famous is Fermat's equation

$$x^n + y^n = z^n. \tag{1}$$

The case $n = 2$ had already been completely understood during the Greek antiquity (our Theorem 1 is due to Diophantus himself—weaker results were known long before, possibly even by Pythagoras), but it was not until some 1400 years later that the next progress was made by Fermat, Leibniz (1646–1716), and Euler, who gave independent proofs of our Theorem 2 and Corollary 2.1, stating that (1) has no solutions with $x, y, z \in \mathbf{Z}$, $x \cdot y \cdot z \neq 0$, for $n = 4$.

It has often been told, but bears repeating, that on Fermat's copy of Diophantus' work (edited by Bachet) one finds a marginal note (presumably from 1637) to the effect that Fermat had found a "truly marvelous proof" of the statement, "(1) is not solvable in non-vanishing integers x, y, z for any integral $n \geq 3$." He added that the proof was too long for insertion in the free space available on that page of Diophantus. We shall refer to this statement as the *Fermat Conjecture*, or the FC.

Since the 17th century, many among the foremost mathematicians have tried in vain to reconstruct the proof that Fermat claimed to possess (or to find another one). The likelihood that Fermat really had a proof may be a tantalizing—but hardly profitable—subject for speculation; those interested in it may want to consult Mordell's beautiful booklet [19].

If the exponent $n > 2$ is not a prime, then it is either a power of 2 or else it is divisible by some odd prime p. In the first case, $n = 4k$ and (1) may be written as $(x^k)^4 + (y^k)^4 = (z^k)^4$. As already mentioned, we have a proof going back to Fermat himself of the fact that the sum of two fourth powers cannot be a fourth power (actually, it cannot even be a perfect square, as we shall see). In the second case $n = pk$, and (1) becomes $(x^k)^p + (y^k)^p = (z^k)^p$. Hence to prove that (1) is not solvable for arbitrary integral powers n, it is sufficient to prove that it is not solvable when $n = p$, an odd prime.

We can simplify the problem still further by observing that if x, y, z are integers satisfying (1) and any two of them are divisible by an integer d, then d also divides the third one, and writing $x = dx_1, y = dy_1, t = dz_1$, (1) shows that also $x_1^n + y_1^n = z_1^n$. Hence it is sufficient to look only for solutions of (1) in integers x, y, z that are coprime in pairs; such solutions are called *primitive solutions*. Finally, in order to obtain a more symmetric formulation, we observe that if p is an odd prime $(-z)^p = -z^p$. This leads us to reformulate the problem as follows:

To prove that if p is an odd prime, then

$$x^p + y^p + z^p = 0 \tag{1'}$$

has no solutions in rational integers x, y, z which are pairwise coprime and with $x \cdot y \cdot z \neq 0$.

It will soon turn out that it is convenient to distinguish between the following two cases

Case I: $p \nmid x \cdot y \cdot z$;

and

Case II: $p \mid x \cdot y \cdot z$.

From $(x, y) = (y, z) = (z, x) = 1$ it follows that in Case II, p divides exactly one of the three integers x, y, or z. It also will appear soon that Case I is the much easier one to deal with—and we can dispose of it almost trivially for small primes.

Not surprisingly, the first case considered successfully was $p = 3$. Incorrect proofs of the insolvability of (1') for $p = 3$ seem to have been proposed already before 1000 A.D. (see [4], Chapter XXI). The first essentially correct (although incomplete) proof for $p = 3$ is due to Euler (1753), while the first complete proof is due to Legendre (after 1800; see [13]). Using ideas similar to those which worked for $p = 3$, Legendre also disposed of the case $p = 5$ (in 1823). This result was also obtained by Dirichlet at almost the same time (see [5]). After that, several other particular cases could be settled (recall Section 10.11), but at the price of increasingly complicated reasonings, and it became clear that different methods were called for if the general case was to be settled.

About 1843, Kummer may have believed he had the proof of the FC in the general case—but Dirichlet, no newcomer to this problem, observed that at one point the argument had a gap, essentially the same one as in Euler's proof for $p = 3$. As we know by now from Chapter 10, the difficulty consists in the fact that the cyclotomic fields in which one is led to operate are not domains of uniqueness of factorization, if $p > 19$. By use of ideals Kummer was able to prove the FC for a large number of primes p—but not for all, not even for infinitely many! Indeed, the machinery needed to deal with the problem increased its complexity and new conditions had to be satisfied. In particular, in order to permit the reasoning to go through, a special condition comes up, which Kummer called "regularity" of the prime (see Section 7). While it is not easy to decide directly the regularity of a given prime p, Kummer also discovered a simple criterion of regularity involving a curious relation between this property and the numerators of the Bernoulli numbers, which were defined in Section 8.7. Other necessary conditions for the solvability of (1') were discovered, and outstanding contributions to the problem were made by Wieferich [27], Mirimanoff [16] and [17], Libri [15], Sophie Germain [13], Dickson [3], Vandiver [23], Hecke [10], and many others. In the last few years, high-speed computers were also used, and by a combination of deep theoretical work, ingenuity, and computers, Rosser (see [22]) and the Lehmers (see [14]) showed that the FC holds in Case I at least up to $p = 253{,}747{,}889$. More recent work by Morishima and Gunderson has raised that bound to 3.10^9, or even to 57.10^9 (see [21] for details), while as already mentioned, Wagstaff [24] proved the FC in Case II (i.e., unconditionally) at least up to $p < 125{,}000$. Recently the already immense literature on Fermat's Conjecture has received two particularly valuable additions. Although their subject matter is ostensibly the same, there is hardly any overlap between them. Both books, by Edwards [6] and by Ribenboim [21], have rapidly become indispensable to any serious student of this topic. Also a third book, by J. M. Gandhi, again unlike either of the two mentioned, has been announced but as of this writing, is not yet available.

In Section 2 we shall easily solve (1) for $n = 2$. In Section 3 we shall (also easily) dispose of the case $n = 4$. All this material was well known before ideals were introduced by Kummer. Also the case $p = 3$, the object of Section 4, can be treated completely without the use of ideals, and had, in fact, been settled by Euler and Legendre, as already mentioned. The last case that will be handled here by essentially elementary methods, namely Case I for $p = 5$, is discussed in Section 5. For general p, Case I is treated as far as this is possible without the theory of ideals in Section 6. There we shall arrive at a dead end. To overcome the basic difficulties of the situation, regular primes are introduced in Section 7, and then by use of the powerful tool of ideal theory, the two cases are taken up again separately and solved, at least for regular primes, in Sections 8 and 9, respectively.

2   SOLUTION OF FERMAT'S EQUATION FOR $n = 2$

Clearly, (1) is not solvable in Case I, because if x, y, z are all odd, $x^2 + y^2$ is even while z^2 is odd. Considering next Case II, $2 | x \cdot y \cdot z$, but $(x, y) =$

$(y, z) = (z, x) = 1$, because we are only interested in primitive solutions; this implies that at most one (hence exactly one) of the three integers x, y, z is even, the other two being odd. Also, if $x = 2m + 1$, $y = 2n + 1$, $z = 2k$, then $x^2 + y^2 \equiv 2 \pmod 4$, $z^2 \equiv 0 \pmod 4$, which is impossible. Therefore, the even one is either x or y, and without loss of generality we may assume that $x = 2q$, y and z being odd. This makes $z + y$ and $z - y$ even, say $z + y = 2m$, $z - y = 2n$. Equation (1) may now be written as $x^2 = z^2 - y^2 = (z + y)(z - y)$ or $4q^2 = 2m \cdot 2n$, so that $q^2 = m \cdot n$. The integers $m = \frac{1}{2}(z + y)$ and $n = \frac{1}{2}(z - y)$ are coprime; otherwise if $(m, n) = d > 1$, then $d \,|\, m + n = z$, $d \,|\, m - n = y$, contrary to the assumption $(y, z) = 1$. The integers m and n are coprime and their product is a perfect square; hence each of them is a perfect square (the reader is invited to prove this), $m = a^2$, $n = b^2$. From $(m, n) = 1$ follows of course that also $(a, b) = 1$. We have obtained so far that $z = m + n = a^2 + b^2$, $y = m - n = a^2 - b^2$ and $x^2 = 4q^2 = 4mn = 4a^2b^2$ or $x = 2ab$, with $(a, b) = 1$. From $(a, b) = 1$ it is clear that a and b cannot both be even, but they cannot both be odd either, because a and b odd would force y and z to be even, which is false. Hence exactly one of a and b is even, and the other is odd. We therefore proved: Every primitive solution of (1) with $n = 2$ is necessarily of the form $x = 2ab$, $y = a^2 - b^2$, $z = a^2 + b^2$, with $(a, b) = 1$, a, b of opposite parity, except for the possibility of interchanging x and y. Conversely, if x, y, z are of this form, one obviously has $x^2 + y^2 = 4a^2b^2 + (a^2 - b^2)^2 = a^4 + 2a^2b^2 + b^4 = (a^2 + b^2)^2 = z^2$, and (1) holds. This completes the proof of

Theorem 1. *All primitive solutions of equation* (1) *with* $n = 2$ *are of the form* $x = 2ab$, $y = a^2 - b^2$, $z = a^2 + b^2$ *or* $x = a^2 - b^2$, $y = 2ab$, $z = a^2 + b^2$ *with* $(a, b) = 1$, *and exactly one of* a, b *even.*

While it is true that the FC has been neither proven nor disproven, it should be observed that $n = 2$ is the only exponent for which solutions are known.

3　PROOF OF THE FC FOR $n = 4$

Fermat proved a statement which is slightly stronger than, and implies the truth of the FC for $n = 4$, namely:

Theorem 2. *The Diophantine equation*

$$x^4 + y^4 = z^2 \tag{2}$$

has no solution in integers.

The proof of Fermat is based on the "method of descent," which he invented and also used successfully in other problems. To disprove the existence of solutions we assume the contrary, namely that a solution exists; next we show that from a given solution we can construct another one with a smaller value of some integral parameter, which, however, has to stay positive. This, of course, is a self-contradictory statement, because starting from a given positive value, in a finite number of descending steps, every integral-valued parameter reaches the value zero or becomes negative. This proves the desired result, namely that it was false in the first place to assume the existence of a solution.

Before proving Theorem 2, let us observe that it implies:

Theorem 2'.   *Equation* (1) *has no solutions for* $n = 4$.

PROOF.   If $x^4 + y^4 = z^4$, then setting $z^2 = z_1$, $x^4 + y^4 = z_1^2$, contrary to Theorem 2.

PROOF OF THEOREM 2.   As already observed, it is sufficient to find a primitive solution, that is a solution with x, y, z coprime in pairs; this will be tacitly understood throughout this chapter, without further mention.

Let us then assume that there exists a primitive solution x, y, z of (2). As in the case $n = 2$, either x or y (but not both) must be even, and z is odd. Let x be even. From Section 3 it follows that if $(x^2)^2 + (y^2)^2 = z^2$ and x is even, then $x^2 = 2ab$, $y^2 = a^2 - b^2$, $z = a^2 + b^2$ with $(a, b) = 1$, a, b of opposite parity. Actually, here it is easy to decide that a is odd and b even, which leads to $y^2 = a^2 - b^2 \equiv 1 \pmod 4$, while a even, b odd would lead to $y^2 \equiv 3 \pmod 4$, which is impossible. Hence set $b = 2c$, $(a, c) = 1$ so that $x^2 = 4ac$, and as in Section 3 we conclude that $a = z_1^2$, $c = e^2$, $(z_1, e) = 1$. It follows that $y^2 = a^2 - b^2 = z_1^4 - 4e^4$ or $(2e^2)^2 + y^2 = (z_1^2)^2$, with $2e$, y, z_1 coprime in pairs. This equation is of the same form as (1) with $n = 2$; hence according to Theorem 1, $2e^2 = 2ml$, $z_1^2 = m^2 + l^2$, $y = m^2 - l^2$, $(m, l) = 1$, m, l of opposite parity. From $e^2 = ml$ and $(m, l) = 1$ follows that $l = x_1^2$, $m = y_1^2$, so that $z_1^2 = l^2 + m^2 = x_1^4 + y_1^4$. Consequently, assuming that a primitive solution x, y, z of (2) exists, we have constructed another primitive solution x_1, y_1, z_1. (Why is x_1, y_1, z_1 a *primitive* solution?) But clearly $z > a^2 = z_1^4 \geq z_1$; hence from any given primitive solution x, y, z we can construct another one, x_1, y_1, z_1 with $z_1 < z$. After a finite number of steps, $z_1 \leq 0$; yet $z_1^2 = x_1^4 + y_1^4 > 0$, unless $x_1 = y_1 = 0$, which is contrary to, say $(x_1, y_1) = 1$. This contradiction finishes the proof of Theorem 2.

4   PROOF OF THE FC FOR $p = 3$

Instead of proving the FC for $p = 3$ in the way we formulate it in (1'), it turns out that it is actually easier to prove a stronger statement, namely:

Theorem 3.   *Let* $\omega = e^{2\pi i/3} = \frac{1}{2}(-1 + i\sqrt{3})$ *and consider the field* $\mathbf{K} = \mathbf{Q}(\omega)$ $= \{A + B\omega | A, B \in \mathbf{Q}\}$, *with the ring of integers* $\mathbf{I} = \mathbf{I}(\omega) = \{a + b\omega | a, b \in$

Z}. *Then the equation*

$$\xi^3 + \eta^3 + \zeta^3 = 0 \tag{3}$$

has no solution in integers $\xi, \eta, \zeta \in \mathbf{I}$ *with* $\xi \cdot \eta \cdot \zeta \neq 0$.

K is a cyclotomic field because $\omega = e^{2\pi i/3}$, so that $\omega^3 = e^{2\pi i} = 1$. It is a field of degree 2. Indeed ω satisfies the equation $x^3 - 1 = (x-1)(x^2 + x + 1) = 0$; but $\omega \neq 1$, so that ω is one of the roots of the quadratic equation $\omega^2 + \omega + 1 = 0$, the other root being its complex conjugate $\bar{\omega} = \frac{1}{2}(-1 - i\sqrt{3})$. One also observes that $\omega^2 = \bar{\omega} = -(1 + \omega)$.

We recall from Chapter 10 that the norm of $\alpha = a + b\omega$ is $N\alpha = (a + b\omega)(a + b\bar{\omega}) = a^2 + ab(\omega + \bar{\omega}) + b^2\omega\bar{\omega} = a^2 - ab + b^2$, where use has been made of $\omega + \bar{\omega} = \omega + \omega^2 = -1$ and of $\omega\bar{\omega} = \omega \cdot \omega^2 = \omega^3 = 1$.

From Theorem 10.7 follows:

Lemma 1.　*There are exactly six units in* **K**, *namely* ± 1, $\pm \omega$, *and* $\pm \omega^2$.

All rational integers are also integers of **K** (because they are of the form $a + b\omega$ with $0 = b \in \mathbf{Z}$); hence Theorem 3 implies:

Corollary 3.1.　*If* $p = 3$, *equation* (1′) *has no solution in rational integers* x, y, z *with* $x \cdot y \cdot z \neq 0$.

Before actually proving Theorem 3, it may be worthwhile to show that (1′) has no solutions for $p = 3$ in Case I. Indeed if $x \cdot y \cdot z \not\equiv 0 \pmod 3$, then x^3, y^3, and z^3 are all congruent to either $+1$ or to $-1 \pmod 9$. Hence their sum is congruent to $-3, -1, +1$, or $+3 \pmod 9$, which is impossible if $x^3 + y^3 + z^3 = 0$. We have proven:

Theorem 3′.　*The FC holds for* $p = 3$ *in Case 1.*

In the proof of Theorem 3, we shall need a few Lemmas.

Lemma 2′.　$\lambda = 1 - \omega$ *is a prime in* **K**.

PROOF.　See Section 10.11.

Lemma 3′.　*There exist exactly three residue classes* mod λ *and* $-1, 0, +1$ *form a complete set of residues.*

PROOF.　The fact that $-1, 0, +1$ are incongruent follows by verifying that λ does not divide any of their differences. Clearly, $\lambda \nmid \pm 1$, because λ is not a unit; also $\lambda \nmid \pm 2$ because

$$\frac{2}{\lambda} = \frac{2}{1 - \omega} \frac{1 - \bar{\omega}}{1 - \bar{\omega}} = \tfrac{1}{3}(2 - 2\omega^2) = \tfrac{1}{3}(2 + 2 + 2\omega) = 1 + \tfrac{1}{3}(1 + 2\omega) \notin \mathbf{I}.$$

The fact that $-1, 0, +1$ actually forms a complete set of residues follows trivially from a general theorem that identifies the number of residue classes with the absolute value of the norm (here $N\lambda = 3$); we may, however, also verify it directly as follows: If $\alpha \in I$, then $\alpha = a + b\omega = a + b(1 - \lambda) \equiv a + b$ (mod λ). From $\lambda\bar{\lambda} = 3$ follows $\lambda|3$; hence if $\beta \equiv \gamma$ (mod 3), *a fortiori*, $\beta \equiv \gamma$(mod λ). But $a + b$ may have only the three least residues $-1, 0, +1$ (mod 3); consequently the same holds for α (mod λ), and as these residues are incongruent (mod λ), they form a complete set of residues; the Lemma is proven.

Lemma 4'. $\alpha \in I, \lambda \nmid \alpha \Rightarrow \alpha^3 \equiv \pm 1$ (mod λ^4).

PROOF. By Lemma 3', $\lambda \nmid \alpha \Rightarrow \alpha \equiv \pm 1$ (mod λ). For concreteness, let $\alpha \equiv 1$ (mod λ), that is $\alpha = \beta\lambda + 1$. Then

$$
\begin{aligned}
\alpha^3 - 1 &= (\alpha - 1)(\alpha^2 + \alpha + 1) = (\alpha - 1)(\alpha - \omega)(\alpha - \omega^2) \\
&= \beta\lambda(\beta\lambda + 1 - \omega)(\beta\lambda + 1 - \omega^2) \\
&= \beta\lambda(\beta\lambda + \lambda)(\beta\lambda + (1 - \omega)(1 + \omega)) \\
&= \beta\lambda^2(\beta + 1)(\beta\lambda + \lambda(1 + \omega)) \\
&= \beta\lambda^3(\beta + 1)(\beta + 1 + \omega) = \beta\lambda^3(\beta + 1)(\beta + 2 - \lambda) \\
&= \lambda^3\beta(\beta + 1)(\beta - 1 + (3 - \lambda)).
\end{aligned}
$$

But $\lambda\bar{\lambda} = 3$ so that $\lambda|3 - \lambda$ and $\beta - 1 + (3 - \lambda) \equiv \beta - 1$ (mod λ); hence using Lemma 3', the three factors $(\beta + 1), \beta, (\beta - 1)$, being incongruent mod λ, are congruent in some order to -1, 0, and $+1$; therefore, $\lambda|(\beta - 1)\beta(\beta + 1)$ and $\alpha^3 - 1 = \lambda^4\gamma, \gamma \in I$, so that $\alpha^3 \equiv 1$ (mod λ^4). The proof that $\alpha \equiv -1$ (mod λ) $\Rightarrow \alpha^3 \equiv -1$ (mod λ^4) is similar.

Lemma 5. *Factorization in I is unique.*

This is the key to the success of the present proof of Theorem 3. The statement itself is contained in Remark 10.3, and follows from the existence of a Euclidean algorithm; hence Lemma 5 will follow from:

Lemma 6 (Euclidean algorithm in I).

$$\alpha, \beta \in I \Rightarrow \exists \gamma, \rho \in I \ni \alpha = \beta\gamma + \rho, \ N\rho < N\beta.$$

PROOF. As the existence of a Euclidean algorithm was only stated but not proved in Chapter 10, we now proceed to verify that claim (compare with Section 10.6).

$$
\begin{aligned}
\frac{\alpha}{\beta} &= \frac{a_1 + a_2\omega}{b_1 + b_2\omega} = \frac{(a_1 + a_2\omega)(b_1 + b_2\bar{\omega})}{(b_1 + b_2\omega)(b_1 + b_2\bar{\omega})} \\
&= \frac{a_1b_1 - a_1b_2 + a_2b_2 + \omega(a_2b_1 - a_1b_2)}{b_1^2 - b_1b_2 + b_2^2} \\
&= C + D\omega, C, D \in \mathbf{Q},
\end{aligned}
$$

because $a_1, a_2, b_1, b_2 \in \mathbf{Z}$. Now let c and d be rational integers closest to C and D, respectively, so that $C - c = F, D - d = G$, with

$$|F| \leq \tfrac{1}{2}, \qquad |G| \leq \tfrac{1}{2}. \tag{4}$$

Then $C + D\omega = c + d\omega + F + G\omega$. Clearly $\gamma = c + d\omega \in \mathbf{I}$ and $F + G\omega \in \mathbf{K}$. Hence $\alpha = \beta\gamma + \rho$, $\rho = \alpha - \beta\gamma$ is an integer and $\rho = \beta(F + G\omega)$. Hence $N\rho = N\beta \cdot N(F + G\omega)$, and $0 \leq N(F + G\omega) = F^2 - FG + G^2 \leq F^2 + |FG| + G^2$. Hence using (4), $0 \leq N(F + G\omega) \leq \tfrac{3}{4} < 1$, and $N\rho \leq \tfrac{3}{4}N\beta < N\beta$; the Lemma is proven, and with it also Lemma 5.

We are now in a position to prove the following version of the statement that (3) has no solution in Case I:

Theorem 3″.   *Equation (3) has no solutions with $\xi, \eta, \zeta \in \mathbf{I}$ and $\lambda \nmid \xi \cdot \eta \cdot \zeta$.*

PROOF.   It is sufficient to show that (3) has no primitive solutions of the stated kind. By Lemma 4′, $\sigma = \xi^3 + \eta^3 + \zeta^3 \equiv \pm 1 \pm 1 \pm 1 \pmod{\lambda^4}$. If all signs are the same, $\sigma \equiv \pm 3 \pmod{\lambda^4}$, otherwise $\sigma \equiv \pm 1 \pmod{\lambda^4}$. If (3) holds, then $\sigma = 0$ so that either $\lambda^4 | 3$ or $\lambda^4 | 1$. The last alternative is obviously impossible, because λ is not a unit. Also $\lambda = 1 - \omega$, $\lambda^2 = 1 - 2\omega + \omega^2 = 1 + \omega + \omega^2 - 3\omega = -3\omega$, $\lambda^4 = 9\omega^2$ so that (remembering that ω^2 is a unit) λ^4 is associated with 9 and $\lambda^4 | 3 \Leftrightarrow 9 | 3$, which is clearly false; Theorem 3″ is proven.

We conclude, just as previously for $n = 2$, that it is sufficient to consider only Case II, which for (3) means that $\lambda | \xi\eta\zeta$. Remembering also that we are looking only for a primitive solution, it follows that λ divides exactly one among the three integers ξ, η, ζ of $\mathbf{I}$. By the symmetry of (3) it is immaterial which one we assume to be divisible by λ; therefore we are free to let $\lambda | \xi$, which will permit us to keep up the analogy with case $n = 2$. If k is the highest power of λ which divides ξ, we may set $\xi = \lambda^k \alpha$, $(\alpha, \lambda) = 1$, and (3) becomes $\lambda^{3k}\alpha^3 + \eta^3 + \zeta^3 = 0$. Once more it turns out that it is easier to prove a slightly more general result (which, however, is in some sense a more natural one, as the experienced reader will presumably agree), namely:

Theorem 4.   *The equation*

$$\varepsilon\lambda^{3k}\alpha^3 + \eta^3 + \zeta^3 = 0 \tag{5}$$

has no solutions with $\alpha, \eta, \zeta \in \mathbf{I}$, $\lambda \nmid \alpha \cdot \eta \cdot \zeta, 1 \leq k \in \mathbf{Z}$ and ε a unit in $\mathbf{I}$.

A primitive solution of (5) (and this is of course the only kind we are interested in) which also satisfies the other conditions of Theorem 4 will be called simply an *admissible* solution. It is clear that Theorem 4 contains in particular Theorem 3 in Case II; hence in view of Theorem 3″, the proof of Theorem 4 will also complete the proof of Theorem 3.

PROOF OF THEOREM 4. This proof gives us another opportunity to use Fermat's method of descent; this time, the descent will operate on the exponent k. More precisely, we shall show

  (i) for every admissible solution of (5), $k \geq 2$; and
  (ii) from every admissible solution of (5) with a $k = k_0 \geq 2$, we can construct another admissible solution with $k = k_0 - 1$.

These two assertions are of course contradictory, because starting with any integer k_0 after a finite number of steps, all possible by (ii), we end up with an admissible solution with $k \leq 1$, and this is impossible by (i).

Assertion (i) is almost obvious; indeed from $\lambda \nmid \eta \cdot \zeta$ and Lemma 4 it follows that $\varepsilon \lambda^{3k} \alpha^3 \equiv -\eta^3 - \zeta^3 \equiv \begin{cases} -2 \\ 0 \\ +2 \end{cases} \pmod{\lambda^4}$. The alternatives ± 2 are impossible even mod λ, because $\varepsilon \lambda^{3k} \alpha^3 \equiv 0 \pmod{\lambda}$ and $\lambda \nmid 2$. Next $\lambda \nmid \alpha^3$, so that $\varepsilon \lambda^{3k} \alpha^3 \equiv 0 \pmod{\lambda^4} \Rightarrow \lambda^4 | \lambda^{3k} \Rightarrow k \geq 4/3 \Rightarrow k \geq 2$, as claimed.

The hardest part is the proof of (ii). We may try to follow the successful pattern of the procedure for $n = 2$, and assuming that there exists an admissible solution with $k \geq 2$, we write (5) in the form $-\varepsilon \lambda^{3k} \alpha^3 = \eta^3 + \zeta^3 = (\eta + \zeta)(\eta + \omega \zeta)(\lambda + \omega^2 \zeta)$. We have $\lambda = 1 - \omega$, so that $\omega \equiv 1 \pmod{\lambda}$ and also $\omega^2 \equiv 1 \pmod{\lambda}$. It follows that $\eta + \zeta \equiv \eta + \omega \zeta \equiv \eta + \omega^2 \zeta \pmod{\lambda}$ and either all, or none of these three integers of **I** are divisible by λ; but their product is $-\varepsilon \lambda^{3k} \alpha^3$ $(k \geq 2)$, so that all three are divisible by λ. Therefore, $\phi = (\eta + \zeta)/\lambda$, $\psi = (\eta + \omega \zeta)/\lambda$, $\chi = (\eta + \omega^2 \zeta)/\lambda$ are all integers in **I**. We claim that (ϕ, ψ), (ψ, χ) and (χ, ϕ) are all units. (The reader may reflect upon the fact that this statement and its proof rely heavily upon Lemma 5.) Indeed assume, for instance, that $(\chi, \phi) = \delta$, δ not a unit. Then $\delta | \phi - \chi$; but $\phi - \chi = \zeta(1 - \omega^2)/\lambda = \zeta(1 + \omega) = -\omega^2 \zeta$, so that $\delta | \zeta$. Also, $\delta | \omega^2 \phi - \chi$ and $\omega^2 \phi - \chi = \eta(\omega^2 - 1)/\lambda = -(\omega + 1)\eta = \omega^2 \eta$ and $\delta | \eta$. If δ is not a unit, then ζ and η are not coprime, and the solution was not primitive, contrary to our assumption. In the same way one shows that (ϕ, ψ) and (ψ, χ) are units.

Equation (5) now reads $-\varepsilon \lambda^{3k} \alpha^3 = \lambda \phi \cdot \lambda \psi \cdot \lambda \chi$, or

$$-\varepsilon \lambda^{3(k-1)} \alpha^3 = \phi \psi \chi, \qquad (\phi, \psi) = (\psi, \chi) = (\chi, \phi) = 1. \qquad (6)$$

From (6) and $k \geq 2$ follows that $\lambda | \phi \psi \chi$, and from the pairwise coprimality that λ divides exactly one of the three factors. Because of the symmetry in ϕ, ψ, χ, we may assume for instance that $\lambda | \phi$, $\lambda \nmid \psi \chi$. Then $\lambda^{3(k-1)} | \phi$ or $\phi = \lambda^{3(k-1)} \theta$, $\theta \in$ **I**, and (6) becomes

$$-\varepsilon \alpha^3 = \theta \psi \chi, \qquad (\theta, \psi) = (\psi, \chi) = (\chi, \theta) = 1. \qquad (7)$$

On account of Lemma 5, θ, ψ, and χ must each be associated with perfect cubes. It is worthwhile to stress the fact that if we had no Lemma 5 at our disposal the last conclusion would not be warranted, and soon we shall be faced with the consequences of this distressing reality! But now we may proceed, and we know that we are permitted to write $\theta = \varepsilon_1\alpha_1^3$, $\psi = \varepsilon_2\beta^3$, $\chi = \varepsilon_3\zeta_1^3$, with $\alpha_1, \beta, \zeta_1 \in I$ and $\varepsilon_1, \varepsilon_2, \varepsilon_3$ units of I. Substituting these values we obtain that $\eta + \zeta = \lambda\phi = \lambda \cdot \lambda^{3(k-1)}\theta = \lambda^{3k-2}\varepsilon_1\alpha_1^3$, and in the same way we compute $\eta + \omega\zeta$ and $\eta + \omega^2\zeta$, obtaining

$$\eta + \zeta = \varepsilon_1\lambda^{3k-2}\alpha_1^3, \eta + \omega\zeta = \varepsilon_2\lambda\beta^3, \eta + \omega^2\zeta = \varepsilon_3\lambda\zeta_1^3, \qquad (8)$$

$$\lambda \nmid \alpha_1 \cdot \beta \cdot \zeta_1 \qquad (\alpha_1, \beta) = (\beta, \zeta_1) = (\zeta_1, \alpha_1) = 1,$$

$$\varepsilon_1, \varepsilon_2, \varepsilon_3 \text{ units of } I.$$

Let us consider now the sum $\sigma = (\eta + \zeta) + \omega(\eta + \omega\zeta) + \omega^2(\eta + \omega^2\zeta)$; on the one hand, it is clear that $\sigma = \eta(1 + \omega + \omega^2) + \zeta(1 + \omega^2 + \omega^4) = 0$; on the other hand, replacing the brackets in σ by their values from (8), we obtain $\sigma = \varepsilon_1\lambda^{3k-2}\alpha_1^3 + \omega\varepsilon_2\lambda\beta^3 + \omega^2\varepsilon_3\lambda\zeta_1^3$. Equating these two values and dividing by $\varepsilon_3\lambda(\neq 0)$, we obtain

$$\varepsilon_4\lambda^{3(k-1)}\alpha_1^3 + \varepsilon_5\beta^3 + \zeta_1^3 = 0, \qquad \text{with} \qquad (5')$$

$$(\alpha_1, \beta) = (\beta, \zeta_1) = (\zeta_1, \alpha_1) = 1,$$

ε_4 and ε_5 units of I. Equation (5') is almost of the same form as (5), with α_1, β, ζ_1 instead of α, η, ζ, and with k replaced by $k - 1$. Were it not for the (so far not determined) unit ε_5, the proof of (ii)—and hence of Theorem 4—would be complete. It remains to show that we actually may take $\varepsilon_5 = 1$. Indeed $k \geq 2$, so that $\lambda^3 | \lambda^{3(k-1)}\varepsilon_4\alpha_1^3$. Also $\lambda \nmid \beta \cdot \zeta_1$, so that by Lemma 4', $\beta^3 \equiv \pm 1$ (mod λ^4), $\zeta_1^3 \equiv \pm 1$ (mod λ^4). Consequently, $\varepsilon_5\beta^3 + \zeta_1^3 \equiv \pm\varepsilon_5 \pm 1$ (mod λ^4), and the congruence holds, *a fortiori*, to the modulus λ^3. We obtain, therefore, that $0 \equiv -\lambda^{3(k-1)}\varepsilon_4\alpha_1^3 \equiv \varepsilon_5\beta^3 + \zeta_1^3 \equiv \pm\varepsilon_5 \pm 1$ (mod λ^3). However $\pm\omega \pm 1$ and $\pm\omega^2 \pm 1$ are not divisible by λ^3 (not even by λ^2; why?); hence $0 \equiv \pm 1\varepsilon_5 \pm 1$ (mod λ^3) is possible only if $\varepsilon_5 = \pm 1$ and (5') becomes $\varepsilon_4\lambda^{3(k-1)}\alpha_1^3 + (\pm\beta)^3 + \zeta_1^3 = 0$, or setting $\pm\beta = \eta_1$, $\varepsilon_4\lambda^{3(k-1)}\alpha_1^3 + \eta_1^3 + \zeta_1^3 = 0$ with $\alpha_1, \eta_1, \zeta_1 \in I$, $\lambda \nmid \alpha_1 \cdot \eta_1 \cdot \zeta_1$; $\alpha_1, \eta_1, \zeta_1$ coprime in pairs and ε_4 a unit of I.

In other words, starting with an admissible solution $\xi = \lambda^k\alpha, \eta, \zeta$, we have constructed another admissible solution $\xi_1 = \lambda^{k-1}\alpha_1, \eta_1, \zeta_1$, with a smaller exponent k. This finishes the proof of (ii), hence that of Theorem 4 and that of Theorem 3.

5   CASE I FOR $p = 5$

While already the proof of the FC for $p = 3$ is not precisely trivial, the general case is still much more difficult. Therefore in order to warm up, let us consider one last case that can be handled with elementary methods, namely Case I for $p = 5$. We propose to prove:

Theorem 5. *The equation*

$$\xi^5 + \eta^5 + \zeta^5 = 0 \tag{9}$$

has no solutions with ξ, η, ζ integers of the cyclotomic field $\mathbf{Q}(\omega)$, $\omega = e^{2\pi i/5}$, none of which is divisible by $\lambda = 1 - \omega$.

For simplicity in the present section we shall denote the field $\mathbf{Q}(e^{2\pi i/5})$ by $\mathbf{K}_5$ and its ring of integers by $\mathbf{I}_5$. Also, in the present section a primitive solution of (9) satisfying all conditions of Theorem 5 will be called for short an *admissible solution*.

Lemma 2″. $\lambda = 1 - \omega$ *is a prime in* $\mathbf{I}_5$.

PROOF.   See Section 10.11.

Corollary 2″.1. $\lambda | 5$.

PROOF.   Obvious.

Corollary 2″.2. λ^4 *and 5 are associates.*

PROOF.   We verify that $\lambda^4 = 5\,(\lambda^3 - 2\lambda^2 + 2\lambda - 1)$ and that the bracket is a unit (why?).

Corollary 2″.3. $x \in \mathbf{Z}$ *and* $\lambda | x \Rightarrow 5 | x$.

PROOF. $\lambda | x \Rightarrow \lambda^4 | x^4 \Rightarrow 5 | x^4$ by Corollary 2″.2. But $x \in \mathbf{Z}$ and 5 is a prime; hence $5 | x$.

It is clear that $\mathbf{Z} \subset \mathbf{I}_5$; hence Theorem 5 and Corollary 2″.1 have as consequence (why?) that for $p = 5$, (1′) has no solutions $x, y, z \in \mathbf{Z}$ with $5 \nmid x \cdot y \cdot z$, so that for $p = 5$ the FC holds at least in Case I. It also is clear that $\lambda \nmid \xi \cdot \eta \cdot \zeta$ is the proper formulation of the Case I restriction, in the present slightly more general setting.

Lemma 3″. *There exist exactly 5 residue classes* mod λ *and* $-2, -1, 0, 1, 2$ *form a complete set of residues* mod λ.

PROOF.   By Section 10.11,

$$\alpha \in \mathbf{I}_5 \Rightarrow \alpha = a_0 + a_1\omega + a_2\omega^2 + a_3\omega^3 + a_4\omega^4 \qquad \text{with } a_m \in \mathbf{Z};$$

hence

$$\alpha = a_0 + a_1(1 - \lambda) + \cdots + a_4(1 - \lambda)^4 \equiv a_0 + a_1 + \cdots + a_4 \,(\text{mod}\,\lambda).$$

However $\sum_{j=0}^4 a_j \equiv \pm 2, \pm 1, 0 \,(\text{mod}\,5)$, and hence using Corollary 2″.1, the same holds *a fortiori* mod λ. The proof is completed as in the case $p = 3$ by checking that these five residues are incongruent not only mod 5, but also mod λ.

Lemma 4″. $\alpha, \beta \in \mathbf{I}_5, \alpha \equiv \beta \pmod{\lambda} \Rightarrow \alpha^5 \equiv \beta^5 \pmod{\lambda^5}$

PROOF. $\alpha^5 - \beta^5 = (\alpha - \beta)(\alpha - \omega\beta)(\alpha - \omega^2\beta)(\alpha - \omega^3\beta)(\alpha - \omega^4\beta)$; but $\omega \equiv 1 \pmod{\lambda} \Rightarrow \omega^k \equiv 1 \pmod{\lambda}$ so that $\alpha - \omega^k\beta \equiv \alpha - \beta \pmod{\lambda}$. Hence if $\lambda | \alpha - \beta$, then $\lambda | \alpha - \omega^k\beta$, each of the 5 factors is divisible by λ, and $\lambda^5 | \alpha^5 - \beta^5$ or $\alpha^5 \equiv \beta^5 \pmod{\lambda^5}$. We also could prove the analog of Lemma 5, stating that factorization into primes is unique in $\mathbf{I}_5$, but we shall not need this result, already stated in Section 10.11.

PROOF OF THEOREM 5.   By assumption $\lambda \nmid \xi \cdot \eta \cdot \zeta$; hence by Lemma 4″, if $\xi \equiv \pm 1 \pmod{\lambda}$, then $\xi^5 \equiv \pm 1 \pmod{\lambda^5}$; if $\xi \equiv \pm 2 \pmod{\lambda}$, then $\xi^5 \equiv \pm 32 \pmod{\lambda^5}$, and similarly for η and ζ. Consequently, $\xi^5 + \eta^5 + \zeta^5 \equiv a + b + c \pmod{\lambda^5}$, where each of a, b, and c may take only one of the four values $\pm 1, \pm 32$. If ξ, η, ζ are an admissible solution of (9) and $a + b + c = f$, then $0 = \xi^5 + \eta^5 + \zeta^5 \equiv f \pmod{\lambda^5}$. Using Corollaries 2″.2 and 2″.3, $f \equiv 0 \pmod{\lambda^5} \Leftrightarrow \lambda^5 | f \Leftrightarrow 5\lambda | f \Rightarrow \lambda | (f/5) \Rightarrow 5 | (f/5) \Rightarrow 5^2 | f$ so that $f \equiv 0 \pmod{25}$. If we now take for a, b, c any of the values ± 1 or ± 32 ($\equiv \pm 7 \pmod{25}$), no combination of three of them leads to zero (the sum with same signs is at most $\equiv 3 \cdot 7 \equiv 21 \pmod{25}$); hence $f \equiv 0 \pmod{25}$ is impossible and Theorem 5 is proven.

6   CASE I FOR AN ARBITRARY PRIME *p*

Let us set $\omega = e^{2\pi i/p}$, and for simplicity of writing let us denote (throughout this section) by **K** the cyclotomic field $\mathbf{Q}(\omega)$ generated by ω (that is, the set $\sum_{j=0}^{p-2} A_j \omega^j, A_j \in \mathbf{Q}$) and by **I** the ring of integers of **K**, $\mathbf{I} = \{\sum_{j=0}^{p-2} a_j \omega^j | a_j \in \mathbf{Z}\}$; also let $\lambda = 1 - \omega$. Exactly as in the case $p = 3$ and $p = 5$ one has (see Section 10.11):

Lemma 2. $\lambda = 1 - \omega$ *is a prime in* **I** *and* $N\lambda = p$.

Corollary 2.1. λ^{p-1} *and p are associates.*

PROOF.   In $(x^p - 1)/(x - 1) = x^{p-1} + x^{p-2} + \cdots + x + 1 = \prod_{j=1}^{p-1}(x - \omega^j)$ set $x = 1$, obtaining $p = (1 - \omega)(1 - \omega^2) \cdots (1 - \omega^{p-1})$. For every $k \not\equiv 0 \pmod{p}$, $1 - \omega$ and $1 - \omega^k$ are associates. Indeed

$$1 - \omega^k = (1 - \omega)(1 + \omega + \cdots + \omega^{k-1}),$$

so that $1 - \omega | 1 - \omega^k$. Next one can solve the congruence $km \equiv 1 \pmod{p}$, because $(k, p) = 1$; let m be the smallest positive solution. Then

$$1 - \omega^{km} = 1 - \omega^{np+1} = 1 - \omega;$$

also

$$1 - \omega^{km} = (1 - \omega^k)(1 + \omega^k + \omega^{2k} + \cdots + \omega^{(m-1)k}),$$

so that

$$1 - \omega = (1 - \omega^k)(1 + \omega^k + \omega^{2k} + \cdots + \omega^{(m-1)k})$$

and $1 - \omega^k | 1 - \omega$. Consequently $1 - \omega$ and $1 - \omega^k$ are indeed associates, $(1 - \omega)(1 - \omega^2) \cdots (1 - \omega^{p-1}) = \varepsilon(1 - \omega)^{p-1} = \varepsilon\lambda^{p-1}$, and the Corollary is proven.

Corollary 2.2. *If* $(k, p) = 1$ *and* $mk \equiv 1 \pmod{p}$, *then* $1 + \omega^k + \omega^{2k} + \cdots + \omega^{(m-1)k}$ *is a unit of* **I**.

PROOF. $1 - \omega$ and $1 - \omega^k$ are associates by Corollary 2.1; hence their ratio $(1 - \omega)/(1 - \omega^k) = (1 - \omega^{mk})/(1 - \omega^k) = 1 + \omega^k + \cdots + \omega^{(m-1)k}$ is a unit.

Corollary 2.3. *For every* $m \not\equiv 0 \pmod{p}$, $1 + \omega + \omega^2 + \cdots + \omega^{m-1}$ *is a unit.*

PROOF. For $p \nmid m$, $1 - \omega^m$ and $1 - \omega$ are associates and $(1 - \omega^m)/(1 - \omega) = 1 + \omega + \cdots + \omega^{m-1}$.

Lemma 3. *There exist exactly* p *residue classes* $\mod \lambda$ *and* $0, \pm 1, \pm 2, \ldots, \pm \frac{1}{2}(p - 1)$ *form a complete set of residues.*

PROOF. Same (*mutatis mutandis*) as for $p = 3$ or $p = 5$.

Lemma 4. $\alpha, \beta \in \mathbf{I}, \alpha \equiv \beta \pmod{\lambda} \Rightarrow \alpha^p \equiv \beta^p \pmod{\lambda^p}$.

PROOF. Same (*mutatis mutandis*) as for $p = 5$.

One would now try to prove the generalization of Lemma 5. As will be seen presently, this would enable us to prove the FC easily, at least in Case I. However, the generalization of the previous proof for $p = 3$ does not go through! It may be worthwhile to observe, though, that when one tries to prove the uniqueness of factorization for $\mathbf{Q}(e^{2\pi i/p})$ with $p = 5, p = 7$, and so on, one seems to succeed each time. Yet each time some new, specific reasoning, some trick that works just once for that particular p seems to be needed, and any attempt at a general proof fails. The reason for this failure is of course that the factorization actually *is not unique* in a general cyclotomic field—yet the first case of failure occurs only for $p = 23$, and this was originally pointed out by Cauchy in 1847 (see [1]), as we already know.

Deprived of Lemma 5 in the general case, let us try, nevertheless, and see what we can achieve with only Lemmas 2, 3, and 4 at our disposal. Equation

(1') may be written as

$$-x^p = y^p + z^p = (y + z)(y + \omega z) \cdots (y + \omega^{p-1} z),$$

with $\alpha_k \underset{\text{def}}{=} y + \omega^k z \in \mathbf{I}$, provided that $y, z \in \mathbf{Z}$. As until now, we are interested only in primitive solutions with x, y, z coprime in pairs.

Restricting ourselves to Case I, that is to solutions x, y, z such that $p \nmid x \cdot y \cdot z$, we claim that the factors α_k with different subscripts are coprime. Indeed if π is a prime, $\pi | y + \omega^k z$ and $\pi | y + \omega^m z$ with, say $k > m$, then $\pi | (y + \omega^k z) - (y + \omega^m z) \Rightarrow \pi | \omega^k (1 - \omega^{m-k}) z$ or $\pi | \lambda \varepsilon z$, with ε a unit of $\mathbf{I}$, by Corollary 2.3. If $\pi | \lambda$, then π and λ are associates; also $\pi | y + \omega^k z \Rightarrow \pi | y^p + z^p \Rightarrow \pi | - x^p \Rightarrow \lambda | x^p \Rightarrow N\lambda | x \Rightarrow p | x$, contrary to our assumption, hence $\pi | z$. This together with $\pi | y + \omega^k z \Rightarrow \pi | y$; consequently, taking norms we show that $N\pi | y$, $N\pi | z$, contrary to the assumption that y and z are coprime.

We can formulate the results obtained so far as follows: If x, y, z are a primitive solution of (1') with $p \nmid x \cdot y \cdot z$, then

$$-x^p = \alpha_0 \alpha_1 \cdots \alpha_{p-1} \qquad \text{with} \qquad \alpha_i \in \mathbf{I} (0 \le i \le p - 1), (\alpha_i, \alpha_j) = \varepsilon_{ij},$$

a unit of $\mathbf{I}$ if $i \ne j$. We would like to continue as in the particular case $p = 3$ and invoke the uniqueness of factorization to the effect that each α_j is the associate of an exact pth power, that is, $\alpha_j = \varepsilon_j \xi_j^p$, ε_j a unit, ξ_j an integer of $\mathbf{I}$. But this conclusion is warranted only if $\mathbf{I}$ is a domain of unique factorization. Let us assume for a moment that this is the case.

In order to continue our reasoning, we still need some more Lemmas, which, however, have nothing to do with the uniqueness of factorization.

Lemma 7. *If ε is a unit of $\mathbf{I}$, then there exists an exponent $k (\in \mathbf{Z})$ such that $\varepsilon = \omega^k \eta$, with real η.*

REMARK 6. $\eta = \varepsilon \omega^{-k}$ is a unit because ε and ω^{-k} are units.

The *Proof of Lemma* 7 is long but otherwise rather routine. It may be found in [12] pp. 225–227 or in [20] pp. 118–120, and will not be reproduced here. It seems worthwhile to mention that the proof of Lemma 7 is based on the following Lemma, of independent interest.

Lemma 8. $\gamma \in \mathbf{I} \Rightarrow \gamma^p \equiv a \pmod{\lambda^p}$ *for some $a \in \mathbf{Z}$.*

PROOF.   By Lemma 3 each residue class mod λ contains some $c \in \mathbf{Z}$ (even some c with $|c| \le \frac{1}{2}(p - 1)$); hence $\gamma \in \mathbf{I} \Rightarrow \exists c \in \mathbf{Z} \ni \gamma \equiv c \pmod{\lambda}$. Using Lemma 4, $\gamma^p \equiv c^p \pmod{\lambda^p}$, and the Lemma is proven with $a = c^p$ (which is somewhat more than claimed).

Corollary 8.1. $\gamma \in \mathbf{I} \Rightarrow \exists a \in \mathbf{Z} \ni \gamma^p \equiv a \pmod{p}$.

PROOF.   By Lemma 8, there exists an integer $a \in \mathbf{Z}$ such that $\lambda^p | \gamma^p - a$; a fortiori, $\lambda^{p-1} | \gamma^p - a$. Yet by Corollary 2.1, $p = \varepsilon \lambda^{p-1}$, and the result follows.

Returning now to $\alpha = \alpha_1 = y + \omega z$, we assume uniqueness of factorization, hence that $\alpha = \varepsilon \xi^p$. Using Lemma 7, we may replace ε by $\omega^r \eta$ (η real), so that $\alpha = \eta \omega^r \xi^p$; next by Corollary 8.1, we can determine $a \in \mathbf{Z} \ni \xi^p \equiv a \pmod{p}$ and obtain $\alpha \equiv \eta \omega^r a \pmod{p}$ or, setting $\eta a = b$, $\alpha \equiv b \omega^r \pmod{p}$ with real $b \in \mathbf{I}$. Hence $b \equiv \omega^{-r} \alpha \pmod{p}$; being real, $b = \bar{b}$, its complex conjugate; that is, $b = \bar{b} \equiv \omega^r \bar{\alpha} \pmod{p}$. We conclude that $\omega^r \bar{\alpha} \equiv \omega^{-r} \alpha \pmod{p}$. Replacing α by $y + \omega z$ and $\bar{\alpha}$ by $y + \bar{\omega} z$, the last congruence becomes

$$y \omega^{-r} + z \omega^{1-r} - y \omega^r - z \omega^{r-1} \equiv 0 \pmod{p}. \tag{10}$$

If $r \equiv 0 \pmod{p}$, then (10) reduces to $y + z\omega - y - z\omega^{-1} \equiv 0 \pmod{p}$; but $p \nmid z$; hence $p | \omega - \omega^{-1}$ or $p | 1 - \omega^2$, that is, $\lambda^{p-1} | (1 - \omega)(1 + \omega)$ or $\lambda^{p-2} | 1 + \omega$, which is false because $p \geq 3$ and $1 + \omega$ is a unit. If $r \equiv 1 \pmod{p}$ one similarly obtains $y \omega^{-1} + z - y\omega - z \equiv 0 \pmod{p}$ with the same conclusion, because $p \nmid y$. Consequently $r \not\equiv 0$, $r \not\equiv 1 \pmod{p}$, and (10) is equivalent to an equation of the form

$$y \omega^{2r} + z \omega^{2r-1} - z\omega - y + pf(\omega) = 0 \tag{10'}$$

with $f(\omega) \in \mathbf{Z}[\omega]$ and where $2r$ may be taken less than p (because $\omega^p = 1$). There are two possibilities. The first is that the terms in (10) mutually cancel out, and then (10') holds with $f(\omega)$ identically zero. In the other alternative, the polynomial in (10'), like any polynomial with root ω, must be a multiple of the irreducible (cyclotomic) polynomial

$$\omega^{p-1} + \omega^{p-2} + \cdots + \omega + 1. \tag{11}$$

A glance at (10') shows that this is not possible. Indeed discarding the cases $p = 3$ and $p = 5$ (for which Case I has been settled), (11) contains at least 7 terms; hence if the polynomial in (10') is a multiple of (11), $f(\omega)$ does not vanish identically. The coefficients of all terms of $pf(\omega)$ are multiples of p; hence if the polynomial in (10') is a multiple of (11), the other coefficients in (10'), namely y and z, also have to be multiples of p, contrary to the assumption $p \nmid x \cdot y \cdot z$. It follows that the first alternative holds. Clearly $2r \equiv 2r - 1 \pmod{p}$ is not possible and $2r \equiv 0 \pmod{p}$ has been ruled out. The only possibility for mutual cancellation of the 4 terms in (10) is therefore $2r \equiv 1 \pmod{p}$, and (10) becomes

$$\omega^{-r}(y + z\omega - y\omega^{2r} - z\omega^{2r-1}) = \omega^{-r}(y + z\omega - y\omega - z) \equiv 0 \pmod{p}$$

or, ω being a unit, $y(1 - \omega) + z(\omega - 1) = (y - z)(1 - \omega) = \lambda(y - z) \equiv 0 \pmod{p}$. p being associated with λ^{p-1}, $\lambda^{p-2} | y - z$; hence as already seen several times, $p | y - z$. By symmetry, starting in (1') from $-z^p = x^p + y^p$, it follows in exactly the same way that $p | x - y$. Hence, $x \equiv y \equiv z \pmod{p}$, $x^p + y^p + x^p \equiv 3x^p \pmod{p}$. If x, y, z are a solution of (1'), the left sum vanishes, so that $p | 3x^p$. For $p > 3$ this means $p | x$, which contradicts the condition $p \nmid x \cdot y \cdot z$ of Case I. We may state the result obtained as follows:

Theorem 6.   *If for some p the cyclotomic field* $\mathbf{K} = \mathbf{Q}(e^{2\pi i/p})$ *has the property of uniqueness of factorization of its integers, then there exists no solution of* (1') *in integers* $x, y, z \in \mathbf{Z}$, *with* $p \nmid x \cdot y \cdot z$.

7   REGULAR PRIMES

In view of our experience with $p = 3$, we are tempted to try to prove a statement like the following: "Let p be an odd rational prime; then the equation

$$\xi^p + \eta^p + \zeta^p = 0 \tag{12}$$

has no solutions in integers $\xi, \eta, \zeta \in \mathbf{I}$ satisfying $\xi\eta\zeta \neq 0$." This, however, would imply the truth of the FC and up to now this statement has been neither proven nor disproven; not even the easier Case I has been settled. However, Kummer succeeded in proving the FC for a certain class of primes, which he called *regular primes*.

DEFINITION 1.   Let p be a rational prime, and denote by h the class number of the cyclotomic field $\mathbf{K} = \mathbf{Q}(\omega)$; then p is said to be a *regular prime* if $p \nmid h$.

REMARK 1.   The small primes are all regular (37 is the first irregular one) but among larger primes many are not regular. How large the class of regular primes is, one does not know; one does not even know today whether there are infinitely many regular primes, or not.

In what follows we shall see a proof of the FC in both Case I and Case II for regular primes. The proof of Case I seems to be the longer and more difficult one, but that is not really the case. Indeed, the proof of Case I is given essentially in full, while in Case II we shall make use of a Theorem of Kummer, which we shall not prove. Its statement is very simple, but the proof takes over 30 pages (in [12]).

8   PROOF OF THE FC FOR REGULAR PRIMES IN CASE I

Theorem 7.   *Let p be a regular prime. Then* (12) *has no primitive solution in integers* $\xi, \eta, \zeta \in \mathbf{I}$, *none divisible by* $\lambda = 1 - \omega$.

REMARK 2.   The conditions $\lambda \nmid \xi, \lambda \nmid \eta, \lambda \nmid \zeta$ are the natural generalization of the original condition $p \nmid x \cdot y \cdot z$ that characterizes Case I. For want of a better expression, any solution satisfying all conditions of Theorem 7 will be called an *admissible solution*.

On account of Theorem 11.32, the relevance of the condition that p be regular is clear. In the present particular case of $\mathbf{K} = \mathbf{Q}(\omega)$ it implies:

Theorem 8.   *If p is a regular prime, then $\mathfrak{a}^p \sim \mathfrak{b}^p \Rightarrow \mathfrak{a} \sim \mathfrak{b}$ for all ideals of $\mathbf{K}$.*

Corollary 8.1.   *If p is a regular prime and $\mathfrak{a}^p$ is a principal ideal of $\mathbf{K}$, then so is $\mathfrak{a}$.*

PROOF.   Take $\mathfrak{b} = (1)$ in Theorem 8.

Finally, the proof of Theorem 7 is made easier if we introduce one more definition.

DEFINITION 2.   An integer $\alpha \in \mathbf{I}$ is said to be *primary* if

(i) $\lambda \nmid \alpha$; and

(ii) $\alpha \equiv a \pmod{\lambda^2}$, $a \in \mathbf{Z}$.

REMARK 3.   This is Landau's ([12] p. 227) terminology; for most modern authors *primary* means a stronger property, which we will not need and the present property is called semiprimary.

The following theorem gives criteria by which one can identify primary integers.

Theorem 9.   *If $\alpha \in \mathbf{I}$, $\lambda \nmid \alpha$, then*

(i) *α^p is always primary;*

(ii) *$\exists k \in \mathbf{Z} \ni \omega^k \alpha$ is primary;*

(iii) *$\alpha \in \mathbf{R} \Rightarrow \alpha$ primary.*

PROOF.   By Lemma 8 $\exists a \in \mathbf{Z} \ni \alpha^p \equiv a \pmod{\lambda^p}$; for $p \geq 2$ this implies $\alpha^p \equiv a \pmod{\lambda^2}$, hence (i).

By Lemma 3 we know that there are exactly p residue classes mod λ each containing a rational integer; therefore $\exists a \in \mathbf{Z} \ni \alpha \equiv a \pmod{\lambda}$ or $\alpha = a + \mu\lambda$, $\mu \in \mathbf{I}$. For the same reason, $\mu \equiv m \pmod{\lambda}$ or $\mu = m + \nu\lambda$, $\nu \in \mathbf{I}$, so that $\alpha = a + m\lambda + \nu\lambda^2$ or $\alpha \equiv a + m\lambda \pmod{\lambda^2}$. From $\lambda \nmid \alpha$ follows $\lambda \nmid a$ and, *a fortiori*, $p \nmid a$. Therefore we can determine $k \in \mathbf{Z}^+ \ni ka \equiv m \pmod{p}$. Observing that

$$\omega^k = (1 - \lambda)^k = 1 - k\lambda + \binom{k}{2}\lambda^2 \cdots \equiv 1 - k\lambda \pmod{\lambda^2},$$

it follows that

$$\omega^k \alpha \equiv (1 - k\lambda)(a + m\lambda) \pmod{\lambda^2} \text{ or } \omega^k \alpha \equiv a + \lambda(m - ka) \equiv a \pmod{\lambda^2}$$

(the last congruence holds because $p \mid m - ka \Rightarrow \lambda \mid m - ka$), and (ii) is proven.

Finally, remembering that $\lambda = 1 - \omega$, from $\alpha \equiv a + m\lambda \pmod{\lambda^2}$ follows that $\alpha \equiv a + m - m\omega \pmod{\lambda^2}$. Also, taking complex conjugates α being real and $\bar{\lambda}$ being associate to λ (indeed $\bar{\lambda}(-\omega) = (\bar{\omega} - 1)\omega = 1 - \omega = \lambda$), $\bar{\alpha} = \alpha \equiv a + m - m\bar{\omega} \pmod{\lambda^2}$; therefore, subtracting,

$$m(\omega - \bar{\omega}) = m(\omega - \omega^{-1}) \equiv 0 \pmod{\lambda^2}$$

or

$$\omega^{-1}m(\omega^2 - 1) = \omega^{-1}m(\omega + 1)(\omega - 1) = -\omega^{-1}(\omega + 1)m\lambda \equiv 0 \pmod{\lambda^2}.$$

Here -1, ω^{-1}, and $\omega + 1$ are units; hence the last congruence becomes $m\lambda \equiv 0 \pmod{\lambda^2}$ and $\alpha = a + m\lambda \pmod{\lambda^2}$ reduces to $\alpha \equiv a \pmod{\lambda^2}$, proving (iii).

After this preparation we are now ready to start the

PROOF OF THEOREM 7. If (12) has any admissible solutions, then it also has solutions with ξ, η, ζ primary. Indeed if, say ξ is not primary, then by Theorem 3, $\xi_1 = \xi\omega^k$ is primary for some $k \in \mathbf{Z}^+$; also $\xi_1^p = \xi^p\omega^{pk} = \xi^p$ and $\lambda \nmid \xi \Leftrightarrow \lambda \nmid \xi_1$. Hence in (12) ξ may be replaced by ξ_1, and similarly, η and ζ may be replaced by η_1, ζ_1, both primary, with

$$\xi^p + \eta^p + \zeta^p = \xi_1^p + \eta_1^p + \zeta_1^p, \quad \lambda \nmid \xi_1, \lambda \nmid \eta_1, \lambda \nmid \zeta_1.$$

Therefore without loss of generality we may, and shall, assume that in any admissible solution of (12)

$$\xi \equiv a, \eta \equiv b, \zeta \equiv c \pmod{\lambda^2}, \quad \text{with } a, b, c \in \mathbf{Z}.$$

Proceeding as in Section 4, (12) may be written as $-\xi^p = \eta^p + \zeta^p = \prod_{j=0}^{p-1}(\eta + \omega^j\zeta)$. In terms of ideals, this is equivalent to

$$(\xi)^p = \prod_{j=0}^{p-1}(\eta + \omega^j\zeta), \tag{13}$$

where $(\eta + \omega^j\zeta)$ now stands for the principal ideal of multiples of $\eta + \omega^j\zeta$; clearly, $\lambda \nmid \xi \Rightarrow (\lambda) \nmid (\eta + \omega^j\zeta)$ $(0 \leq j \leq p - 1)$. We can no longer conclude that if the solution is primitive, then ξ, η, ζ have to be coprime in pairs, because the ideal (η, ζ), for instance, need not be a principal ideal. But if $(\eta, \zeta) = \mathfrak{d} \neq (1)$, then

(i) $\mathfrak{d}$ divides every factor $(\eta + \omega^j\zeta)$, which is rather obvious; and

(ii) $\mathfrak{d}$ is actually the g.c.d. of any two such factors, which may not be completely obvious. To convince ourselves, let $\mathfrak{f}$ be any common factor of $(\eta + \omega^j\zeta)$ and $(\eta + \omega^{j'}\zeta)$, $j < j'$. We have to prove that $\mathfrak{f} | \mathfrak{d}$.

We observe that $\mathfrak{f}$ divides the difference $(\eta + \omega^j\zeta) - (\eta + \omega^{j'}\zeta)$ or $\mathfrak{f} | \omega^j(1 - \omega^{j'-j})\zeta$. Suppressing units, it follows that $\mathfrak{f} | (1 - \omega)\zeta = \lambda\zeta$. I claim that actually $\mathfrak{f} | \zeta$. To conclude this it is sufficient to show that the prime ideal (λ) does not divide $\mathfrak{f}$. Indeed $\mathfrak{f} \nmid (\eta + \omega^j \zeta)$ and $(\eta + \omega^j \zeta) \mid (\xi)^p$; therefore $\mathfrak{f} \nmid (\xi)^p$. Hence if $(\lambda) \mid \mathfrak{f}$, then $(\lambda) \mid (\xi)^p \Rightarrow \lambda \mid \xi$, contradicting one of our assumptions. We have therefore proven that $\mathfrak{f} \mid \zeta$. In exactly the same way one shows that $\mathfrak{f} \mid \eta$; hence $\mathfrak{f} \mid (\eta, \zeta) = \mathfrak{d}$, as claimed. We now set $\eta + \omega^i \zeta = \mathfrak{d} \mathfrak{c}_i$, and (12) becomes $(\xi)^p = \mathfrak{d}^p \prod_{j=0}^{p-1} \mathfrak{c}_j$; hence $\mathfrak{d} \mid (\xi)$, and if we set $(\xi) \mathfrak{d}^{-1} = \mathfrak{g}$, then $\mathfrak{g}^p = \prod_{j=0}^{p-1} \mathfrak{c}_j$. By the uniqueness of factorization of ideals it now follows that each $\mathfrak{c}_j$ is the pth power of an ideal in $\mathbf{K}$, say $\mathfrak{c}_j = \mathfrak{e}_j^p$ and $(\eta + \omega^j \zeta) = \mathfrak{d} \mathfrak{e}_j^p$, $(\lambda) \nmid \mathfrak{e}_j$. One observes that all ideals $\mathfrak{d} \mathfrak{e}_j^p$ are principal, and consequently, equivalent; by Lemma 11.1 and Theorem 11.10 it now easily follows that $\mathfrak{e}_j^p \sim \mathfrak{e}_i^p$, or by Theorem 8 that $\mathfrak{e}_j \sim \mathfrak{e}_i$ for all $i, j, 0 \le i, j \le p - 1$. It should be emphasized that this crucial conclusion makes essential use of Theorem 8, which in turn was proven only for regular primes.

From the definition of equivalence we know that there exist principal ideals (α_j) and (β_j) such that $(\alpha_j) \mathfrak{e}_j = (\beta_j) \mathfrak{e}_0$. Remembering that $(\lambda) \nmid \mathfrak{e}_j$, it is clear that if $\lambda^k \mid \alpha_j$, then $\lambda^k \mid \beta_j$. Dividing out, if necessary, this highest power of λ, we obtain an equality of the form $(\gamma_j) \mathfrak{e}_j = (\delta_j) \mathfrak{e}_0$, $\lambda \nmid \gamma_j$, $\lambda \nmid \delta_j$, whence $(\gamma_j^p) \mathfrak{e}_j^p = (\delta_j^p) \mathfrak{e}_0^p$ or $(\gamma_j^p) \mathfrak{e}_j^p \mathfrak{d} = (\delta_j^p) \mathfrak{e}_0^p \mathfrak{d}$, that is,

$$\left(\gamma_j^p \right) \left(\eta + \omega^j \zeta \right) = \left(\delta_j^p \right) (\eta + \zeta). \tag{14}$$

This equality (14) involves only principal ideals; therefore on account of Theorem 11.2, we may just as well replace them by their respective generators, that is, by integers of $\mathbf{I}$, provided we also introduce some as yet not further determined unit of the ring $\mathbf{I}$. As a matter of fact, we have used the ideals so far to be able to factor without ambiguity and to replace coprime factors whose product was a pth power by the pth power of some other factor. The ideals have done their duty, but from here on we return to integers of $\mathbf{I}$; in particular, equation (14) is equivalent to $\gamma_j^p (\eta + \omega^j \zeta) = \delta_j^p (\eta + \zeta) \varepsilon_j'$, with ε_j' a unit. We now invoke again Lemma 7, according to which $\varepsilon_j' = \omega^{k_j} \varepsilon_j$, with ε_j real and $k_j \in \mathbf{Z}$. Taking congruences mod λ^p we may (see Lemma 8) also replace γ_j^p and δ_j^p by rational integers, say c_j and d_j, and obtain

$$c_j (\eta + \omega^j \zeta) \equiv d_j (\eta + \zeta) \omega^{k_j} \varepsilon_j \; (\mathrm{mod} \; \lambda^p), \qquad \lambda \nmid c_j d_j. \tag{15}$$

We now determine a rational integer g_j by the condition $g_j c_j \equiv d_j \; (\mathrm{mod} \; p^2)$. Any such congruence will hold, *a fortiori*, mod λ^p. Multiplying (15) by g_j and simplifying, we obtain $\eta + \omega^j \zeta \equiv (\eta + \zeta) \omega^{k_j} g_j \varepsilon_j \; (\mathrm{mod} \; \lambda^p)$, $\lambda \nmid g_j$. Denote $g_j \varepsilon_j$ by β_j; then β_j is real and $\lambda \nmid \beta_j$ so that (15) becomes

$$\eta + \omega^j \zeta \equiv \beta_j \omega^{k_j} (\eta + \zeta) \; (\mathrm{mod} \; \lambda^p). \tag{16}$$

By Theorem 9, β_j is primary because it is real, so that $\beta_j \equiv a_j \; (\mathrm{mod} \; \lambda^2)$, $a_j \in \mathbf{Z}$. Actually more is true, namely $a_j = 1$ for all $j (0 \le j \le p - 1)$. In order to prove this assertion, we recall that η and ζ are themselves primary, so that $\eta \equiv b$ (mod λ^2), $\zeta \equiv c$ (mod λ^2), $b, c \in \mathbf{Z}$. Also, as already observed in the proof of Theorem 9, $\omega^k = (1 - \lambda)^k \equiv 1 - k\lambda$ (mod λ^2), so that $\eta + \omega^j\zeta \equiv b + (1 - j\lambda)c$ (mod λ^2), $\beta_j\omega^{k_j}(\eta + \zeta) \equiv \beta_j(1 - k_j\lambda)(b + c)$(mod λ^2), and (16) becomes $b + (1 - j\lambda)c \equiv \beta_j(b + c)(1 - k_j\lambda)$(mod λ^2). Simplifying and taking the congruences only mod λ, we obtain $b + c \equiv \beta_j(b + c)$(mod λ); from $\lambda \nmid b + c$ (see Problem 23) now follows $\beta_j \equiv 1$ (mod λ). On the other hand, β_j being primary, $\beta_j \equiv a_j$ (mod λ^2). Comparing the last two congruences it is clear that $\beta_j \equiv 1$ (mod λ^2), as claimed. Using this result and taking congruences mod λ^2 rather than mod λ^p, (16) now becomes $b + (1 - j\lambda)c \equiv (b + c)(1 - k_j\lambda)$(mod λ^2), which simplifies to $-j\lambda c \equiv -(b + c)k_j\lambda$ (mod λ^2) or $jc \equiv (b + c)k_j$ (mod λ). Here both sides are rational integers; hence if they are congruent mod λ, then they are also congruent mod p (why? Hint: See the proof of Theorem 5; also Problem 8), so that

$$jc \equiv (b + c)k_j \;(\text{mod } p). \tag{17}$$

We now determine $r \in \mathbf{Z} \ni r(b + c) \equiv c$ (mod p); this is always possible, because $p \nmid b + c$. Multiplying (17) by r, we obtain

$$rjc \equiv r(b + c)k_j \equiv ck_j \;(\text{mod } p) \text{ or } rj \equiv k_j \;(\text{mod } p).$$

We can now replace the (so far unknown) k_j in (16) and obtain (remark that now β_j reappears; it is congruent to one only mod λ^2, but not necessarily mod λ^p, $p > 2$)

$$\eta + \omega^j\zeta \equiv \beta_j\omega^{rj}(\eta + \zeta) \;(\text{mod } \lambda^p).$$

From here on we may proceed essentially as in Section 6. First, taking complex conjugates, $\bar{\eta} + \omega^{-j}\bar{\zeta} \equiv \beta_j\omega^{-rj}(\bar{\eta} + \bar{\zeta})$ (mod $\bar{\lambda}^p$), because β_j is real. However, as seen, $\lambda = 1 - \omega$ and $\bar{\lambda} = 1 - \bar{\omega} = 1 - \omega^{p-1}$ are associates; hence the last congruence also holds mod λ^p. Next we eliminate the unknown β_j between these two congruences. For that, we multiply the first by $(\bar{\eta} + \bar{\zeta})\omega^{-rj}$ and the second by $(\eta + \zeta)\omega^{rj}$; then both second members become $\beta_j(\eta + \zeta)(\bar{\eta} + \bar{\zeta})$, and the corresponding first members have to be congruent mod λ^p; that is,

$$(\bar{\eta} + \bar{\zeta})\omega^{-rj}(\eta + \omega^j\zeta) \equiv (\eta + \zeta)\omega^{rj}(\bar{\eta} + \omega^{-j}\bar{\zeta}) \;(\text{mod } \lambda^p). \tag{18}$$

Before proceeding, let us observe that again, as in Section 6, both $r \equiv 0$, (mod p) and $r \equiv 1$ (mod p) have to be ruled out. Indeed if $r \equiv 0$ (mod p), then

$$r(b + c) \equiv c \;(\text{mod } p) \Rightarrow p | c,$$

which is false, and if $r \equiv 1 \pmod{p}$, then

$$r(b + c) \equiv c \pmod{p} \Rightarrow p \mid b,$$

which is equally false.

Simplifying and rearranging the terms, (18) may be written as

$$(\bar{\eta} + \bar{\zeta})\eta + \omega^j(\bar{\eta} + \bar{\zeta})\zeta - \omega^{(2r-1)j}(\eta + \zeta)\bar{\zeta} - \omega^{2rj}(\eta + \zeta)\bar{\eta} \equiv 0 \pmod{\lambda^p},$$

$$(19)$$

and this congruence has to hold for $j = 0, 1, 2, \ldots, p - 1$. It is conceivable that changing j into $p - j$, we obtain congruences that are not essentially distinct; but in any case, we seem to get $(p + 1)/2$ distinct congruences, by taking, for instance, $j = 0, 1, 2, \ldots, ((p - 1)/2)$. Case I has already been settled for $p = 3$ and $p = 5$; therefore, we may assume that $p \geq 7$, $(p + 1)/2 \geq 4$, so that η and ζ have to satisfy at least 4 congruences like (19). These are nonlinear, which is rather unpleasant. However if we take as "unknowns" the four expressions $\phi = (\bar{\eta} + \bar{\zeta})\eta$, $\psi = (\bar{\eta} + \bar{\zeta})\zeta$, $\bar{\phi} = (\eta + \zeta)\bar{\eta} = \chi$, and $\bar{\psi} = (\eta + \zeta)\bar{\zeta} = \theta$, then the congruences become linear and homogeneous in ϕ, ψ, χ, θ, namely

$$\phi + \omega^j \psi - \omega^{(2r-1)j}\theta - \omega^{2rj}\chi \equiv 0 \pmod{\lambda^p}.$$

$$(20)$$

Exactly as in the case of linear equations, a system of m linear homogeneous congruences in t variables, say, $\phi_1, \phi_2, \ldots, \phi_t$, whose coefficients are integers in some number field $\mathbf{K}$ and with more congruences modulo an ideal $\mathfrak{a}$ of $\mathbf{K}$ than variables, has, in general, no solutions. Even if the number of congruences is the same as the number of variables, that is $m = t$, there are, in general, no solutions except the trivial ones; by this we understand solutions $\phi_1, \phi_2, \ldots, \phi_m$ such that $((\phi_j), \mathfrak{a}) \neq \mathfrak{i}$ for all $j(1 \leq j \leq m)$. The system has nontrivial solutions, that is solutions where $((\phi_j), \mathfrak{a}) = \mathfrak{i}$ holds at least for one $j(1 \leq j \leq m)$, only if the determinant formed with the coefficients is congruent to zero mod $\mathfrak{a}$. The proof follows step by step that of Cramer's rules for linear equations, and is left as an exercise to the reader (see Problem 10).

In our present situation, $\lambda \nmid \eta$, $\lambda \nmid \zeta$; but also $\lambda \nmid \eta + \zeta$, because $\eta + \zeta$ is a factor of ξ^p, so that $\lambda \mid \eta + \zeta \Rightarrow \lambda \mid \xi$, contrary to our assumptions. It follows that the required solutions of (20) are nontrivial, and restricting ourselves only to the first 4 congruences of (20) (which always exist for $p \geq 7$), the determinant δ of the coefficients has to satisfy $\delta \equiv 0 \pmod{\lambda^p}$. One easily computes

$$\delta = \begin{vmatrix} 1 & 1 & 1 & 1 \\ 1 & \omega & \omega^{2r-1} & \omega^{2r} \\ 1 & \omega^2 & \omega^{4r-2} & \omega^{4r} \\ 1 & \omega^3 & \omega^{6r-3} & \omega^{6r} \end{vmatrix}$$

because this is a Vandermonde determinant. We find

$$\delta = (1 - \omega)(1 - \omega^{2r-1})(1 - \omega^{2r})(\omega - \omega^{2r-1})(\omega - \omega^{2r})(\omega^{2r-1} - \omega^{2r}).$$

Remembering that $r \not\equiv 0, 1$, one sees that the factors $1 - \omega^{2r}$ and $\omega - \omega^{2r-1}$ cannot vanish; hence

$$1 - \omega^{2r} = \lambda(1 + \omega + \cdots + \omega^{2r-1}) = \lambda\varepsilon_1,$$

$$\omega - \omega^{2r-1} = \omega(1 - \omega^{2r-2}) = \omega\lambda(1 + \omega + \cdots + \omega^{2r-3}) = \lambda\varepsilon_2,$$

with ε_1 and ε_2 units. Consequently, also replacing $1 - \omega$ by λ and $\omega^{2r-1} - \omega^{2r}$ by $\omega^{2r-1}\lambda$, $\delta = \lambda^4\varepsilon_0\varepsilon_1\varepsilon_2(1 - \omega^{2r-1})^2$, ε_0 unit. We now verify that $2r \equiv 1$ (mod p); indeed if $2r \not\equiv 1$ (mod p), then $1 - \omega^{2r-1} = \lambda(1 + \omega + \cdots + \omega^{2r-2}) = \lambda\varepsilon_3$, $\delta = \lambda^6\varepsilon_0\varepsilon_1\varepsilon_2\varepsilon_3^2$, or defining the unit ε by $\varepsilon\varepsilon_0\varepsilon_1\varepsilon_2\varepsilon_3^2 = 1$, $\varepsilon\delta = \lambda^6$. Consequently, $\delta \equiv 0$ (mod λ^p) implies $\lambda^6 \equiv 0$ (mod λ^p) or $\lambda^p | \lambda^6$, obviously impossible for $p \geq 7$. It follows that indeed $2r \equiv 1$ (mod p). Now, r was defined by $c \equiv r(b + c)$ (mod p); hence $2c \equiv 2r(b + c) \equiv b + c$ (mod p) or $b \equiv c$ (mod p), and, *a fortiori*, $b \equiv c$ (mod λ). Consequently, $\eta \equiv \zeta$ (mod λ). Proceeding in the same way but writing (12) as $-\zeta^p = \xi^p + \eta^p$, we obtain $\xi \equiv \eta$ (mod λ). Consequently $\xi \equiv \eta \equiv \zeta$ (mod λ), $\xi^p + \eta^p + \zeta^p \equiv 3\xi^p$ (mod λ). From $p > 3$ follows $\lambda \nmid 3$; hence if (12) holds, then $\lambda | \xi$, contrary to our assumption. So far we were able to avoid threatening contradictions by proper choices of parameters. This is no longer possible. The assumption that (12) has solutions in **I** and that p is regular leads to $\lambda | \xi$; hence there are no solutions in Case I. The reader may even have obtained the impression that we actually proved more. Indeed from $\xi \equiv \eta \equiv \zeta$ (mod λ) and (12) follows $0 = \xi^p + \eta^p + \zeta^p \equiv 3\xi^p \equiv 3\eta^p \equiv 3\zeta^p$ (mod λ); hence from $\lambda \nmid 3$ follows $\lambda | \xi$, $\lambda | \eta$, $\lambda | \zeta$, and the solution is not even primitive. This suggests the conclusion that (12) cannot have any admissible solutions (regardless of Case I or Case II) because it cannot have primitive solutions. This reasoning however, is not warranted, because in the present proof we have repeatedly made use of the assumption $\lambda \nmid \xi$, $\lambda \nmid \eta$, $\lambda \nmid \zeta$; consequently the conclusion is valid only in Case I. In any case, Theorem 7 is proven as stated. The remaining Case II will be the object of the following section.

9   PROOF OF THE FC FOR REGULAR PRIMES IN CASE II

Let us once more take up equation (12), keeping the assumption that p is a regular prime and as always restricting our attention to primitive solutions (in the sense that ξ, η, ζ are not simultaneously divisible by any *integer* $\alpha \in$ **I**, α not a unit, but not necessarily in the sense that $(\xi, \eta) = (\eta, \zeta) = (\zeta, \xi) = 1$) which is clearly sufficient. However, we want to drop the restriction that

$\lambda = 1 - \omega$ may not divide any of the integers ξ, η, ζ. In other words, our aim in the present section is to prove:

Theorem 10. *Let p be a regular prime; then (12) has no solutions in integers $\xi, \eta, \zeta \in \mathbf{I}, \xi \cdot \eta \cdot \zeta \neq 0$.*

This statement is already much more satisfactory than that of Theorem 6, although it still falls short of the goal because of the restriction on p to be regular, coupled with our present knowledge of the fact that there are infinitely many primes that are not regular (see [11]).

As already mentioned, we shall use without proof the following result due to Kummer.

Theorem 11. *If p is a regular prime and ε is a unit of $\mathbf{I}$ satisfying $\varepsilon \equiv a$ (mod λ^p) for some $a \in \mathbf{Z}$, then there exists in $\mathbf{I}$ a unit ε_1 such that $\varepsilon = \varepsilon_1^p$.*

Besides the already mentioned proof in [12] pp. 240–270, one may find some others—for instance in [25] or [7]—but all are rather long and cumbersome, and it seems preferable at a first study of the subject to accept this result without proof, in order to get a clearer picture of the scheme of the proof of Theorem 10.

Some further preliminary remarks are in order: First if $\xi, \eta, \zeta \in \mathbf{I}$ and are a primitive solution of (12), then λ now divides one but not all of ξ, η, ζ. If it divides any two, then by (12) it follows that it also divides the third, which is ruled out for primitive solutions. Consequently, it divides exactly one of the integers ξ, η, ζ. On account of the symmetry of (12) in ξ, η, ζ, it does not matter which we select to be divisible by λ, and in order to maintain the notational analogy with the cases $p = 2$ and $p = 3$, let $\lambda | \xi$. If $\xi = \lambda^k \alpha, \lambda \nmid \alpha$, then (12) becomes $\lambda^{kp} \alpha^p + \eta^p + \zeta^p = 0$. Secondly, we remember that while the FC has been proven directly for $p = 5$ in Case I (see Theorem 5), Case II for $p = 5$ has not been previously discussed; therefore the present section ought to (and shall) also take care of $p = 5$. Finally just as in the case $p = 3$, it turns out that it is easier to prove a slightly stronger statement, namely:

Theorem 12. *If p is a regular prime, then the equation*

$$\varepsilon \lambda^{kp} \alpha^p + \eta^p + \zeta^p = 0 \tag{21}$$

has no solutions in integers $\alpha, \eta, \zeta \in \mathbf{I}$, none divisible by λ, for any unit ε of $\mathbf{I}$ and any rational integer $k > 0$.

Theorem 12 contains in particular (namely for $\varepsilon = 1$) the statement that for p regular the equation $\xi^p + \eta^p + \zeta^p = 0$ has no solutions in integers $\xi, \eta, \zeta \in \mathbf{I}, \lambda | \xi, \lambda \nmid \eta, \lambda \nmid \zeta$. Combining this result wth Theorem 7, we immediately obtain Theorem 10. It only remains to prove Theorem 12. Any primitive solution of (21) satisfying all conditions of Theorem 12 will be called simply an *admissible solution*.

PROOF OF THEOREM 12.   We shall proceed, as in the case $p = 3$, in two steps:

  (i) show that in (21) k has to satisfy $k \geq 2$; and
  (ii) show that from every admissible solution α, η, ζ of (21), with some $k = k_0 \geq 2$, we can construct another admissible solution with $k \leq k_0 - 1$.

These two statements, being plainly contradictory, prove that there cannot exist an admissible solution of (21); hence Theorem 12 holds.

As in the study of Case I, there is no loss of generality in assuming ξ, η, and ζ to be primary,

$$\xi \equiv a, \eta \equiv b, \zeta \equiv c \,(\text{mod } \lambda^2). \tag{22}$$

THE PROOF OF (i) is almost as easy as in the particular case $p = 3$. Indeed (21) may be written as

$$-\varepsilon\lambda^{kp}\alpha^p = \eta^p + \zeta^p = \prod_{j=0}^{p-1} \left(\eta + \omega^j\zeta\right),$$

or passing from integers to ideals,

$$(\lambda)^{kp}(\alpha)^p = \prod_{j=0}^{p-1} \left(\eta + \omega^j\zeta\right), \tag{23}$$

where $(\eta + \omega^j\zeta)$ now stands for the principal ideal generated by the integer $\beta_j = \eta + \omega^j\zeta$ of **I**. Now $\beta_0 - \beta_j = \zeta(1 - \omega^j) = \zeta(1 + \omega + \cdots + \omega^{j-1})(1 - \omega) = \zeta\varepsilon\lambda$, ($\varepsilon$ a unit, but this is irrelevant here). Consequently $\beta_0 \equiv \beta_j \,(\text{mod } \lambda)$ for all $j(0 \leq j \leq p - 1)$, and either all or none of the principal ideals (β_j) are divisible by (λ). Their product being divisible, all are divisible.

In particular, using (22) and $\lambda|\beta_0$, $\beta_0 = \eta + \zeta \equiv b + c \,(\text{mod } \lambda^2) \Rightarrow \lambda|b + c \Rightarrow p|b + c \Leftrightarrow \varepsilon\lambda^{p-1}|b + c \Rightarrow \lambda^{p-1}|b + c$; consequently $0 \equiv b + c \equiv \eta + \zeta \,(\text{mod } \lambda^2)$. It follows that the product $\prod_{j=0}^{p-1}(\eta + \omega^j\zeta)$ contains the ideal factor (λ) at least $p + 1$ times; but $(\lambda) \nmid (\alpha)$, so that on account of (23), $kp \geq p + 1$, or $k \geq 1 + (1/p) > 1$, that is $k \geq 2$ because k is a rational integer.

This proves Part (i); but again as in the particular case $p = 3$, the real difficulty lies in:

THE PROOF OF (ii). Let $(\eta, \zeta) = \mathfrak{d}$ (unfortunately, as already mentioned, from the fact that no integer of **I** is a common factor of η and ζ, one cannot infer that η and ζ are coprime; they may have as greatest common divisor a nonprincipal ideal $\mathfrak{d}$ of **K**). From $\lambda \nmid \eta$ follows that $(\lambda) \nmid \mathfrak{d}$. Also $\mathfrak{d}|\eta$, $\mathfrak{d}|\zeta \Rightarrow \mathfrak{d}|(\eta + \omega^j\zeta)$ for every $j(0 \leq j \leq p - 1)$. Hence remembering also that

$(\lambda)|(\beta_j)(0 \le j \le p - 1)$, it follows that $(\lambda)\mathfrak{d}$ is a common divisor of the β_js. It is in fact the greatest common ideal divisor of the β_js, even taken in pairs. To prove this it is sufficient to show that every common ideal divisor $\mathfrak{c}$ of β_j and β_m $(m > j)$ also divides $(\lambda)\mathfrak{d}$.

From $\mathfrak{c}|\beta_j, \mathfrak{c}|\beta_m$ and $(\beta_j - \beta_m) = (\omega^j(1 - \omega^{m-j})\zeta) = (\lambda)(\zeta)$ follows $\mathfrak{c}|(\lambda)(\zeta)$; similarly (or directly, by considerations of symmetry) $\mathfrak{c}|(\lambda)(\eta)$ and $\mathfrak{c}|(\lambda)(\eta, \zeta)$, or $\mathfrak{c}|(\lambda)\mathfrak{d}$, as claimed.

In particular it also follows that (λ) cannot divide any (β_j), $j \ne 0$ to higher than the first power, because we saw that $(\lambda)^2|(\beta_0)$, while $(\lambda) \nmid \mathfrak{d}$. We may therefore set $(\beta_j) = (\lambda)\mathfrak{d}\mathfrak{b}_j(j \ne 0)$, $(\beta_0) = (\lambda)^{kp-p+1}\mathfrak{d}\mathfrak{b}_0$, with pairwise coprime ideals $\mathfrak{b}_j$ $(j = 0, 1, \ldots, p - 1)$; also $(\lambda) \nmid \mathfrak{b}_j$. Equation (23) now becomes

$$(\lambda)^{kp}(\alpha)^p = \prod_{j=0}^{p-1} (\beta_j) = (\lambda)^{kp}\mathfrak{d}^p \prod_{j=0}^{p-1} \mathfrak{b}_j,$$

or by Theorem 11.12, $(\alpha)^p = \mathfrak{d}^p \prod_{j=0}^{p-1}\mathfrak{b}_j$. From $\mathfrak{d}^p|(\alpha)^p$ it follows that $(\alpha) = \mathfrak{d}\mathfrak{f}$, so that $\prod_{j=0}^{p-1}\mathfrak{b}_j = \mathfrak{f}^p$, and by a now already familiar reasoning that each ideal $\mathfrak{b}_j$ is the pth power of some ideal $\mathfrak{q}_j$ in $\mathbf{K}$, $\mathfrak{b}_j = \mathfrak{q}_j^p$, say. Clearly $(\lambda) \nmid \mathfrak{b}_j \Rightarrow (\lambda) \nmid \mathfrak{q}_j$. Returning to the β_js, $(\beta_0) = (\lambda)^{kp-p+1}\mathfrak{d}\mathfrak{q}_0^p$ and $(\beta_j) = (\lambda)\mathfrak{d}\mathfrak{q}_j^p(j \ne 0)$. From the fact that all (β_j) and (λ) are principal ideals it now follows that all $\mathfrak{d}\mathfrak{q}_j^p$ belong to the same class (actually all are principal ideals), say that of $\mathfrak{d}\mathfrak{q}_1^p$. By Lemma 11.1 and Theorem 11.10 it follows further that $\mathfrak{q}_j^p \sim \mathfrak{q}_1^p$.

So far we have made no use of our assumption that p is a regular prime; that is, the results obtained up to this point are valid for all primes p, regular or not. The next step, however, consists in inferring that all ideals $\mathfrak{q}_j$ belong to the same class, and for that we have to invoke Theorem 8, which assumes the regularity of p. Therefore, reluctantly, we restrict ourselves from here on to the class of regular primes, use Theorem 8, and infer from $\mathfrak{q}_j^p \sim \mathfrak{q}_1^p$ that $\mathfrak{q}_j \sim \mathfrak{q}_1$. This means, by the definition of equivalence, that there exist integers $\mu_j, \nu_j \in \mathbf{I}$ such that $(\mu_j)\mathfrak{q}_j = (\nu_j)\mathfrak{q}_1$. Remembering that $(\lambda) \nmid \mathfrak{q}_j$, it is clearly possible to select that μ_js and ν_js so that $\lambda \nmid \mu_j, \lambda \nmid \nu_j$ $(0 \le j \le p - 1)$.

Of the $p - 1$ equalities just proven, we select two, say for $j = 0$ and $j = 2$, and have a closer look at them. The first one, $(\mu_0)\mathfrak{q}_0 = (\nu_0)\mathfrak{q}_1$, implies

$$\mathfrak{d}(\lambda)^{(k-1)p+1}(\mu_0^p)\mathfrak{q}_0^p = \mathfrak{d}(\lambda)^{(k-1)p}(\lambda)(\nu_0^p)\mathfrak{q}_1^p,$$

that is, $(\eta + \zeta)(\mu_0^p) = (\lambda)^{(k-1)p}(\eta + \omega\zeta)(\nu_0^p)$. Similarly the second one, $(\mu_2)\mathfrak{q}_2 = (\nu_2)\mathfrak{q}_1$, leads to $(\eta + \omega^2\zeta)(\mu_2^p) = (\eta + \omega\zeta)(\nu_2^p)$.

Once more we have used the theory of ideals in order to be able to factor without ambiguity, but have managed to end up with products involving only principal ideals. We may therefore pass back to integers of $\mathbf{I}$, by introducing appropriate units, and obtain:

$$(\eta + \zeta)\mu_0^p = \varepsilon_1 \lambda^{(k-1)p}(\eta + \omega\zeta)\nu_0^p$$

and

$$(\eta + \omega^2\zeta)\mu_2^p = \varepsilon_2(\eta + \omega\zeta)\nu_2^p,$$

respectively. If we multiply the first equation by $\omega\mu_2^p$, and the second by μ_0^p and add, we obtain

$$\mu_0^p\mu_2^p\{\omega(\eta + \zeta) + \eta + \omega^2\zeta\} = \{\varepsilon_1\lambda^{(k-1)p}\omega\nu_0^p\mu_2^p + \varepsilon_2\nu_2^p\mu_0^p\}(\eta + \omega\zeta).$$

In the first member,

$$\omega\eta + \omega\zeta + \eta + \omega^2\zeta = \eta(1 + \omega) + \omega\zeta(1 + \omega) = (\eta + \omega\zeta)(1 + \omega).$$

Replacing the bracket by this value and simplifying by $\eta + \omega\zeta(\neq 0)$, one obtains $\mu_0^p\mu_2^p(1 + \omega) = \varepsilon_1\lambda^{(k-1)p}\omega\nu_0^p\mu_2^p + \varepsilon_2\nu_2^p\mu_0^p$, or dividing by the unit $1 + \omega$ and setting $\alpha_1 = \nu_0\mu_2$, $\gamma_1 = \nu_2\mu_0$, $\zeta_1 = -\mu_0\mu_2$,

$$\varepsilon_3\lambda^{(k-1)p}\alpha_1^p + \varepsilon_4\gamma_1^p + \zeta_1^p = 0. \tag{24}$$

From $\lambda \nmid \mu_j$, $\lambda \nmid \nu_j$ follows, of course, that $\lambda \nmid \alpha_1$, $\lambda \nmid \gamma_1$, $\lambda \nmid \zeta_1$; hence we have almost proven Part (ii). Indeed (24) very much resembles (21) and has the factor $\lambda^{(k-1)p}$ instead of λ^{kp}. However in (24) we have the extra unit ε_4. In the particular case $p = 3$, when we reached this point we could simply try out all six units in $\mathbf{Q}(e^{2\pi i/3})$ and check that all except $\varepsilon_4 = \pm 1$ led to contradictions. In the present general case, such a procedure is clearly not feasible. Yet actually it is not even necessary to show that $\varepsilon_4 = \pm 1$; all one would need to know is that in $\mathbf{I}$ there exists a unit ε_5 such that $\varepsilon_4 = \varepsilon_5^p$. Then indeed $\varepsilon_4\gamma_1^p = \varepsilon_5^p\gamma_1^p = (\varepsilon_5\gamma_1)^p = \eta_1^p$ with $\eta_1 = \varepsilon_5\gamma_1 \in \mathbf{I}$, $\lambda \nmid \eta_1$, and (24) becomes $\varepsilon_3\lambda^{(k-1)p}\alpha_1^p + \eta_1^p + \zeta_1^p = 0$, $\lambda \nmid \alpha_1$, $\lambda \nmid \eta_1$, $\lambda \nmid \zeta_1$, which would finish the proof of Part (ii), and hence that of Theorem 12. The proof of the existence of ε_5 is actually the most difficult point in the whole proof of Theorem 10. However, having Theorem 11 at our disposal, all we have to do to obtain the desired result is to show that $\varepsilon_4 \equiv a \pmod{\lambda^p}$ for some $a \in \mathbf{Z}$. This we achieve as follows.

By Lemma 8, there exist $s, t \in \mathbf{Z} \ni \gamma_1^p \equiv s$, $\zeta_1^p \equiv t \pmod{\lambda^p}$. We now look at (24) modulo λ^p, and observing that $k \geq 2$, we obtain $\varepsilon_4\gamma_1^p + \zeta_1^p \equiv \varepsilon_4 s + t \equiv 0 \pmod{\lambda^p}$. From $\lambda \nmid \gamma_1$ follows $\lambda \nmid s$, and because $s \in \mathbf{Z}$, that $p \nmid s$; consequently one can solve the congruence $sx \equiv 1 \pmod{p^2}$. This congruence will hold, a fortiori, mod λ^p. From $\varepsilon_4 s + t \equiv 0 \pmod{\lambda^p}$ now follows $\varepsilon_4 sx + tx \equiv 0 \pmod{\lambda^p}$ or $\varepsilon_4 + tx \equiv 0 \pmod{\lambda^p}$; but $a = -tx \in \mathbf{Z}$, so that Theorem 11 is indeed applicable. This means that a unit ε_5 exists in $\mathbf{I}$ such that $\varepsilon_5^p = \varepsilon_4$, and setting $\eta_1 = \varepsilon_5\gamma_1$, (24) now becomes

$$\varepsilon_3\lambda^{(k-1)p}\alpha_1^p + \eta_1^p + \zeta_1^p = 0, \quad \lambda \nmid \alpha_1, \lambda \nmid \eta_1, \lambda \nmid \zeta_1;$$

the proof of Theorem 12 is complete, and with Theorem 12, Theorem 10 is also proven.

10   SOME FINAL REMARKS

The reader whose patience has let him reach this point will have convinced himself that it is hardly profitable to try to penetrate deeper into the problem of the FC without a solid study of algebraic number fields and of related topics. Therefore, the following completions are given for the information of the curious, rather than for the edification of the thorough student. For the latter, there is no shortcut. It is recommended that he study thoroughly the theory of algebraic numbers (see, e.g., [9], [1], or [26]) and then continue with the more specialized work on the FC, for instance, by Dickson [3], Vandiver [23], Furtwängler [8], Landau [12], and others, among whom I would like to mention again the very recent and extremely valuable books by Edwards [6] and by Ribenboim [21].

For the former, however, the following may be of some interest. First, as already mentioned in Section 1, Kummer found a criterion which permits us to decide whether a prime p is regular or not without the need to compute the class number $h = h(\mathbf{K})$ of $\mathbf{K} = \mathbf{Q}(e^{2\pi i/p})$. This criterion states that p is regular if it does not divide the numerators of any of the first $(p - 3)/2$ Bernoulli numbers of even index.

In 1909, Wieferich [27] showed that (12) has no solutions in Case I unless $2^{p-1} \equiv 1 \pmod{p^2}$. It may be observed that $p = 1093$ is the first prime for which this congruence holds. Shortly afterwards (1910; see [16] and [18]) Mirimanoff showed that (12) can have no solutions in Case I unless $3^{p-1} \equiv 1 \pmod{p^2}$ also holds. It seems likely that the existence of a solution of (12) in Case I would actually imply *all* congruences $q^{p-1} \equiv 1 \pmod{p^2}$ for q any prime different from p. This would essentially rule out solutions of (12) in Case I. Apparently the best result known at present in this direction is due to Furtwängler (1912; see [8]), and states that if $x_1^p + x_2^p + x_3^p = 0$ with $x_1, x_2, x_3 \in \mathbf{Z}$ is a primitive solution, then $q^{p-1} \equiv 1 \pmod{p^2}$ holds for every q which is a factor of those among the integers x_i $(i = 1, 2, 3)$ that are not divisible by p (there are at least two of them); the congruence also holds for the factors q of $x_i + x_j$ and $x_i - x_j$, provided i and j are such that $p \nmid x_i^2 - x_j^2$ (again, at least two such couples i, j exist, because the solution is primitive). This criterion is not sufficient to settle the problem, not even in the easier Case I. However, the many congruences of the type $q^{p-1} \equiv 1 \pmod{p^2}$, each hard to satisfy by itself, have permitted us to reach the result quoted earlier (see [14]) that the FC in Case I holds at least for $p < 3.10^9$. In Case II much more delicate procedures are needed, and only much more recently has it been possible to show that the FC holds unconditionally for all primes at least up to $p \le 125,000$ (see [24]).

This discussion of the FC would be incomplete without the mention of two nonmathematical items. The first refers to the fact that in 1908, P. Wolfskehl left the Academy of Sciences of Göttingen (Gesellschaft der Wissenschaften zu Göttingen) a legacy of 100,000 Mark to be paid as a prize for the first complete proof of the FC. This enormous amount of money shrank to insignificance during the inflation following the first World War. But from 1908 to the early 1920s it represented a strong temptation—for nonmathematicians. It seems (see [4] p. 764) that between 1908 and 1912 over 1,000 false proofs were published. Most of these contain such gross errors that one can spot them at a glance and no serious journal printed them. But their authors could not be convinced (often they were unable to understand even how a factorization in **I** could be anything but unique) and went ahead anyhow with private printings. Incidentally, Wolfskehl himself, by all appearances a wealthy man, was a very competent mathematician. Presumably he himself had tried to prove the FC but without success—at least he does not seem to have published anything on this subject. He did work, however, and successfully too, on the class number of certain fields (see [28]) and was fully equipped to appreciate the difficulty of the task.

This brings us to the last item; it had been mentioned already in Chapter 1 and we shall conclude with it. It is simply a reflection upon the fact, which no student of the last few chapters could have failed to notice, that the abundance of deep fertile ideas generated by the attempt to find a proof for a conjecture made by the Judge Pierre de Fermat far exceeds in value the relevance of the conjecture itself.

PROBLEMS

1. Prove: $m, n, q \in \mathbf{Z}$, $(m, n) = 1$, $m \cdot n = q^2 \Rightarrow \exists a, b \in \mathbf{Z} \ni m = a^2$, $n = b^2$.

2. Show that the "solution" x_1, y_1, z_1 constructed in the proof of Theorem 2 is primitive.

3. Prove: In $\mathbf{Q}(e^{2\pi i/3})$ all norms are non-negative and $N\alpha = 0 \Leftrightarrow \alpha = 0$.

4. If $\alpha, \beta \in \mathbf{I}(e^{2\pi i/3})$ and $N\beta = m$, show that the total number k of steps in the Euclidean algorithm starting with $\alpha = \beta\gamma_1 + \rho_1$ satisfies $k \le m$.

5. In (8), compute $\eta + \omega\zeta$ and $\eta + \omega^2\zeta$ as functions of α_1, β, ζ_1.

6. Show that in (8) α_1, β, ζ_1 are coprime in pairs.

7. If $\omega = e^{2\pi i/3}$, show that
$$1 + \omega^r + \omega^{2r} = \begin{cases} 3 & \text{if } r \equiv 0 \ (\text{mod } 3), \\ 0 & \text{otherwise.} \end{cases}$$

8. Generalize the result of Problem 7 to $\omega = e^{2\pi i/p}$.

9. Let $\lambda = 1 - \omega$, $\omega = e^{2\pi i/3}$; prove that $\lambda^2 \nmid \omega \pm 1$ and $\lambda^2 \nmid \omega^2 \pm 1$.

10. Let $\omega = e^{2\pi i/5}$ and set $\lambda = 1 - \omega$. Show that λ satisfies the equation $\lambda^4 - 5\lambda^3 + 10\lambda^2 - 10\lambda + 5 = 0$ and that this equation is irreducible over $\mathbf{Q}$; conclude from this that λ is a prime in $\mathbf{I}(\omega)$. (Hint: use the uniqueness of factorization in $\mathbf{I}(\omega)$ and consider $N\lambda$.)

11. Prove that $\lambda^3 - 2\lambda^2 + 2\lambda - 1 (\lambda = 1 - e^{2\pi i/5})$ is a unit in $\mathbf{I}_5$.

12. Prove that $\pm 2, \pm 1, 0$ are incongruent mod $\lambda (\lambda = 1 - e^{2\pi i/5})$.

13. Can Lemma 4 be strenghtened to $\alpha \equiv \beta \ (\text{mod } \lambda) \Rightarrow \alpha^p \equiv \beta^p$ $(\text{mod } \lambda^{p+1})$, so that it should generalize Lemma 4' and reduce to it for $p = 3$, $\lambda \nmid \alpha$?

14. Let $\omega = e^{2\pi i/p}$, $\lambda = 1 - \omega$; prove that $N\lambda = p$.

15. Give the details of the proof of Lemma 3.

16. Give the details of the proof of Lemma 4.

17. Let π be a prime in $\mathbf{I}(e^{2\pi i/p})$, let $a \in \mathbf{Z}$, and assume that $\pi | a$; show that there exists a rational prime q uniquely defined by π and such that $q | a$. (Hint: $N\pi = q^k$, $1 \le k \le p - 1$.)

18. Let $\eta, \zeta, \alpha \in \mathbf{I}$, and set $(\eta, \zeta) = \delta$; prove that $\delta | (\eta + \alpha\zeta)$.

19. Prove that if the ideal $\mathfrak{f}$ divides two principal ideals (α) and (β), then it divides the principal ideal of their difference, $\mathfrak{f} | (\alpha - \beta)$.

20. Let $\lambda, \xi \in \mathbf{I}$, with λ indecomposable in $\mathbf{I}$, and assume that $(\lambda) | (\xi)^p$; prove that $\lambda | \xi$.

21. Let $\mathfrak{a}_j (1 \le j \le k)$ and $\mathfrak{b}$ be ideals in $\mathbf{K}$, with $\mathfrak{a}_1 \mathfrak{a}_2 \cdots \mathfrak{a}_k = \mathfrak{b}^p$; if $(\mathfrak{a}_i, \mathfrak{a}_j) = (1)$ for all $i \ne j$, then show that there exist ideals $\mathfrak{c}_j$ in $\mathbf{K}$ such that $\mathfrak{a}_j = \mathfrak{c}_j^p (1 \le j \le k)$.

22. (a) If $c, d \in \mathbf{Z}$ and $\lambda \nmid c$, why is it always possible to determine $g \in \mathbf{Z}$ such that $gc \equiv d \ (\text{mod } p^2)$?

    (b) If $gc \equiv d \ (\text{mod } p^2)$, why does $gc \equiv d \ (\text{mod } \lambda^p)$ also hold?

23. In the proof of Theorem 7 an auxiliary result was that $\beta_j \equiv 1 \ (\text{mod } \lambda^2)$. While proving it, use was made of the fact that $\lambda \nmid b + c$ where $\eta \equiv b$, $\zeta \equiv c \ (\text{mod } \lambda^2)$; justify this fact!

24. From $\beta \equiv 1 \ (\text{mod } \lambda)$ and the fact that there exists some $a \in \mathbf{Z}$ such that $\beta \equiv a \ (\text{mod } \lambda^2)$, prove that $\beta \equiv 1 \ (\text{mod } \lambda^2)$.

25. Prove: If $a \in \mathbf{Z}$, then $\lambda | a \Rightarrow p | a$. (Hint: $\lambda | a \Rightarrow a = \lambda\mu$; now take norms, remembering that $Na = a^n = a^{p-1}$ and $N\lambda = p$.)

26. In (19) is the set of congruences with $j > p/2$ *linearly dependent* on the set with $j < p/2$ or not?

27. Let $\alpha_{kj} \in \mathbf{I}$ and let $\mathfrak{a}$ be an ideal in $\mathbf{K}$; prove that the system of linear homogeneous congruences $\sum_{j=1}^{m} \alpha_{kj}\phi_j \equiv 0 \pmod{\mathfrak{a}}$ $(1 \le k \le m)$ can have nontrivial solutions, that is solutions where at least one ϕ_j satisfies $((\phi_j), \mathfrak{a}) = \mathfrak{i})$, only if the determinant δ of the α_{kj} is divisible by $\mathfrak{a}$. (Hint: Let γ_{kj} be the cofactors of the α_{kj}; then

$$\sum_{k=1}^{m} \alpha_{kj}\gamma_{ki} = \begin{cases} \delta & \text{if } j = i, \\ 0 & \text{if } j \neq i. \end{cases}$$

Hence

$$\sum_{k=1}^{m} \gamma_{ki} \sum_{j=1}^{m} \alpha_{kj}\phi_j = \sum_{j=1}^{m} \phi_j \sum_{k=1}^{m} \alpha_{kj}\gamma_{ki} = \delta\phi_i.$$

If the left member is divisible by $\mathfrak{a}$, then so is the right member, and if $\mathfrak{a} \nmid \delta$, then $((\phi_i), \mathfrak{a}) \neq (1)$ for each i.)

28. Consider the following two meanings of "coprimality" for integers $\xi, \eta \in \mathbf{I}$:

   (a) $\alpha|\xi, \alpha|\eta \Rightarrow \alpha = \varepsilon$, a unit of $\mathbf{I}$; and

   (b) $((\xi),(\eta)) = \mathfrak{i}$.

   Does (a) $\Rightarrow$ (b)? Does (b) $\Rightarrow$ (a)? Or is it true that (a) $\Leftrightarrow$ (b)? Which statement (if either) is stronger?

29. Justify the sequence of implications following (23): $\{\lambda|\beta_0$ and $\beta_0 = \eta + \zeta \equiv b + c \pmod{\lambda^2}\} \Rightarrow \lambda|b + c \Rightarrow p|b + c \Rightarrow \lambda^{p-1}|b + c$, used in the proof of Theorem 12.

30. Prove that $\mathfrak{c}|(\lambda)(\eta), \mathfrak{c}|(\lambda)(\zeta) \Rightarrow \mathfrak{c}|(\lambda)\mathfrak{d}$, where $\mathfrak{d} = (\eta, \zeta)$.

31. Let $(\lambda) \nmid \mathfrak{d}$ and let $(\lambda)\mathfrak{d}$ be the g.c.d. of the ideals $(\beta_0), (\beta_1), \ldots, (\beta_{p-1})$ taken in pairs; show that $(\lambda)^2|(\beta_0) \Rightarrow (\lambda)^2 \nmid (\beta_j)$ for $j \neq 0$.

BIBLIOGRAPHY

1. E. Artin, *Theory of Algebraic Numbers* (Lectures) translated by G. Striker. *Mathem. Institut Göttingen*, 1959.

2. A. L. Cauchy, *Comptes Rendus Acad. Sciences* (Paris) 24 (1847) 578–584.

3. L. E. Dickson, *Quarterly Journal of Mathem.* 40 (1908) 27–45.

4. L. E. Dickson, *History of the Theory of Numbers*, vol. 2. New York: Chelsea, 1952.

5. G. L. Dirichlet, Mémoire read at the Royal Acad. of Sciences, Paris (Institut de France) on July 11, 1825, but not published until 1828 in

the *Journal für die reine und angew. Mathem.* 3 (1828) 354–375; *Werke* 1, pp. 21–46.

6. H. M. Edwards, *Fermat's Last Theorem*. New York: Springer, 1977.

7. R. Fueter, *Synthetische Zahlentheorie*. Berlin: DeGruyter, 1925.

8. P. Furtwängler, *Sitzungsberichte Akad. Wiss. Wien* 121 (1912) 589–592.

9. E. Hecke, *Theorie der algebraischen Zahlen*. Leipzig: Akad. Verlagsgesellschaft, 1923; New York: Chelsea, 1948.

10. E. Hecke, *Nachrichten der Akad. Wiss. Göttingen* II, *Mathem.-Physik. Klasse* 1910, 420–424.

11. K. L. Jensen, *Nyt Tidsskrift for Mathematik, Afdeling, B.* 1915, 73–83.

12. E. Landau, *Vorlesungen über Zahlentheorie*, vol. 3. Leipzig: S. Hirzel, 1927.

13. A. M. Legendre, *Essai sur la Théorie des Nombres*. Paris: Duprat, 1808 (Legendre gives credit for many results to Sophie Germain).

14. D. H. Lehmer and E. Lehmer, *Bull. Amer. Math. Soc.* 47 (1941) 139–142.

15. G. Libri, *Journal f. d. reine u. angw. Mathem.* 9 (1832) 270–275.

16. D. Mirimanoff, *Comptes Rendus Ac. Sci. Paris* 150 (1910) 204–206.

17. D. Mirimanoff, *L'Enseignement Mathématique* 11 (1909) 49–51.

18. D. Mirimanoff, *Journal f. d. reine u. angw. Mathematik* 139 (1911) 309–324.

19. L. J. Mordell, *Three Lectures on Fermat's Last Theorem*. Cambridge: The University Press, 1921.

20. H. Pollard, *The Theory of Algebraic Numbers* (Carus Monograph No. 9). New York: Wiley, 1950.

21. P. Ribenboim, 13 *Lectures on Fermat's Last Theorem*. New York: Springer, 1979.

22. J. B. Rosser, *Bull. Amer. Math. Soc.* 46 (1940) 299–304; 47 (1941) 109–110.

23. H. S. Vandiver, *Amer. Math. Monthly* 53 (1946) 555–578; see also a large number of the papers quoted in this survey article.

24. S. S. Wagstaff, *Mathem. Comput.* 32 (1978) 583–591.

25. H. Weber, *Lehrbuch der Algebra*, vol. 2. Braunschweig: F. Vieweg & Sohn, 1899.

26. H. Weyl, *Algebraic Theory of Numbers* (Annals of Mathem. Studies, No. 1). Princeton, N.J.: Princeton University Press, 1940.

27. A. Wieferich, *Journal f. d. reine u. angw. Mathem.* 136 (1909) 293–302.

28. P. Wolfskehl, *Journal f. d. reine u. angw. Mathem.* 99 (1886) 173–178.

Answers to Selected Problems

CHAPTER 3

1. No   **11.** $d = 9$, $m = 2^4 3^2 7 \cdot 11 = 11{,}088$
13.(a) 441 and 693   **(b)** $21 = p \in \mathbf{H}$, $21 \nmid 9$, $21 \nmid 49$, but $21 | 9 \cdot 49 = 441$
15. $20! + 2, 20! + 3, \ldots, 20! + 21$

CHAPTER 4

3. $(a, 1)$ corresponds to rational integers; $(0, 1)$ to zero; $(1, 1)$ (or more generally (a, a)) to one
7. For instance, $10, 20, 30, 40, 50, 60, 70 (\equiv 3, 6, 2, 5, 1, 4, 0 \pmod 7$, respectively). The problem has no solution for 6, because for even $a \in \mathbf{Z}$, $(10a, 6)$ is even; hence for no integer a can one obtain, e.g., $10a \equiv 1 \pmod 6$
10. The binomial coefficients of $(a + b)^p$ (more generally, the multinomial coefficients of $(a + b + \cdots + m)^p$) are all divisible by p
11. If $3 | m$, then $m^2 \equiv 0 \pmod 3$; otherwise $m^2 \equiv 1 \pmod 3$, while $3n^2 - 1 \equiv -1 \not\equiv 0 \pmod 3$, and also $\not\equiv 1 \pmod 3$
12. $n^2 - 1 = (n - 1)(n + 1)$, where $n - 1, n + 1$ are consecutive even integers; hence one of them is divisible by 4 and the other by 2
13. If $3 | n$, then $27 | n^3$; a fortiori, $9 | n^3$. Otherwise $n = 3m \pm 1$, $n^3 = 27m^3 \pm 27m^2 + 9m \pm 1 \equiv \pm 1 \pmod 9$
14. (a) $2^{2n} - 1 \equiv (-1)^{2n} - 1 \equiv 1 - 1 \equiv 0 \pmod 3$
    (b) $2^{3n} - 1 = (2^3)^n - 1 = 8^n - 1 \equiv 1 - 1 \equiv 0 \pmod 7$
    (c) $2^{4n} - 1 = (2^4)^n - 1 = 16^n - 1 \equiv 1^n - 1 \equiv 0 \pmod{15}$
16. If $m = nk$ with odd $n \neq 1$, then $2^m + 1 = (2^k)^n + 1$ is divisible by $2^k + 1 = r$, say with $1 < r < 2^m + 1$, and $2^m + 1$ is not a prime
19. $240^{37} \equiv 2^{37} \equiv 2^{36} \cdot 2 \equiv 1 \cdot 2 \equiv 2 \pmod 7$

20. 2, if $p = 2$; 1 otherwise

22. $x = 15 + 49k, y = 3 + 10k$ **25.** $x = 8 + 14n, y = -2 + 14n$

CHAPTER 5

3. Quadratic residues modulo 11: $1, 3, 4, 5, 9$; quadratic residues modulo 13: $1, 3, 4, 9, 10, 12$

10. 3 is a quadratic residue for $p \equiv \pm 1 \pmod{12}$, and a quadratic nonresidue for $p \equiv \pm 5 \pmod{12}$. -5 is a quadratic residue for $p \equiv 1, 3, 7, 9 \pmod{20}$ and a quadratic nonresidue for $p \equiv 11, 13, 17, 19 \pmod{20}$

11. $S = 1^2 + 2^2 + \cdots + ((p-1)/2)^2 = ((p-1)/2)((p+1)/2) \cdot p/6 = (p(p^2 - 1))/24 \equiv 0 \pmod{p}$, because for $p > 3$, $24 | (p-1)(p+1)$ (see Problem 4.12)

19. (a) 2;   **(b)** 0

CHAPTER 6

3. 748   **4.** 8, 60, 850   **5.** 5080   **9.** 6000

11. Both expressions equal 2592   **12.** The only exception is $n = 6$

16. 496 and 8128   **17.** Both are equal to -1

CHAPTER 7

1. and 2.

n	1	2	3	4	5	6	7	8	9	10
$p(n)$	1	2	3	5	7	11	15	22	30	42

3. $p_\mathbf{P}(5) = p_\mathbf{P}^{(3)}(5) = 2$, $p_{\mathbf{P},3}(5) = p_\mathbf{P}^{(0)}(5) = p_\mathbf{P}^{(e)}(5) = 1$

4. $\displaystyle\prod_{m=0}^{\infty} \frac{1}{(1 - x^{5m+1})(1 - x^{5m+4})} = \sum_{n=0}^{\infty} p_{\mathbf{R}(5)}(n)x^n$;

$\displaystyle\prod_{m=0}^{\infty} \frac{1}{(1 - x^{5m+2})(1 - x^{5m+3})} = \sum_{n=0}^{\infty} p_{\mathbf{N}(5)}(n)x^n$

14. $\displaystyle\prod_{m=0}^{\infty} (1 - x^{5m+1})(1 - x^{5m+4})(1 - x^{5m+5})$

$\displaystyle = \sum_{n=-\infty}^{\infty} (-1)^n x^{(5n^2 + 3n)/2}$

CHAPTER 8

1. If there are only finitely many primes, then there is a largest one, say q.
Then $\zeta(s) = \prod_{\substack{p=2 \\ p \in \mathbf{P}}}^{q} \dfrac{1}{1 - p^{-s}}$, and for $s = 2$, $\zeta(2) = \prod_{\substack{p=2 \\ p \in \mathbf{P}}}^{q} \dfrac{1}{1 - p^{-2}}$, a rational
number. This contradicts the fact that $\zeta(2) = \pi^2/6$ is not rational and so
no largest prime q can exist

4. $|\zeta(s)| = |\sum_{n=1}^{\infty} n^{-s}| \leq \sum_{n=1}^{\infty} |n^{-s}| = \sum_{n=1}^{\infty} n^{-\sigma} \leq \sum_{n=1}^{\infty} n^{-2} = \zeta(2)$

CHAPTER 9

11. (a) yes   **(b)** no   **(c)** yes   **(d)** no

CHAPTER 10

5. (a) $x = \pm 1, y = 0$ **(b)** has no solutions
   (c) $x = \pm 6, y = 0$ **(d)** $x = \pm 9, y = \pm 1$
6. $\pm 1, (\pm 1 \pm \sqrt{-3})/2$ **9.** $x^4 - 4x^2 + 9 = 0$ and $x^4 + 4x^2 + 49 = 0$
10. (a) $\phi = \sqrt{2 + \sqrt{-5}}$ will do, because $\sqrt{-5} = \phi^2 - 2 \in \mathbf{Q}(\phi)$
   (b) One may take, e.g., $\psi = \sqrt{2 + \sqrt{-5}} + \sqrt{-2 + 3\sqrt{-5}}$
11. $\psi^8 + 8(39\psi^4 + 160\psi^2 + 162) = 0$
17. $\Delta = p^{p-2}$ **20.** $A^3 + B^3 + C^3 - AC^2 - AB^2 + ABC$
21. Same as for Problem 20   **22.** $P(\alpha) = 3A^2 - B^2 + 4AC + C^2 + 3BC$

CHAPTER 12

8.

n	1	2	3	4
χ_0	1	1	1	1
χ_1	1	i	$-i$	-1
χ_2	1	-1	-1	1
χ_3	1	$-i$	i	-1

9.

$$1 + \frac{i}{2} - \frac{i}{3} - \frac{1}{4} + \frac{1}{6} + \frac{i}{7} - \frac{i}{8} - \frac{1}{9} + \frac{1}{11} + \frac{i}{12} - \frac{i}{13} - \frac{1}{14} + \cdots$$

$$= 1 - \frac{1}{4} + \frac{1}{6} - \frac{1}{9} + \frac{1}{11} - \frac{1}{14} + \cdots$$

$$+ i\left(\frac{1}{2} - \frac{1}{3} + \frac{1}{7} - \frac{1}{8} + \frac{1}{12} - \frac{1}{13} + \cdots\right)$$

$$= \sum_{m=0}^{\infty}\left(\frac{1}{5m+1} - \frac{1}{5m+4}\right) + i\sum_{m=0}^{\infty}\left(\frac{1}{5m+2} - \frac{1}{5m+3}\right)$$

$$= .834 \cdots + .194 \cdots i$$

11. $L(1, \chi) = 1 - \frac{1}{3} + \frac{1}{5} - \frac{1}{7} + \cdots = \pi/4$

CHAPTER 13

4. $x = 1/(m^2 - 1)$, $y = m(m^2 - 1)$ for all $m \in \mathbf{Z}$
5. $x = m^2 - 1$, $y = (m - 1)m(m + 1)$ for all $m \in \mathbf{Q}$
6. $x = m^2 - 2$, $y = m^3 + 1$ for all $m \in \mathbf{Q}$
7. By setting $x = x_1 + 1/10$, $y = y_1 + 9/5$, $z = z_1 + 13/10$, one obtains $Q_1(x_1, y_1, z_1) = x_1^2 + 2y_1^2 + 3z_1^2 + x_1 y_1 - y_1 z_1 = 42/5$
8. Let $s_2(n)$ be the number of distinct solutions and $r_2(n)$ be the total number of solutions; then

n	1	2	3	4	5	6	7	8	$\cdots$	18
$r_2(n)$	4	4	0	4	8	0	0	4	$\cdots$	4
$s_2(n)$	1	1	0	1	1	0	0	1	$\cdots$	1

9. $n = 100 = (\pm 10)^2 + 0^2 = 0^2 + (\pm 10)^2 (4 \text{ representations})$

$$= (\pm 8)^2 + (\pm 6)^2 = (\pm 6)^2 + (\pm 8)^2 (8 \text{ representations})$$

for a total of $r_2(100) = 4 + 8 = 12$ representations.
$n = 101 = (\pm 10)^2 + (\pm 1)^2 = (\pm 1)^2 + (\pm 10)^2$, so that $r_2(101) = 8$.
$n = 102 = 2 \cdot 3 \cdot 17$ has no such representations, $r_2(102) = 0$
10. $8 = 0^2 + (\pm 2)^2 + (\pm 2)^2$ with three permutations of the terms, so that $r_3(8) = 3 \cdot 4 = 12$.
$19 = (\pm 1)^2 + (\pm 3)^2 + (\pm 3)^2$ with three permutations of the terms, so that $r_3(19) = 3 \cdot 8 = 24$.
$20 = (\pm 2)^2 + (\pm 4)^2 + 0^2$ with 6 permutations, so that $r_3(20) = 6 \cdot 4 = 24$.
$29 = (\pm 2)^2 + (\pm 5)^2 + 0^2 = (\pm 2)^2 + (\pm 3)^2 + (\pm 4)^2$; the first yields $6 \cdot$

$4 = 24$ representations, the second one $6 \cdot 8 = 48$, for a total of $r_3(29) = 72$. $33 = (\pm 1)^2 + (\pm 4)^2 + (\pm 4)^2 = (\pm 5)^2 + (\pm 2)^2 + (\pm 2)^2$; each of these essentially distinct representations is counted $3 \cdot 8 = 24$ times for a total of $r_3(33) = 48$

11. (i) and (ii) have no nontrivial solutions, (iii) has, e.g., $x = z = 1$, $y = 2$, or $x = y$ an arbitrary integer, $z = 0$, etc.
13. $x = z = 2$, $y = 4$ 14. $r_4(7) = 64$, $r_4(10) = r_4(20) = 144$, $r_4(25) = 248$
15. One may start from $x = y = 1$ 17. The equations have no solutions

CHAPTER 14

26. When we replace j by $p - j$, we obtain the same congruence, with η and ζ replaced by $\bar{\eta}$ and $\bar{\zeta}$, respectively
28. (b) $\Rightarrow$ (a), but (a) $\not\Rightarrow$ (b); hence (b) is stronger

SUBJECT INDEX

Mathematical terms without particular number theoretic relevance (such as *entire functions,* or *residue of a pole*) have not been listed here. It also appeared impractical to list all pages at which certain often used terms (such as *number,* or *divisor*) occur. An effort has been made to list the first occurrence, as well as other important ones. If a certain term occurs frequently after a certain page, this is indicated by the symbol "& seq.".

NAME INDEX